INTRODUCTION TO **SYSTEMS**

ANALYSIS & DESIGN

FIFTH EDITION

IGOR HAWRYSZKIEWYCZ

Prentice
Hall

Pearson Education Australia
Unit 4, Level 2
14 Aquatic Drive
Frenchs Forest NSW 2086

www.pearsoned.com.au

Acquisitions Editor: Cath Godfrey
Senior Project Editor: Jeremy Fisher/Robi van Nooten
Copy Editor: Matthew Stevens
Cover and text design: Ivan Finnegan Design

Typeset by Midland Typesetters, Victoria

Printed in Malaysia, PP
2 3 4 5 03 02 01
ISBN 1 7400 9280 5

National Library of Australia
Cataloguing-in-Publication data

Hawryszkiewycz, I.T. (Igor Titus)
Introduction to systems analysis and design

　　5th edition
　　Bibliography
　　Includes index

　　ISBN 1 7400 9280 5

　　1. Systems analysis. 2. System design. I. Title.

004.21

A division of Pearson Education Australia

Preface   xxi

CHAPTER 1 – **The system development environment**   1

CHAPTER 2 – Work practices and workgroup systems

CHAPTER 3 – **Business systems** 49

CHAPTER 4 – **Requirements analysis** 79

CHAPTER 5 – The development process

CHAPTER 6 – Supporting system development

CHAPTER 7 – Conceptual design

CHAPTER 10 – **Process descriptions** 247

CHAPTER 11 – **Object modeling**   269

CHAPTER 12 – **Object analysis—identifying objects**   303

CHAPTER 13 – Defining the requirements 321

CHAPTER 16 – Program design

CHAPTER 17 – Physical design

The fifth edition of this book is a substantial revision. There is a change in emphasis from a study of information systems as a topic on its own, to closely integrating it with the strategic objectives of the enterprise. Much of the material from chapters on strategic planning has now been included in the earlier chapters and has been closely integrated with business systems development. Chapter 3 now sees information systems as one dimension of a larger picture that also includes strategic and organizational development.

This fifth edition recognizes the significant trends in electronic commerce and electronic business and their impact on information systems design. The impact is addressed early in Chapter 1 and is given considerable emphasis in early chapters, especially in organizational strategy and information systems. New material introduced in Chapter 3 considers how the trends in electronic commerce affect information systems development. In particular, this addresses integration of systems with the Worldwide Web and the kinds of requirements that arise through closer integration with clients and other businesses.

The book also builds on some of the new directions introduced in the fourth edition. It continues with the distinction between the development process, the management process and supporting processes in line with evolving standards. It also elaborates more on communication, making distinctions between the kinds of communication needed in different phases of the development process, stressing the importance of using the best terms at different development phases. A greater range of development processes is now covered, to go beyond prototyping and include the rapid application development often found in the development of electronic business applications.

This edition develops some technical aspects in more detail. The chapters on object modeling have been substantially changed to make the techniques taught consistent with the UML standard. Some new material is also introduced, especially on object-oriented approaches to provide a better balance with conventional structured systems analysis, as well as an introduction to open systems methods. The book's emphasis on modeling continues to stress the way modeling is used for good communication, especially during requirements analysis. It provides a balanced approach of structured analysis and object-oriented methods to stress that different modeling methods are often suitable for different problems. In this context, it introduces some ideas from soft-systems methodologies, especially for use in initial development phases that require a view of the whole enterprise.

Systems analysis and design techniques continue to be introduced in the sequence in which they would be needed during systems analysis and design. The book first outlines the steps followed in analysis and design and then describes each step in detail, introducing the techniques to be used at that point. It then discusses how techniques for describing data, processes and flows can be integrated, and how models developed during analysis can be converted to working systems during design.

The book continues to be one for beginners. It does not assume any knowledge of systems analysis or design. However, it does assume that readers have some basic knowledge about computers and some knowledge of the Internet. For example, readers should know that data is stored on computer disks and that these disks are controlled by a processor, and what a URL is. Readers should also know that computers can communicate between themselves or remote devices via communications links. Some knowledge of programming and algorithms is also useful but not essential.

To assist students' understanding, additional examples have been provided. Also, each chapter is followed by discussion questions and problems to illustrate the techniques described. Four text cases are used throughout the book to illustrate the various methods described in the book.

Igor Hawryszkiewycz

THE SYSTEM DEVELOPMENT ENVIRONMENT

CHAPTER 1

CONTENTS

KEY LEARNING OBJECTIVES

Development, management, and support processes
The importance of specifying systems
The role played by users, analysts, and designers in an information system
The relationship between the various people in building information systems
The role of computers in information systems
The structure of computer-based information systems

INTRODUCTION

Computers are now becoming part of virtually every activity in any organization. The earliest applications that used computers were business applications, such as keeping records of transactions, airline reservations, or the quantity of parts available in a warehouse. There were also early applications for design problems, such as designing a building or setting up a project schedule. Computer use has gradually expanded, and now computers are increasingly used for everyday activities, such as sending messages, or arranging appointments or meetings. They are also increasingly used throughout the whole of society through electronic commerce and the development of electronic communities.

There are not only a variety of uses of computers in systems, but also many different ways to build systems. Sometimes a system just evolves, which often happens with Internet applications. It may start with a very simple program, and more and more functionality is gradually added to it. Sometimes systems are built experimentally to try out new ideas as the system is built. On the other hand, the development of an information system for a business application is often carefully planned and monitored, especially given the complexity of many contemporary systems. Many information systems now use computers for manipulating information and are sometimes called computer-based information systems. This book is about building such computer-based information systems. It describes the different approaches for building them and the use to which each different approach is put.

PROCESSES IN BUILDING SYSTEMS

Development process A set of steps used to build a system.

An number of activities or processes are associated with building systems. Foremost in this book is the **development process**, which is what we actually do to build a system. But other things go on as well. For example, the **management process** is concerned mainly with organizing the development work, ensuring that adequate resources are made available, and monitoring progress. There are also **supporting processes** that provide the necessary development tools to developers, manage documentation, and facilitate communication between people working in teams on projects and between system designers and users.

Management process The tasks required to manage a development process.

Supporting process A process to provide facilities needed by development teams.

ORGANIZATION OF THE BOOK

System specification A precise description of what the system must do.

User requirements What users expect the system to do for them.

This book covers all these processes but concentrates on system development for a wide variety of systems. It begins by describing some typical systems in Chapters 2 and 3 and how they fit into organizations. Chapter 3 also describes how to choose what systems to build and the importance of identifying what people require from the system. This is followed by developing a **system specification**, which correctly specifies these **user requirements**. Chapters 4 and 5 describe ways to collect user requirements and alternative development processes for building computer-based information systems. Chapter 6 then describes project management processes and tools used to support system development.

The next chapters describe in more detail the modeling methods used in analysis and design. Chapters 7 and 12 describe modeling methods that are used during system development. The main goal of modeling is to gain precision and improve communications by avoiding ambiguities that are often found in natural language system descriptions. Modeling achieves precision by using precise modeling constructs and process descriptions to define system requirements. Many different modeling methods are used in practice, some to describe processes, others to describe data, and still others to describe information flows in systems. The book describes methods used in structured systems analysis as well as methods used in object modeling.

The book then continues with ways that are used to construct the specified systems. Chapter 13 begins by developing a specification of a new system and continues by describing the detailed techniques used in system design. Chapter 14 discusses the importance of interfaces to computers and how to design them. Subsequent chapters describe how system specifications are converted to a working system. Chapters 15 and 16 describe how data structures and programs are designed in detail, followed by Chapter 17, which describes the conversion of designs into databases and programs. Precision is again emphasized in design. We need to precisely convert the proposed model to a working system and not go through a process of interpretation, thus losing some precision. Chapters 19 and 20 then describe some techniques used in interviewing and assuring system quality.

The book uses several cases to illustrate the modeling techniques and design ideas. These cases are used in a number of chapters to illustrate how systems evolve during information system design. The cases emphasize two important design issues. One is integration of one or more existing or **legacy systems** to meet some new requirement; the other is distribution of processing across a variety of distributed locations. They also cover the important area of designing applications based on the Internet. From this you will see how complex some of these systems are and what must be done to build them. You will also see why a book about analyzing, designing, and building systems is needed.

Legacy system An existing working computer system that is to be used in a new business process.

DEVELOPMENT PROCESSES

Ideally one should plan what the system is to do and then build the planned system. How this is done depends on the type of system. Methods of building systems vary from almost *ad hoc* to highly structured. *Ad hoc* approaches are often used to build simple personal support systems—for example, a spreadsheet to keep the budget for a manager or a personal system to keep addresses. More formal ways are needed to develop larger business systems, which are becoming increasingly complex but at the same time must be *correct* and do what is expected of them. A formal rather than *ad hoc* approach is required to achieve correctness. Development must begin by defining exactly what the system must do and specify its requirements before actual construction begins. Otherwise, how can we say that the system is correct? As a result there is a lot of similarity between building computer systems and building other kinds of systems. For example, if you look at Figure 1.1, at the way a house is built, you will see that there are a number of important activities. First, there is the concept, the dream house, which is generally reduced to a more concrete specification, taking into consideration the needs of the owners, their available

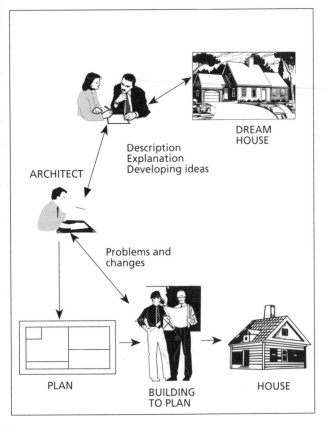
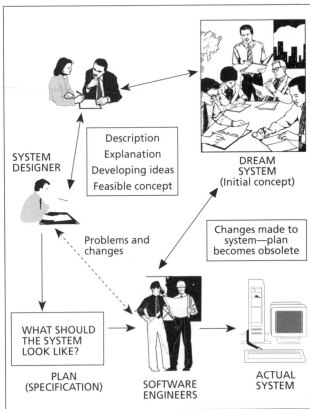

Figure 1.1 *The building process*

funds, the rules that govern building, and any other constraints. Such specifications are usually developed by an architect, who has extensive knowledge of the building trade. The result is a plan that specifies the requirements of a system, and this plan must be followed by the builder to produce the required house.

The same principles apply when building computer systems. First, there is the concept of a dream system to solve a business problem. This is then discussed and analyzed to clearly identify user requirements while taking into account the various needs and constraints. The requirements are then used to produce a system specification, which corresponds to the house plan, and defines the future information system to satisfy user requirements. It is important here to distinguish between the initial concept and the resulting system specification. Both play an important role in the development process. The first comes up with an idea, whereas the second is the practical realization of that idea. In between are other activities, like carrying out a detailed analysis of the system and identifying detailed user requirements. These detailed development activities are described in Chapter 5.

There are other ways of building systems—for example, we may have a grand plan but build it piece by piece, just as we may build a part of a house but leave scope for later extensions. There are also experimental ways of building systems. We may build an initial prototype, experiment with it, discover new requirements, amend the system, and so on. This process, often used in engineering, is to add new features to the system at each step and test them in use before proceeding to the next step. Furthermore, not all projects start by building a totally new system. Thus, one part of the system may be changed or a new component may be added, just like improving a house through upgrading a kitchen or bathroom or opening up a wall between two rooms.

There are, however, some differences that distinguish the building of information systems from the building of other systems. An important one is the lack of universal standards for building information systems. Thus, for example, house plans are often checked by local councils and must adhere to building codes. But there are no such codes for information systems. Information systems requirements also tend to change as development proceeds. Furthermore, by adhering to tight deadlines, people often fail to record changes in the original plan, which can subsequently lead to the eventual breakdown of the whole process. In contrast, when building houses, any change must be approved by the relevant authorities, again ensuring that standards are met. There is, however, increasing emphasis on developing universal standards for building information systems and managing change.

METHODS AND TOOLS

An engineering approach for building systems requires methods and tools to ensure that systems are built in the most effective way. One term often found here is the development **process**. The methodology defines the set of steps followed to build the system. It also provides a variety of supporting methods and tools. **Modeling methods** produce models, which help us to understand the system and its requirements and then to develop system specifications. These models are used primarily in analysis. There are also **productivity tools** to help people develop the models and convert them to working systems.

A large number of system development methodologies and tools are used to build computer-based information systems. Different methodologies use different sets of steps or processes to build systems as well as different productivity tools. Good methodologies must include a way to determine system requirements. The methodologies must also provide supporting methods and tools, which ensure that a system is built in the quickest and cheapest way and is acceptable to users. The term *quality* is often used to describe systems that work well and satisfy all user criteria. We must thus choose the right method to build a quality system. We will come back to this term many times in this book.

The usual sequence of steps in methodologies is to propose the solution, develop a system specification, and then design and build the system. Analysis is used during these activities to gain an understanding of an existing system and what is required of it. Usually analysis produces a system description and a set of user requirements for a new system. These requirements lead to the specification for the new system. Design, which follows, proposes the new system that meets these requirements. Once the design is approved, the system is built. A new system may be built afresh or by changing the existing system. There is some advantage to using suitable existing modules to build a system by putting them together in a way that satisfies user requirements. This reduces the cost of building a system and the time needed to put it into place.

Process A component of a DFD that describes how input data is converted to output data.

Modeling method A method used to construct a model of a system.

Productivity tools Software systems that assist analysts and designers to build computer-based information systems.

MANAGEMENT PROCESSES

A myriad of things must be done to build or change a system. One of the most important activities is to choose the right problem to solve, propose a feasible way to solve it, and describe this way in a system specification. The chosen system must also fit into an existing business environment and must be very easy to use. It is often necessary to spend considerable time gaining a thorough understanding of the system and its changing business environment. It is only after developing a good understanding of the business that it becomes possible to propose changes that will produce a better system with little risk of causing unforeseen effects.

Identifying what is to be done is one of the major tasks in management. As shown in Figure 1.2, management identifies the tasks needed to realize a project goal and assigns people to these tasks, making sure that people with the right expertise are chosen for the tasks. These tasks include choosing equipment, designing new business processes and databases, and writing new programs to support these processes on the equipment. The systems must be tested, and users must be shown how to use them.

SUPPORTING PROCESSES

A number of supporting activities are needed in information systems development. One of the most obvious is to ensure that the necessary equipment is provided for building the systems. But there are other needs here, in particular, providing support for people to work together on teams, and documenting the work.

THE IMPORTANCE OF TEAMWORK AND COMMUNICATION

Information systems are built by teams of people. These teams must include people with the expertise needed to build the system and organize the teams in ways that suit the development process. Chapter 5 outlines different development processes and the team structures to support them. It describes how to organize teams and assign team members with the right expertise to each task. Such teams must be supported by communication services that ensure that all team members are aware of each other's activity. Such communication is needed to avoid overlap and unnecessary delays and to ensure that everyone is working towards the same goal.

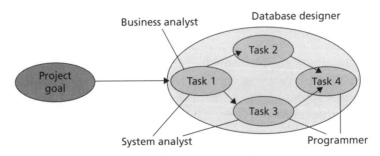

Figure 1.2 *Defining development tasks*

In most organizations this now means access to e-mail and the ability to exchange documents. It also means providing ways to discuss issues, whether through workshops, video- or teleconferencing, and ensuring that everyone has the most recent information about the status of a project.

KEEPING TRACK OF DESIGN DOCUMENTS

Documentation plays a major role in communication processes. Many kinds of documents must be kept during system development. These include proposals, records of interviews, and decisions, which require word processing. There is also the need to develop presentations using systems such as Microsoft PowerPoint.

Project plans and budgets require spreadsheets to keep track of financial details, and various project management tools, such as Gantt charts, to keep track of project progress. Documents that describe the status of tasks must be quickly distributed to those team members affected by these tasks.

Then there are the documents that describe the system itself. These include the specifications themselves. These are then followed by other documents such as designs of specific parts, test specifications, and the actual programs. In addition, there are the various system models that are developed during the life of the project. Computer Assisted Software Engineering (CASE) tools often maintain these. Finally, there are the system programs themselves, as well as the instructions on how to use them.

All these documents are related, but they can also change during the development process. They are used by team members to build different parts that eventually make up the whole system. In large systems these documents may well number in the hundreds. Special care must be taken to keep track of the documentation and ensure that team members always have access to the latest version of any document. The term **configuration management** is used to describe systems that manage documents. This is described in Chapter 6.

Configuration management Managing a configuration.

WHAT ARE THE KINDS OF SYSTEMS?

Systems analysts are continually challenged with the kinds of systems they have to develop and the continuously changing environment. They must continually learn about such trends, their effect on business, and how to build the systems. The remainder of this chapter describes some of the kinds of systems that analysts work with. It is also important to remember that the kind of systems change over time, and analysts must always keep abreast of such changes.

Information systems have evolved significantly over the last few years. This evolutionary trend is broadly illustrated in Figure 1.3. It shows personal and corporate systems gradually coming together as corporate networks and electronic commerce. Early corporate information systems supported **transaction processing systems** that followed routine procedures, such as processing account transactions, personnel records, and so on. Each of these procedures was often supported by a different system. The next generation of corporate information systems integrated these functional systems to support **integrated business processes**, which include more than one functional area. For example, the integration of accounting with

Transaction processing system A computer system that manages transactions.

Integrated business process A system that uses more than one business function.

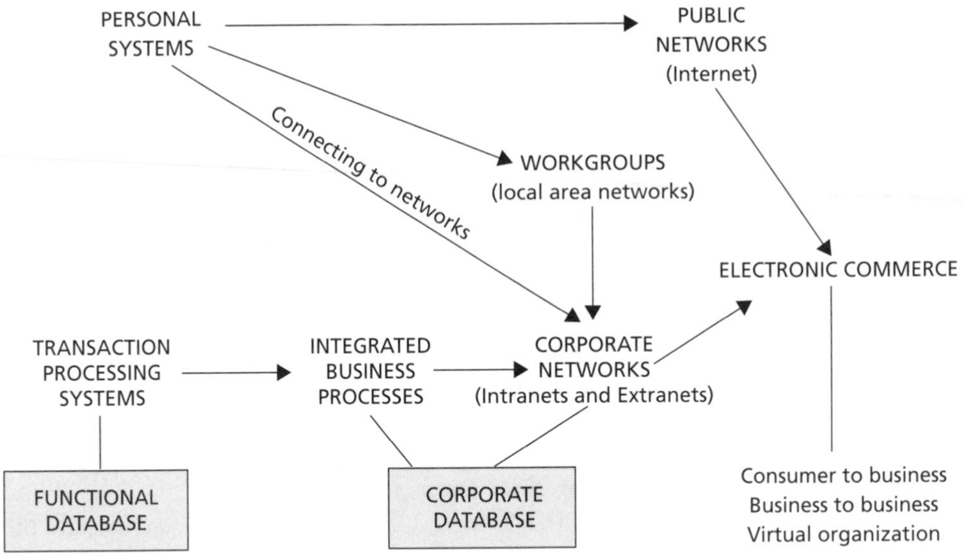

Figure 1.3 *Trend in applications*

delivery systems to directly initiate invoicing and payments. The next extension is to corporate networks, where not only systems but people are networked through intranets and extranets. In parallel, there has been a growth of people using personal computers. This began with keeping simple records, but personal computers are now increasingly connected into networks, either public networks such as the Internet, or to corporate networks. Increasingly, people are also being connected into **workgroups** within corporate networks to solve organization-wide problems. Corporate systems themselves are being networked, connecting people into the corporate networks. In addition, the corporate networks are now beginning to open up to the external world, leading to what is now commonly called electronic commerce or e-business.

You may then ask whether such change is likely to continue. The answer is yes, and even more rapidly. A recent paper by Truex (1999) describes many of the reasons. Globalization is one driving factor for such change, leading to the Internet's being used for collaboration across distance. Information technology will be increasingly used in everyday work practices such as sending messages and documents between people, who will increasingly work at a distance from each other. Furthermore, networks are now going beyond the organization and increasingly support electronic commerce by allowing external customers to place orders and get services from an organization. They are also increasingly used to support business alliances and trade. This wide range of applications poses many challenges to designers looking at how information systems can improve business operations. System analysts and designers will have to be able to analyze these ever more complex systems and provide innovative solutions within them. Many writers suggest that virtual organizations are around the corner. Anyone planning to enter this field must be prepared to continue to learn and adapt to new working environments.

Workgroup A group of people working to a common goal.

SYSTEM PERSPECTIVES

Different designers approach systems design from different perspectives, depending on what they see as the purpose of the system. One question that people often ask when discussing system analysis and design is: what is a system? What is it that we are generally analyzing and designing? There are a number of ways of answering this question, depending on the general approach that is being used in design. Some see them from the more abstract perspective, others as simply computer systems, still others as a combination of people, process, and technology.

One view of the system is shown in Figure 1.4. It simply shows the system as a set of connected components. The system in Figure 1.4 is made up of three components: a computer, a file, and a workstation that provides an interface to the user. Figure 1.4 also shows the connections between these three components.

The view of the system shown in Figure 1.4 is too constraining for what we are doing in this book. It simply emphasizes information technology and not how it is used. Our aim is to study systems that use computers to solve problems for users. We will not use the view in Figure 1.4 in this book, because we are considering systems with a wider scope than simply computers. We will study systems that include computers to solve some wider problem.

Another way to view systems is shown in Figure 1.5. This is a more theoretical or abstract view that concerns the fundamental systems issues.

Theoretical approaches to systems have introduced many generalized principles. *Goal setting* is one such principle. It defines exactly what the system is supposed to do. Then there are principles concerned with system structure and behavior. One such principle is the **system boundary**. This defines the components that make up the system. Anything outside the system boundary is known as the **system environment**. A system can be made up of any number of **subsystems**.

System boundary The set of system components that can be changed during system design.

System environment Things outside the system study that can affect system behavior.

Subsystem A part of a system.

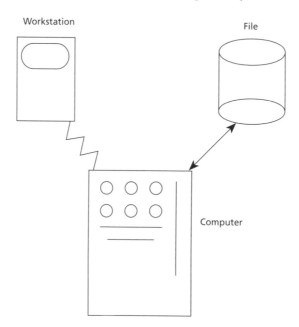

Figure 1.4 *The computer as a system*

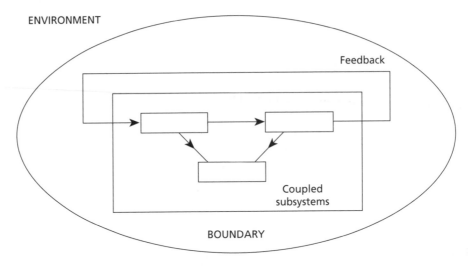

Figure 1.5 *A theoretical view of systems*

Each subsystem carries out part of the system function. Subsystems are important because they can help handle systems complexity and thus improve the understanding of a system. Each subsystem carries out some part of the system goal. The subsystems communicate by passing messages between themselves. A good system will be made up of highly independent subsystems with minimal flows between them. Minimizing flows in turn minimizes complexity and simplifies the system. It also makes it easier to change a part of a system.

A number of principles are also concerned with system behavior. One such important principle is **feedback**. Feedback is the idea of monitoring the current system output and comparing it with the system goal. Any variations from the goal are then fed back into the system and used to adjust it to ensure that it meets its goal. To do this it is necessary to **monitor** the system to see if it is meeting its goal.

We will not emphasize the theoretical approach in the book but will nevertheless use many principles from it. For example, the idea of subsystems is very important. Any information system contains many subsystems. Some may use computers, others may be manual processes. Goals are also important in information systems. Each such system has a purpose and must work toward that purpose. Feedback also arises in information systems. It provides the outputs that are used to monitor system performance and possibly change some activities in the system.

In this book we place greater emphasis on the components that make up a computer-based information system, but at the same time we use the general ideas about systems. Our approach, however, goes beyond the simple system shown in Figure 1.4 and includes people as well as their work processes.

Thus the general view of the system used in this book is shown in Figure 1.6. Figure 1.6 views systems as collections of *people* using *information technology* and *processes* that define how people carry out their work. In addition, there are the more **informal interactions** that take place in any organization and are now increasingly supported by computers through systems such as electronic mail.

Feedback Using variations from a system goal to change system behavior.

Monitoring a system Checks made to see if a system is meeting its goal.

Informal interaction Working together without a set of prearranged rules.

budget reports, personnel movements, and so on. The format of such reports is often required to change because of changing management needs. Analysts must ensure that any systems that are built can accommodate such changing needs.

DATA WAREHOUSING AND DATA MINING

Many large organizations keep a central repository of records in their centralized system. For example, banks, insurance companies, and airlines need to maintain records of their customers. A popular term used for this kind of system is **data warehousing**. Here, all companywide information is stored on one large database, which is made available to users throughout the whole organization. At the same time, the data stored in a large database contains a large amount of undiscovered information. This information may be in the form of patterns of events that on their own have little significance but when combined yield interesting insights. For example, it may be possible to find out about the correlation of purchasing habits and economic factors. Similarly, emerging patterns of insurance claims may be detected, with obvious impact on premiums. This search for such information is now commonly known as **data mining**. Designers here are required to choose the best structure for the data so that it can be used across many applications. Security of data is also high on the design agenda.

Designers here are often called upon to design the **expert systems** needed to carry out these searches.

Data warehousing The storage of large volumes of data for organizational use.

Data mining Looking for patterns in databases.

Expert system A system for helping people make decisions.

DISTRIBUTED COMPUTER SYSTEMS

Typically, however, computer systems are now becoming more distributed, and most organizations have more than one computer system connected together to form the **computer network**. In most organizations, the computer networks now support virtually all the organization's functions, including their financial data, production outputs and plans, current inventory, sales, and personnel records. Furthermore, as workstations become more and more powerful, it is becoming more common to also carry out a significant amount of computation on the workstation itself. Such distribution is supported by client–server systems.

Computer network A set of computers connected by communication lines.

CLIENT–SERVER SYSTEMS

In a **client–server** network, the workstation is known as the client, and the computer system is known as the server. The server stores the data commonly used by its connected clients, as well as common programs for the users. When some computation is needed, both the data and program can be sent from the server to the client workstation, the computation is completed on the workstation, and then the data is returned to the server.

A network can be made up of many servers, and each server can support a different database. One such system is shown in Figure 1.10. Here, server 2 supports the financial database and server 3 the inventory database. Server 1 provides some local services to the workstations—for example, printer services.

People in the network can access any of the servers through their workstations. Links between the servers allow a workstation connected to one server to access data in other servers and reports to be generated using data from all servers. In

Client–server process A process that describes how a server provides a service to a client.

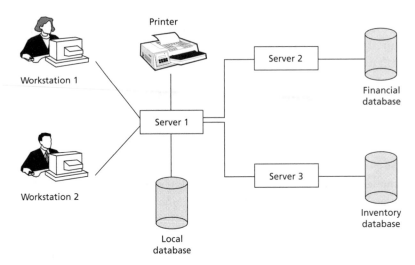

Figure 1.10 *A computer network*

some cases, a server can support a workgroup, whereas connections between servers support connections between workgroups.

Computer networks introduce their own design issues. One, for example, is how to distribute the database. Another is how to make the distribution **transparent** to a user. All the user has to do is request an action in the system and obtain a response but not be concerned with which machine carries out the action, or to access data as if it were on one machine, with the system itself taking care of the distribution.

CONNECTING CORPORATE SYSTEMS

Many organizations, however, have computers built by different manufacturers with programs written with different software. Such systems were often developed in different parts of the organization. Special care is needed to connect systems into networks. Such networks are known as **heterogeneous** networks, compared with **homogeneous** networks, where all computer systems use the same equipment and software. A problem arises when organizations wish to develop business processes that can span a number of different systems and must thus construct heterogeneous networks. They must then decide whether to totally redesign their systems using homogeneous software and hardware or to come up with structures that support the connection of heterogeneous systems.

An important aspect of using heterogeneous systems is how to gain access to data in these systems. There are two options in **integrating** databases in more than one system. One is to integrate them by placing them under one controller. The other is to have a **federated** system, where databases in different systems are individually managed but can be accessed in a unified way to solve a particular problem. Figure 1.11 shows one way of federating databases.

The nodes in a heterogeneous network, shown in Figure 1.11, are accessed through common software that can exchange messages between heterogeneous systems. Application programs that use data from more than one node must access the network nodes through this software. Such integration is becoming more

Transparent access Accessing data from a network independently of its location in the network.

Heterogeneous network A computer network made up of different computers and software.

Homogeneous network A computer network made up of the same computers and network.

Integrated databases A set of databases managed by a single controlling system.

Federated database system A set of databases managed independently but accessible in a unified way.

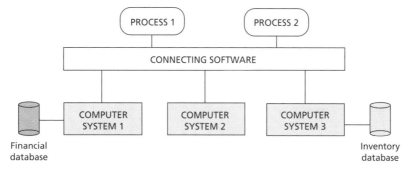

Figure 1.11 *Connecting heterogeneous systems*

important as more and more new business procedures are required to use information from more than one existing system. A large amount of the designer's time here is spent on defining standards so that data can be shared between systems.

WORKGROUP SYSTEMS

There is an increasing tendency for people to work in groups in organizations for specific and often transient tasks. People are increasingly connected into workgroups and supported by an **office information system**. Such subsystems can help people in the same office share documents and other information, and they can support distributed groups. People in workgroups share documents, change these documents, and access common background information. Their purpose may be to make a major decision, provide a report to clients, or evaluate a proposal. Workgroups are supported by software often known as **groupware**, which is described in the next chapter. Combinations of groupware can be used to support the activities commonly found in offices. Systems that support such workgroups are sometimes known as **group support systems**, especially in cases where they support groups to make a decision.

Designers of workgroup systems select the groupware needed by the workgroup and install it on workstations used by the workgroup members. They must also provide ways to store the large variety of information needed by the workgroup.

Office information system A system that supports office operation.

Groupware Software that assists workgroups.

Group support system A system that supports decision making by groups.

NETWORKING

Corporate networks are usually restricted to processing transactions and recording them against the corporate database. However, now networking is beginning to take on a new meaning. It is being extended to include the networking of people, driven mainly by the increasing use of the Internet.

THE INTERNET

Networking is now going beyond the organization. The Internet has now become almost a household name, with millions of people connected to its services. Each network user has their own unique electronic address, and all messages with that address are directed to that user. The user can send electronic mail and read a variety of news groups on specialist topics. The danger with e-mail is that people may suddenly find themselves flooded with messages, in much the same way as many letterboxes are flooded with advertising brochures. There is also a limit to what can be done with e-mail. Most networks support information exchange that goes beyond simple electronic messages. This can include:

* attaching files to messages
* using a filter, where a receiver can specify the characteristics of received messages
* broadcasting a message to a whole group or collecting information from a group
* setting up bulletin boards and news services; here members direct all their messages to a bulletin board where they are posted, and other members can look up the bulletin board at their convenience
* setting up file libraries using FTP (File Transfer Protocol) sites
* using Internet search engines to find information based on specified keywords

Many of these services are presented to users by software that makes the facilities easy to use. For example, Figure 1.12 shows how a product known as Eudora supports electronic mail.

Most of the early electronic mail systems were restricted to textual information. Since the early 1990s, the Worldwide Web has become part of the Internet and can be used to store and distribute information in any medium.

THE WORLDWIDE WEB

The Worldwide Web (WWW) came into use in the early 1990s and became quickly accepted, because it could be used to store multimedia data and had means to easily search through this data. It also has a natural presentation to users, mixing images

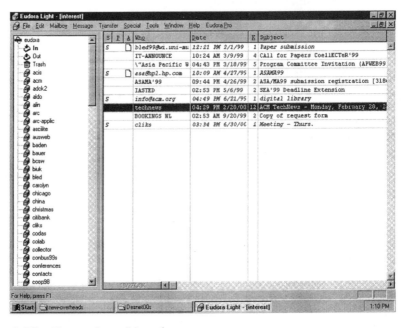

Figure 1.12 *Electronic mail interface*

and text on the one screen. Information on the WWW is stored as pages, with each page having a unique address, known as the URL (uniform resource locator) address. Usually a site has a large number of interconnected pages. The site has a home address, which is the starting point for any information search. The home page then provides links to other pages at the site. Such systems are now increasingly used by businesses and individuals to establish a presence on the WWW and to engage in electronic commerce.

ELECTRONIC COMMERCE

Electronic commerce generally means doing business across the Internet, especially the WWW. Electronic commerce is now developing in many directions. Most organizations take up electronic commerce in an evolutionary way. A typical scenario is where organizations:

- establish an electronic profile through which an enterprise can advertise its presence on the Web
- extend their Web site to trade with their consumers, most often allowing them to place orders using the WWW
- engage in business-to-business operations
- evolve to virtual organizations.

Electronic commerce
Conducting business between organizations and consumers using the Internet.

ESTABLISHING AN ELECTRONIC PRESENCE

Any organization that participates in electronic commerce must maintain a WWW site. Anyone who wishes to find out about the organization and its services must know the URL of the organization's home page. Once they get to the address they can follow links to other pages that provide various services.

Organizations can provide a variety of information services and information across the WWW. The most common pages on most sites include:

- a brief description of the organization's mission
- news items about recent events and successes in the organization
- frequently asked questions about the organization
- descriptions of the organization's products and services
- contact points and addresses
- in an increasing number of cases, the ability to place orders directly with the organization.

Perhaps the most common early use of the WWW is for organizations to publicize their activities. A home page in this case includes a brief description of the organization's mission, usually in the form of products and services that it provides. Some sites now provide information such as product catalogs and help facilities for customers having problems with the organization's products. Designers must design these pages and organize them into linked structures.

A typical WWW home page is shown in Figure 1.13, the home page for the Australian airline Qantas. The site provides information on Qantas's services to any Internet user. At the top of the page you will find links to its schedules as well as arrival and departure information. On the right are links to a page on holiday specials, a news room that contains news releases about recent activities, interesting

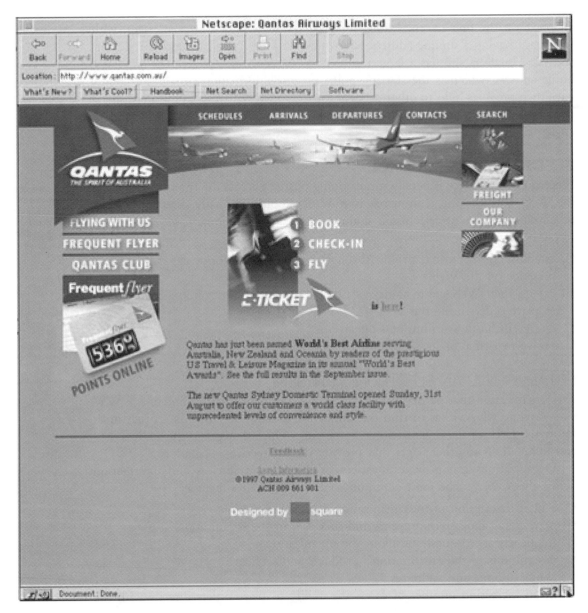

Figure 1.13 *The home page for Qantas*

articles, and even the entertainment guide for the current month. There is also information about inflight shopping and freight.

This is typical of many other sites that provide information about their products and contact points.

CONSUMER TO BUSINESS TRADE

There is also an increase in the number of sites that allow direct ordering of services, although these have to take special precautions to ensure the security of financial transactions. Such sites also increasingly support what are commonly known as client loyalty services. Thus, on the left of the Qantas home page you will find a link for frequent fliers, how to make reservations, and any specials currently on offer. There are also airline schedules for future dates that give the best route for a given start and end location, and there is arrival and departure information for the current day.

In fact, this leads to the possibility of corporate systems interacting with their customers in a variety of ways and engaging in what is commonly known as electronic commerce. Thus, sites allow customers to place orders for the products or services. Many bank sites allow customers to operate on their accounts.

Management of such financial transactions introduces a number of problems, especially in ensuring the security and privacy of transactions and in preventing any fraudulent activities.

BUSINESS-TO-BUSINESS TRADE

The next generation of electronic commerce systems will place more emphasis on **business-to-business networking**. Business-to-business commerce has many possibilities. One is change to supply chains. Many businesses use parts produced by another business in their production process. Other businesses distribute goods to customers. Such flow of goods can be improved by tracking them through business-to-business networks and matching them to the work processes within the businesses. Advantages here include reduced inventory levels, wider distribution networks, and reduced delivery times. Ravi Kalakota and Marcia Robinson describe some of these processes in more detail in their book (see Bibliography).

Business-to-business network Conducting business across the Internet.

Other possibilities are joint ventures and establishment or trading hubs. Joint ventures require networked applications where information is readily exchanged between businesses. Trading hubs allow suppliers to register their services while buyers register their needs. The hub then provides tools to match buyer needs with supplier offers.

An extension of trading hubs is virtual organizations. Once businesses develop trust in the trading hubs, they may not need to maintain links with the same organizations but establish more transient relationships formed temporarily to meet some specialized need. Such enterprises are known as virtual organizations. These are collections of individuals or businesses set up temporarily to meet some need with a network.

THE CORPORATE INTRANET—BRINGING CORPORATE AND PERSONAL SYSTEMS TOGETHER

Just as the Internet connects people across organizations, it is also possible to use the same technology to connect people within organizations. Initially, networks within organizations were known as local area networks, where they usually supported one organizational unit. Now increasingly, **intranets** are being set up to support communication across the whole organization. Intranets use the same technology as the Internet, but access is restricted to people within the organization. Perhaps the initial impact of intranets is to give people access to organizational working documents. Their more important contribution is to make it easier for people to communicate and work together in organizations, especially when these

Intranet A network supporting information exchange within an organization.

are globally distributed. Ways of doing this are described in detail in Chapter 2.

Intranets are becoming almost a must for **knowledge-intensive organizations**. These are organizations that deal predominantly with knowledge. They include consultancy companies, financial advisers, and similar organizations. They require easy access to both stored information and to people who possess any expert knowledge needed to meet a client's need. They must then provide services that enable the users to collaboratively develop ideas or solve client problems.

Knowledge-intensive organization An organization that primarily produces knowledge.

EXTENDING TO THE CLIENT

Most current developments are to extend corporate intranets to customers or to support alliances between organizations. Clients, for example, can be given limited access to the database to monitor their personal accounts. Such networks are now often called **extranets**.

Extranet A network supporting exchanges between clients and an organization.

Privacy Ensuring that information remains accessible to one or a selected number of people.

Security Ensuring that computer system faults do not destroy the information.

ISSUES IN NETWORK DESIGN

Networking introduces a number of other issues. One important issue is **privacy** or **security**. Security must be provided to prevent information from being lost or destroyed. Privacy is needed to ensure that users connected to such systems do not either deliberately or inadvertently access or, worse, change information that does not belong to them. Thus, a system like our interactive marketing system must be built in such a way that producers cannot gain orders placed with other producers and consumers cannot gain access to orders placed by other consumers.

TEXT CASE A:
Interactive Marketing

SITUATION

A cooperative organization for growers of perishable goods has been formed to help the producers sell their products. The producers are all located close to one another. The cooperative's objectives are to advise producers on the products they should sell and to help them sell their products. The cooperative includes a large number of producers of perishable goods and a large number of consumers or sales outlets. Each consumer usually buys a small proportion of the goods. For this reason, the producers, or the cooperative on their behalf, must advertise available products widely as soon as they are available, almost on a daily basis. A distribution network is also needed to get the goods to the consumers as quickly as possible to prevent loss through wastage. The problem is further complicated by the fact that there are a range of goods and a large number of consumers each with a different preference. For example, the goods may be different kinds of fish, such as tuna, shark, and so on. Each kind of fish or catch can be of a given size or quality, and catches can vary over time. Consumers may be restaurants, supermarkets, or small shops. Each may have specialized needs for some of the products.

The cooperative has proposed that an interactive marketing system, or trading hub, be developed to connect the producers and consumers. The proposed system would look like that shown in Figure 1.14.

The proposed system would support a number of activities that are needed to get the product to the consumer. The activities are shown in Figure 1.14. First,

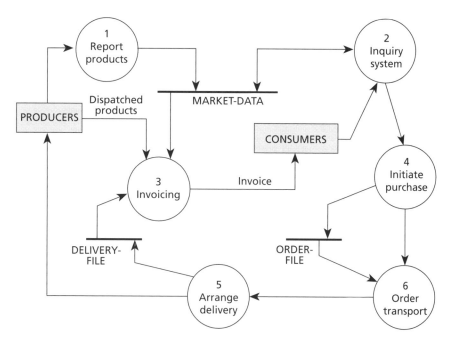

Figure 1.14 *An interactive marketing system*

consumers must be informed about the availability of the product. To do this, the system would allow producers to record their available products and look up these products in an inquiry system. These functions are illustrated by functions 1 and 2 in Figure 1.14. The information about products is stored in a file, named MARKET-DATA. Then subsystems must be provided to allow product orders to be placed and deliveries of the products to be arranged. Finally, an invoicing subsystem is needed to arrange payments.

To ensure that deliveries are made promptly, the proposed system also allows consumers to place orders directly on the system and arranges deliveries of these orders. It is proposed to add automatic billing later to the system. The proposed system includes an ORDER-FILE to store all orders placed in the system and a DELIVERY-FILE to keep a record of deliveries. A study is now needed to determine whether it is feasible to use computers to provide support for such a system.

ISSUES

One important issue identified here is how to educate a large number of users, most of whom have little contact with computers, to use computer systems in their everyday work. Another issue is who will maintain the proposed system, which will be distributed in a manner very similar to that shown in Figure 1.14. Here the producers and consumers have a terminal connected to a central computer with access to data on the computer database, but there is an option of actually distributing the processing across a number of computers. There has also been a suggestion that the computer store and display pictures of the products offered for sale.

There are also some shortfalls in the system described so far. It does not support direct contact between producers and consumers, but allows them only to access data on the same database. Thus, an important human element is removed, preventing activities such as negotiating a price. It is also difficult to keep track of the progress of orders and answer questions on expected delivery times. We will deal with some of these issues in Chapter 2.

WHO IS INVOLVED IN BUILDING INFORMATION SYSTEMS?

Programmer
A person who writes computer programs.

System analyst
A person who analyzes the way the system works and its problems.

Business analyst
A person analyzing a business at the subject level.

Network manager
A person reponsible for a computer network.

Network programmer
A person developing programs to support networking.

Banking business analyst A person analyzing banking systems at the subject level.

Web designer A person developing Web sites.

User requirements What users expect the system to do for them.

Many skills are needed to build today's information system. Consequently, a large number of different people are involved in building a system. The kinds of skills needed often depend on the kinds of systems being built. If you look at Figure 1.3, the traditional skills have been those needed to build transaction processing systems and integrate business systems. The kinds of roles needed to build these systems were **programmers**, **systems analysts**, **business analysts**, and **network managers**. As systems evolve, people with many additional skills and new specialist roles are becoming evident. These include different kinds of programmers, for example **network programmers**, to build and maintain the corporate and public networks. Analysts with specialized skills, such as **banking business analysts**, are needed to build the increasingly specialized systems required. Electronic commerce is now calling for additional people such as **Web designers**, who are involved in building today's electronic commerce systems. It is also increasingly found that people take a mix of these roles, especially in smaller projects or organizations. In that case, people must develop a mix of skills needed for the type of application they are building.

Perhaps the most important people in any system are its users. *Users* are people who have a stake in the information system because they need the system for their work within the organization. Such users are called system users or end users. The term 'end user' is often reserved for users who not only use the system but also develop some of the computer parts of the system. Users play a very important role in systems analysis. They define what new systems must do and specify the **user requirements** for the new system in the ways described in Chapter 4.

The organization's *management* is also involved in system development. Managers clearly define project business goals, estimate the resources needed, and set the completion times. They then monitor project progress to see if these goals are being achieved. Many systems analysts find management reporting demands a nuisance. However, it must be remembered that building an information system is only one of the activities of the organization. As an alternative, the organization could use funds expended on a new system to expand its sales force instead. Thus, management must at all times be aware of what is going on to ensure that it uses the organization's resources in the best possible way. It is important to remember that management may also be making decisions at a number of levels. It may, for example, be deciding whether to open a new store and thus increase sales outlets or to install a new computer to increase efficiency. New computer developments in that context will have to be justified in terms of overall organizational strategy rather than by simply choosing a computer system for a particular activity.

ORGANIZING THE PEOPLE

In most organizations, people are allocated to departments. These departments carry out the routine work of the organization, such as hiring people, building the organization's products, and selling them. Most organizations also have a department known as the Information Systems (IS) department. In the early days of computing, all computer systems were developed by the people in the IS department. The IS department was responsible for all computing equipment and was the only department with people with the skills needed to build computer-based systems. The IS department was important because it controlled scarce resources, namely computers and the people who develop systems for them and decided what systems to build. The IS department liaised with user departments to determine their computing needs. It also liased with management to set priorities for developing systems and acquiring the equipment needed to develop new systems that met user needs.

Many organizations have now distributed some of the responsibility for developing computer-based information systems to operating departments or specialized units, such as those responsible for developing electronic commerce interfaces with users. The IS department is now often responsible for developing systems that support the data warehouses. It is also responsible for building the core business processes that are used throughout the organization. Operating departments develop systems specific to their local tasks. They often set standards for software and hardware and manage the increasingly complex computer network. They also provide specialized consultancy services and advice to end users.

In addition, many organizations have adopted the policy of **outsourcing**. This means that they create a separate organization out of their IS department. The IS department must run as an independent entity, providing service to the operating department in competition with other possible sources of this service.

Outsourcing
Arranging for computer processing to be done outside the organization.

WHAT IS INVOLVED IN SYSTEMS ANALYSIS?

Systems analysis is an important activity that takes place when new information systems are being built or existing ones are changed. It is necessary in any development process, irrespective of whether it is a core process managed by the IS department, a system being built for an operating unit, or an electronic commerce application. The most crucial role of systems analysis is to define user requirements. But why are special skills, such as systems analysis, needed to build good information systems? Why don't these things happen as a matter of course in an organization?

Systems analysis, now often called business systems analysis to emphasize its importance in business, is needed in the first instance to clearly identify the business goals of the new system and to specify how it will be built and how it will work. This includes gathering the necessary data and developing models and plans for new systems. This is not an easy task, because in large systems many people need to be satisfied and many conflicts must be resolved. In that context, a systems analyst must play many roles. Primarily, systems analysts help people solve their business problems by defining what new systems can do to improve ways of doing business. In that role, they must understand the problems and suggest solutions and ways of implementing them. Often they must help resolve conflicts, as different people in the organization may have different needs, and the analyst has to justify the solutions to different classes of users. Often such solutions must be justified in terms of the whole organization, and thus systems analysts are expected to see their work as being relevant to the whole organization. It is not always an easy job, as the analyst is often an agent of change where some people may not want change.

Thus, as a rule, systems analysis is difficult but rewarding work. There are many constraints imposed on the analyst and many people to satisfy. But there is the reward of successfully implementing a new system. A systems analyst must spend a lot of time talking to users and finding out how they use the system, any problems they may have with the system, and what they expect from it. As a result, such work calls for good *analytic and communication skills*. Systems analysts must learn about the system and understand users' expectations of the system. They should be receptive to new ideas but still have concern for users of the existing system and their needs. It is the systems analyst's job to satisfy the requirements of all these users within the constraints imposed by management.

The systems analyst must be able to find out the details about the way the business system works. To do this, the analyst will have to look at such things as forms used in the system, data used by the organization's personnel, contents of computer files, and computer outputs and inputs.

Systems analysts must also be able to work in environments that have considerable ambiguity and uncertainty. In such environments there are often conflicting accounts of what is happening and what is needed, and different users perceive different system problems. The systems analyst must be able to resolve these kinds of conflicts and produce an agreed-upon statement of the system operation and problems.

Systems analysis is a *creative* and *imaginative* activity that produces new solutions to meet user requirements. Systems analysts must have both the knowledge of techniques and devices that exist outside the organization and a good understanding of what is needed inside. They must then put all the pieces together in a creative way to satisfy these requirements. This of course may require more than one try. One usually starts with more than one possible solution. Some of these solutions may be totally unsuitable, whereas others may be acceptable after some change. Such solutions may then be amended and then justified again. Often this is done in an interactive way, and at all times systems analysts must be responsive and creative, using all their knowledge to work toward some effective and acceptable solution.

Systems analysis has become more difficult over the last decade or so. Early systems analysts worked on relatively simple systems, which did only one thing or had only one function. Now the systems tend to be more complex, with many of the systems using data from a large number of sources and a larger variety of equipment than was available with early systems. Consequently, systems analysts need better tools to help them in their analytical work. A major objective of this book is to describe such tools and how analysts would use them to analyze systems.

WHO DOES THE ANALYST INTERACT WITH?

There are now many specialized tasks in building information systems, and analysts are increasingly required to interact with the people who carry out these tasks. Users and management are perhaps the most important. But there are also a number of specialists now in the development area. There are the operating system people, who understand networks, and database designers. In the world of electronic commerce there are Worldwide Web site designers, programmers, and network managers, each with their specialized knowledge and tasks.

It is necessary for the analyst to know what these people do and the terminology they use to be able to discuss problems and possible solutions with them. It is also important for the analyst to know the basic technology they work with and how they contribute to a project. Knowing the potential and limitations of technology will enable the analyst to make realistic proposals.

SUMMARY

This chapter defines what is meant by computer-based information systems and by systems analysis. It then describes what is involved in designing such systems. It describes trends toward integrating processing across a number of distributed sites. Computer-based information systems are made up of people, procedures, and equipment. Systems analysts design computer-based information systems that help the organization's personnel in their work.

We saw that a wide variety of people are involved in a system, and that systems analysts must work with all the people involved. Hence, analysts are required to have good communication skills. It is also necessary to provide systems analysts with good tools to help them in their work.

The chapter also describes the kinds of systems that are now emerging in organizations. It covers both corporate systems and the emerging electronic commerce applications. Analysts will be increasingly required to build these kinds of applications.

DISCUSSION QUESTIONS

1.1   What are the main processes in system development?

1.2   Why is systems analysis necessary?

1.3   What qualities should a systems analyst have? How would you acquire these qualities?

1.4   What is the role of users in system development?

1.5   How does the IS department get involved in systems analysis and design?

1.6   What conflicts are likely to occur in systems analysis and design?

1.7   What is meant by the outsourcing of the IS department?

1.8   What are some important system principles?

1.9   What are the characteristics of personal computer systems?

1.10   How has distribution changed the way systems work?

1.11   What are client-server systems?

1.12   What do you understand by *data warehousing*?

1.13   What are workgroup systems?

1.14   Why are security and privacy important on distributed systems?

1.15   Why is it important to be able to integrate existing systems?

1.16   What do you understand by a *knowledge-intensive organization*?

1.17   Describe the difference between intranets and the Internet.

1.18   What do you understand by the term *virtual organization*?

1.19   What is an extranet?

EXERCISES

1.1   Consider the interactive marketing system. Can you propose some alternative structures for such systems? For example, what would the structure look like if:

   (a)   the producers were distributed across the country rather than within close proximity to each other?

   (b)   some of the functions, for example ordering and delivery, were carried out on different computers?

1.2   Is negotiation between producers and suppliers important? What facilities would have to be included in the system to support negotiation?

BIBLIOGRAPHY

Alter, S. (1999), *Information Systems: A Management Perspective* (3rd edn), Addison-Wesley, Reading, Massachusetts.

Duchessi, P. and Chengalur-Smith, I. (May 1998), 'Client/server benefits, problems, best practices', *Communications of the ACM*, Vol. 41, No. 5, pp. 87–94.

Elmagarmid, A.K. and Pu, C. (eds) (September 1990), *ACM Computing Surveys*, Special Issue on Heterogeneous Databases, Vol. 22, No. 3.

Ferrat, T.W. and Fogel, L. (1998), 'Toward developing global IS specialists', *Proceedings of the Conference on Personnel Research*, New Orleans, April 1998, pp. 96–103.

Heimberger, D. and McLeod, D. (July 1985), 'A federated architecture for information management' *ACM Transactions on Office Information Systems*, Vol. 3, No. 3, pp. 253–78.

Jiang, J.J., Klein, G. and Means, T. (January 1999), 'The missing link between systems analysts, actions and skills', *Information Systems Journal*, Vol. 9, No. 1, pp. 21–34.

Kalakota, R. and Robinson, M. (1999), '*e-Business: Roadmap to Success*', Addison-Wesley, Reading, Massachusetts.

Mueller, M. (ed) (June 1999), 'Emergent Internet Infrastructures', ACM Special Issue.

Mukanata, T. (November 1999), 'Knowledge Discovery', *Special Issue of the Communications of the ACM*.

Parker, C.S. (1992), *Understanding Computers and Information Processing: Today and Tomorrow* (4th edn), The Dryden Press, A Harcourt Brace Jovanovich College Publisher.

Pyle, R. (June 1996), 'Electronic commerce and the Internet', *Communications of the ACM, Special Issue on Electronic Commerce*.

Schultheis, R. and Sumner, M., *Management Information Systems: The Manager's View* (2nd edn), Irwin, Homewood, Illinois.

Truex, D.P., Baskerville, R. and Klein, H. (August 1999), 'Growing Systems in Emergent Organizations', *Communications of the ACM*, Vol. 42, No. 8, pp. 117–23.

WORK PRACTICES AND WORKGROUP SYSTEMS

CONTENTS

KEY LEARNING OBJECTIVES

The importance of analyzing work practices
The kinds of work processes in organizations
How people communicate in organizations
The difference between working as individuals and working in groups
The different kinds of group activities
Ways to support group activities

CHAPTER 2

problems. Such coordination through the management structure is necessary to ensure that management is informed of any changes in resource requirements and that all levels of management are aware of changes at any point of the operational structure. This kind of structure has always assumed a relatively stable environment, simply because a change to such hierarchical structures often requires major organizational changes.

TOWARD FLATTER STRUCTURES

Many organizations are now moving to environments where customer preferences and economic factors are continually changing. As a consequence, organizations must continually change to meet rapidly changing demands in these more volatile environments. In addition, products often require inputs of many skills and must often be customized to particular customer needs. This requires organizations that can quickly bring together people who have such skills and that can make quick changes at the operational level by rearranging both resources and the tasks that people do. We are thus looking for more adaptive structures that make it easier to bring people from different parts of the organization together and to rearrange their activities as customer needs change. Such rearrangement is often difficult if it is to proceed through a number of hierarchical organization levels, each with its own priorities. As a result, there has been a tendency to reduce the number of levels and encourage change by supporting coordination at the operational levels. Computers also make fewer levels possible because they make it easier to reduce information from objective levels to operations levels. Trends to flatter organizations also result in changes to work practices. In hierarchical organizations, each individual is concerned with their individual task, be it inventory control or financial management. In flatter structures, each individual must also have some understanding of the functions carried out by other parts of the organization.

Workgroup A group of people working to a common goal.

Empowerment Giving people additional authority within an organization.

The result is what is sometimes known as the formation of **workgroups** concerned with specific and often limited tasks. Such workgroups are usually **empowered** to make decisions on the use of resources without reference to management, whose main goal in this kind of organization is to provide support to the groups rather than directing them. Workgroups are focused, with specialists from various functions contributing to the task. Close collaboration in workgroups requires individuals to have a wider knowledge of the organization.

Workgroups are also becoming more important in knowledge-intensive organizations. Here people from many organizational units are brought together to solve some problem or develop new products or services. Such a task team can be made up of people from a number of organizational units. Figure 2.2, for example, shows a team of people from the personnel, production, marketing and accounting departments. Such workgroups do things like jointly preparing a document, such as the marketing plan in Figure 2.2. Another workgroup, which may contain some members from the first workgroup, may develop a production plan using the marketing plan as input. Each workgroup may require people with different skills to match the objectives of the workgroup.

Thus, in computer system design there may be a group of analysts and a group of designers. They will meet quite often for discussion to make decisions on their

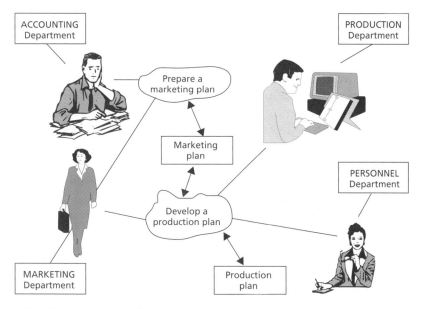

Figure 2.2 *Flat task-oriented structures*

next activity and on what is to be done about designing the system. Then there are various engineering designs, where one may, for example, be designing a building or a bridge.

This view of organizations has been elaborated by Drucker (1988) in what is now a famous paper that suggests that flatter structures are needed because organizations are becoming information rich. This implies that most information resides at the operations level of the enterprise, and work must be organized to make best use of this information. Flatter structures thus enable information that exists at these levels to be coordinated, which leads to better decisions. The flattening of organizational structures changes the way people work. It adds responsibilities at the operational levels, placing more emphasis on control through coordination at operational levels rather than by direction from the management levels. How individuals work and how they interact with each other is also different. One individual may now interact with individuals in many other groups. In hierarchical structures, interaction usually takes place only between individuals at adjacent hierarchical levels. Management in the flatter structures facilitates rather than directs activities.

THE IMPORTANCE OF PROCESS

Considerable emphasis is now placed on processes in organizations. Process fundamentally describes the way we do things and defines the steps we follow. It is important that processes work well. If we do things in a sloppy and inaccurate manner, then we will not have a good outcome. The term quality is often used to describe how we want our processes to work. This word does not have a precise definition. It can mean no errors in the system, serving a customer in a minimum time, or reducing our routine work. It is often up to the organization to define what it means by quality and to set up processes to achieve it. Many people are now looking at the processes in their business and rearranging them to meet quality requirements. The term **business process re-engineering** is often used in this context. It means rearranging the way we do our business. Such rearrangement is not centered on increasing the use of computers. It simply asks how we want things done and then seeing whether computers will support us in doing these things better.

Business process re-engineering
Changing an existing business process.

Processes can be classified in a number of ways. At one end of the spectrum are those processes that can be completely prespecified. At the other are those that arise spontaneously because of some new and unexpected event. Then there are also adaptive processes, where a prespecified process may need to follow an unexpected path because of some unanticipated output.

SUPPORTING PLANNED WORK

Prespecified or structured processes are usually defined as workflows made up of a set of steps. The process is predefined as a sequence of steps, and one person usually carries out each step. For example, Figure 2.3 shows a process used to buy a part. This process starts with a request for a part. The request is approved, purchases are arranged, and a delivery is made. There will be one or more, often different, persons involved at each step. A further important property of interaction here is that each person may be at a different location. Furthermore, each person may carry out his or her tasks at different times. Thus, a person enters a request at some particular time, whereas the approver can look at this request much later.

One way is for persons to simply pass messages between themselves. Thus a requester will pass the request either as a form or by word of mouth to the approver. Another way is for coordination to be achieved through shared files. Each action is recorded by a person on a file. Such records are then used to inform others to carry out some action. Figure 2.3 shows two files. One is a file of requests, the

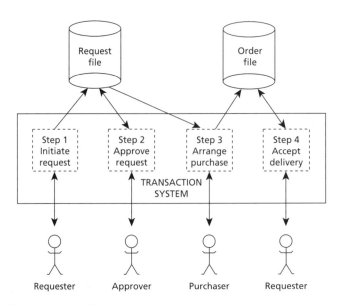

Figure 2.3 *Structured workflows*

REQUEST-FILE, and the other a file of orders, the ORDER-FILE. A request made by a requester is recorded on a file and activates an approver. Once the request is approved (step 2), it goes to the purchaser, who creates an order. The order is placed on the order file and is used to initiate a purchase (step 3). The next step (step 4) is for the requester to accept a delivery when the parts are ready. Each step has some particular task, and a different person may carry out each task. Each task may read and update one or more files.

Systems like that shown in Figure 2.3 are often called **workflow processes**. The term 'workflow' is used because there are a defined flow of information and defined actions to be taken at different points of this flow. Computers can easily support this kind of process. To do this we often use **transaction processing systems**.

TRANSACTION PROCESSING SYSTEMS

The term transaction is used here because it often implies an interaction with the database. Such interactions occur continually in a workflow, as people have to access, change, and store data at defined workflow stages. Thus, each step in Figure 2.3 makes a change to the database.

Most computer systems provide software for processing transactions against a database. These often include a standard computer procedure for making a transaction. One example of this kind of procedure is shown in Figure 2.4. The transaction system first checks the transaction to ensure that no erroneous data is input into the computer. The transaction must then pass through a number of checks. The first check is usually called an edit, which ensures that all the needed data is included in the transaction in the correct format. Thus, for example, we may check to see if an account number has been entered in a bank withdrawal transaction and if this is in the correct format. This means, for example, that numeric data appears in numeric fields and alphanumeric data appears in

Workflow process A process made up of a predefined set of steps.

Transaction processing system A computer system that manages transactions.

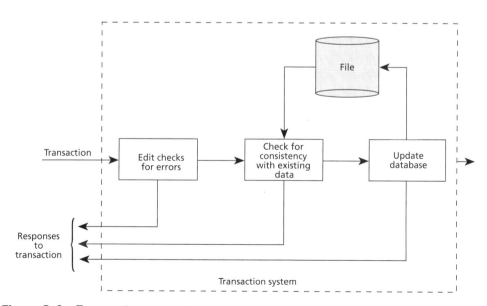

Figure 2.4 *Transaction processing*

TEXT CASE B:
Managing an Agency

SITUATION

The agency manager is required to prepare a feasibility study on building a road between two locations. This requires some preliminary designs to be completed, financial estimates to be made, and consideration of external policy issues such as the environment and the road position. The project must therefore consider both the strategic and social issues centered on building roads and the more operational and managerial tasks of road design and cost.

The manager must form a committee to work on the policy issues. To do this, the manager plans to establish two task-oriented workgroups, one on technical planning and one on financial matters, to report to the committee, propose preliminary designs, and make financial estimates on the cost of building the road. External experts and community group representatives will be invited to the committee to contribute on environmental and social issues. The manager consults the organization's directory and other directories or individuals to obtain people with the right background for the committee. The workgroups are composed mainly of internal agency staff but can include external advisors where necessary.

The manager draws a rough diagram, shown in Figure 2.5(a), to show what happens in the agency. This diagram is based on the idea of a *rich picture*, which comes from Checklands's (1990) Soft-Systems Methodologies. It shows the major activities as clouded shapes and the roles of people as icons associated with each activity. The output documents used and produced by the activities are shown as rectangular boxes. It shows the activities for each of the two workgroups and the

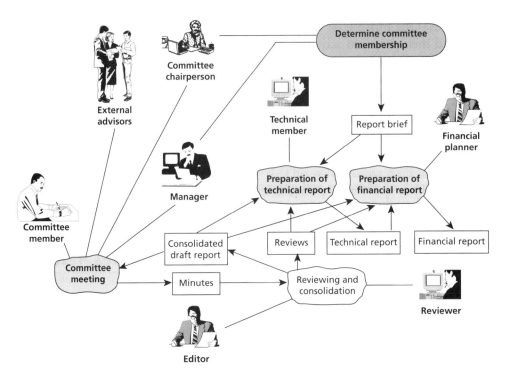

Figure 2.5(a) *A rich picture showing group activities*

committee meetings. The reports produced by the workgroups go to the 'reviewing and consolidation' activity, where they are consolidated by an editor and sent for reviews. Following receipt of the reviews, the consolidated report is forwarded to the committee, or further changes are requested from the workgroups. The deliberation of the committee can also result in minutes that require report changes. Such changes can be carried by the editor or referred back to the task groups. The workgroup members require access to the necessary databases and reports, both financial and design, to do their work. These databases may be distributed and heterogeneous, and some may use a variety of media and design tools.

Figure 2.5(b) is a transition diagram, which illustrates coordination between the activities. Such coordination manages messages sent between activities. Thus, the first activity, 'create committee', starts the activity of the two workgroups through the messages 'initiate financial report' and 'initiate technical report'. Consolidation is initiated with 'report ready' messages.

ISSUES

The manager must now decide how to support the activities shown in Figure 2.5—in particular, how they can be supported by computer systems. The manager must provide support to the workgroups, including coordination between them to ensure that financial estimates use the correct design data. There must be support for joint report preparation, materials distribution, easy consultations and coordination, making decisions, and disseminating information to community groups. Support for formal and informal interchange of messages between workgroup members and for distributing material to committee members is also important. This includes support for committee members to hold their meetings and relate policy issues to any information produced by the workgroups. The committee members must keep track of their meetings and organize the paperwork involved in these meetings. Support is also needed to keep track of the documents and to ensure that all team members always have access to the latest information.

Provision has to be made to easily assign and redistribute tasks among workgroup members and to keep each workgroup participant aware of the

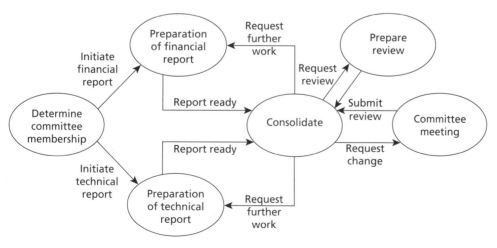

Figure 2.5(b) *Transitions between activities*

progress of both the workgroup and other groups. It should also be possible to change the workgroup structure itself. Examples of the kinds of change supported may be:

- changes to the composition of a committee and redistribution of tasks between other members
- movement of committee members between different locations
- changes to the composition of the workgroups.

We now look at how tools can be put together to support different kinds of activities.

SUPPORTING GROUPS AND TEAMS

Groupware Software that assists workgroups.

CSCW (Computer Supported Cooperative Work) Systems that support groups of people working toward a common goal.

Asynchronous cooperation Cooperation where the participants refer to shared information at different times.

Synchronous cooperation Cooperation where the participants refer to shared information at the same time.

Most people suggest that the best way to provide support to teams is to use commercial software rather than building special systems. There is a growing number of software products often known as **groupware** or **CSCW** (computer-supported cooperative work) to support teams. Such software must be capable of supporting a variety of working environments. The support provided will depend on the relative location of the team members, the type of team, the kind of work carried out by the team, and whether they work at the same or different times. The term **asynchronous** is used where people refer to shared information at different times. The alternative is **synchronous** interaction, where people are in communication at the same time.

Figure 2.6 shows a number of possibilities and also gives an indication of the technology needed in each. One example is where team members interact at the same time but at different locations. The most obvious support here is a video conference meeting. Another example is sharing a database in the same place but at different times. The different time, different place situation could be a mail system such as e-mail, or a workflow management system. Electronic meeting rooms exemplify same time, same place support.

Many other groupware tools are used to support the different kinds of work carried out by teams. One example is discussion systems. These can be used to keep track of offline conversations between people. An example is shown in Figure 2.7. It shows a discussion using the LiveNet system. Here, persons enter statements into

	SAME PLACE	DIFFERENT PLACE
SAME TIME	Electronic meeting Electronic board Shared screen Brainstorming Audience response	Video conferencing Conversation support Cooperative design Group editors
DIFFERENT TIME	Shared files Design tools	Structured workflow Electronic mail Bulletin board

Figure 2.6 *One way to characterize group work*

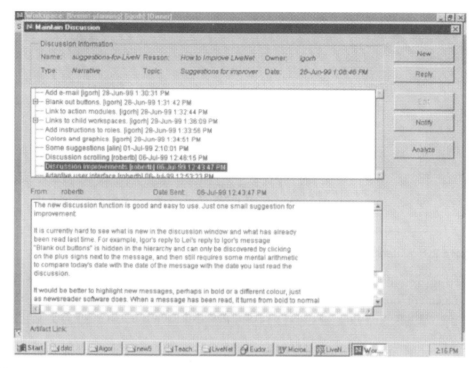

Figure 2.7   *A discussion system*

the discussion. Other people can then reply to these statements any time later. The statements and replies are organized in a way where replies always follow the statements, thus providing a record of the discussion. Chatboxes are another groupware tool that supports such exchange, but synchronously. In chatboxes, responses are entered immediately after a statement is made, thus leading to a more spontaneous conversation.

Other groupware tools are:

- **brainstorming tools**, where again the idea is the same as discussion, but now it is based on many people making entries on some idea; the suggestions are then evaluated later and often voted on

- **appointments systems**, which allow people to keep track of their appointments. This often includes a diary of each person's appointments on the computer system. The appointments system can then be used to find a convenient meeting time for a selected group. To do this it refers to the group members' diaries to find a time when all members are available.

Usually a workgroup requires more than one tool. The usual approach is to identify the tools needed to support each workgroup activity, and then provide the tools through an integrated **workspace** in a seamless way. The goal as shown in Figure 2.8 is to quickly put together the discussion, brainstorming, artifact management, and any other tools into the one workspace. The result should be **seamless**. Seamless means that to the user the tools are seen as integrated, with information moved between each tool with ease and without unnecessary conversions or recourse to operating systems for support. Thus, we can start by

Brainstorming Coming up with new ideas.

Appointments system Keeping track of appointments.

Workspace The space and facilities provided for a user on a screen.

Seamless platform A platform whose services are closely integrated.

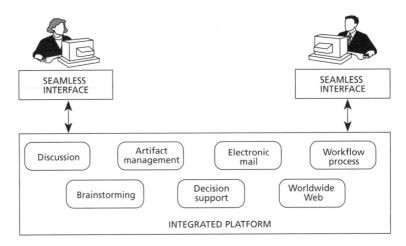

Figure 2.8   *Integrating platform*

selecting a tool to manage the team meeting. We may also select an artifact management module and also tailor it to the needs of the group. Then we must integrate these modules so that we can pass information between them at the interface.

The only alternative to such customization is to specially develop a workspace. This is usually not economical because of the complexity of the systems. Thus, you can be sure that with a trend to group work, there will be more and more emphasis on building groupware tools that can be customized to particular needs. Group situations are naturally dynamic and require tools that can change as group situations change.

The goal is to develop a workspace, like that shown in Figure 2.8, that allows a variety of tools, produced by different manufacturers, to be connected into the one seamless workspace. This is not always easy. The kind of problem that often arises is that groupware does not do exactly what is needed, or that it is difficult to connect two groupware tools intended for related activities within the group. The two approaches here are either to use a core technology that seamlessly supports a large variety of services and allows new services to be added, or to develop **middleware**, which connects different services.

Middleware
Software that connects network services.

We now look at possible tool combinations for different kinds of workgroup applications and the workspaces needed to support them. A rough division of the kind of applications is as follows:

- Support for informal networking simply to interchange information between people, which is usually based on electronic mail or the WWW.
- Support for personal relationships that require a sequence of interchanges with records kept about them, such as in decision making.
- Support for design teams who must create new artifacts and whose work is more situated work rather then planned work.

SUPPORTING MEETINGS

Formal meetings or committees are perhaps the most common group interactions found in organizations. They most often happen face to face at the same location.

Computer support provides the possibility of holding a synchronous meeting with people at different locations. As a matter of interest, let us see which tools are needed to support such meetings. These are illustrated in Figure 2.9. First, each member would need a screen and the same information. Perhaps even the faces of the other members would appear on each screen. There must also be protocol control to define the process to be followed in the meeting. This is where there is a big difference between electronic meetings and face-to-face meetings. In a face-to-face meeting, the tone of voice, gestures, raised voices, all have some meaning and can influence the meeting. This is not the case with electronic meetings. Here we can only input messages into the system, and a process must be devised to control the flow of messages. Such control can also make it difficult for individuals to gain unwarranted priority by raising their voices or because of their position. This leads to considerable debate about the conduct of such a meeting and whether the outcome is better or worse than a face-to-face meeting. Further research is needed to determine whether electronic meetings produce effective results.

Figure 2.9 illustrates one advantage of electronic meetings where members are all at the same place. This is the link to the corporate database. Participants can easily gain access to information in the database and bring it forward for discussion at the meeting. This leads, of course, to the possibility of mixed electronic and face-to-face components of the meeting. Electronic parts are used to display information, whereas actual discussion takes place face to face. Other tools can also be brought into the meeting—for example, support for decision making and argumentation.

DECISION SUPPORT SYSTEMS

Decision support systems help groups to make decisions. There are many ways to make decisions. Some decisions require an optimization algorithm. For example, the goal of scheduling of deliveries is to reduce costs by scheduling deliveries to minimise travel expenses. You cannot simply choose the best delivery routes by entering a number of independent transactions. You would need to set up a number

Decision support system A system that supports decision making.

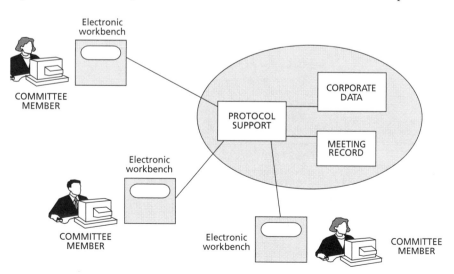

Figure 2.9 *An electronic meeting*

of trial routes, estimate their costs, and gradually improve on them. You may also solve this problem by using a mathematical algorithm. Other decisions are more qualitative in nature. The steps taken in such qualitative decisions are as follows:

- Define an objective.
- Define alternative solutions.
- Make assumptions for each solution.
- Make arguments for and against each solution.
- Evaluate alternatives based on agreed criteria.

For example, different price trends may be input into a portfolio management system, and their effect on a given portfolio is examined. More sophisticated systems may try different price variations to find one that satisfies desired performance criteria. Share portfolio management is a typical decision support system, the question here being which shares to buy to get a balance of capital growth and income return. The user must enter expected trends and then try different portfolios to see how they perform against these trends.

Decision support systems need support to keep track of the alternatives, assumptions, and arguments, especially in asynchronous environments. One approach is to use an amended form of a discussion system that is adapted to an **argumentation structure** or **design rationale** like that shown in Figure 2.10. This provides an argument structure to which users can add new arguments or rationale. It provides simple actions such as entering and deleting alternatives and arguments, but also allows making summaries that may favor a particular decision. Decision support systems may also associate people with particular arguments and activate them whenever a statement related to that argument is made. More advanced support provides references to artifacts used to support the arguments and computations on these artifacts. Each user must also have access to various tools that can be used to analyze the structure in order to make further inputs. Different users can be assigned different roles, with some having the ability to select alternatives as the accepted solution.

Argumentation structure A set of arguments used in making a decision.

Design rationale The reasoning behind making a decision.

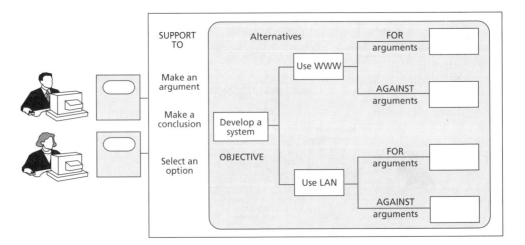

Figure 2.10 *Argumentation structure*

Decision support usually assumes asynchronous interaction, although there can be some advantage in synchronous discussion to resolve conflicts.

DESIGN PROCESSES

There is considerable interest in supporting cooperative design by using computers. The design process often starts by brainstorming ideas for the design. This is a highly creative task where new ideas are proposed. It is often performed with close and spontaneous interaction between a number of designers. Thus, someone may make some informal sketches and ask for a second opinion. The response may include another alternative. Following a response, the first person may make further changes and ask for other changes. This kind of discussion may continue for a while. Questions arise on how much of this discussion must be carried out in the same place and how much can be carried out by people working across distance. The criteria often center on social issues, with the goal of providing the same kind of rapport as exists with face-to-face discussion at the same location.

Usually support is a mix of synchronous and asynchronous work. This recognizes that the nature of design tasks can change at different stages of the design process. We may go from highly creative interaction to more routine evaluations that require fewer interactions. We may also move between synchronous and asynchronous interaction. Correspondingly, interfaces for collaborative design should support random swapping between these different kinds of tasks and interactions.

A possible set of design tasks and support tools for a design workspace is shown in Figure 2.11. The goal is to produce an artifact. The design proceeds in an environment of constant negotiation and division of work by the team members.

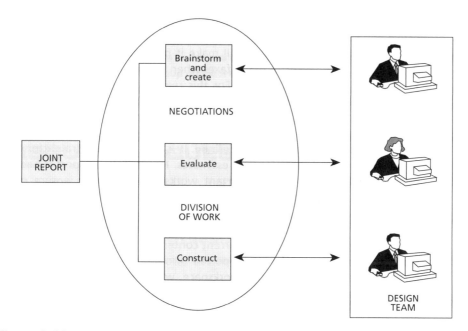

Figure 2.11   *Collaborative design activities*

2.8   Describe the characteristics of some different group support systems.

2.9   Why is it preferable to synthesize group support systems from existing components rather than building the systems from scratch?

2.10   What do you understand by a *discussion system*?

2.11   What kind of support tools would be useful for holding meetings across distance?

EXERCISE

Consider the interactive marketing organization. Do you think any of the kinds of support facilities described in this chapter could be usefully employed in interactive marketing?

BIBLIOGRAPHY

Anthony, R.N. (1965), *Planning and Control Systems: A Framework for Analysis*, Harvard Graduate School of Business Administration, Boston.

Berners-Lee, T., Cailliau, R., Luotonen, A., Nielsen, H. and Secret, A. (1994), 'The World Wide Web', *Communications of the ACM*, Vol. 37, No. 8, pp. 76–82.

Checklands, P.B. and Scholes, J. (1990), *Soft Systems Methodologies in Action*, John Wiley and Sons, Chichester, UK.

Dawson, F. (December 1997), 'Emerging calendaring and scheduling standards', *Computer*.

December, J. and Randall, N. (1995), *The World Wide Web Unleashed* (2nd edn), Sams.net Publishing, Indianapolis.

Drucker, P.F. (January–February 1988), 'The coming of the new organizations', *Harvard Business Review*, pp. 45–53.

Ellis, C.A., Gibbs, S.J. and Rein, G.L. (January 1991), 'Groupware: some issues and experiences', *Communications of the ACM*, Vol. 34, No. 1, pp. 38–58.

Grudin, J. (January 1991), 'Special section on computer supported cooperative work', *Communications of the ACM*, Vol. 34, No. 12.

Hawryszkiewycz, I.I. (1997), *Designing the Networked Enterprise*, Artech House, Boston.

Khoshafian, S. and Buckiewicz, M. (1995), *Groupware, Workflow, and Workgroup Computing*, John Wiley and Sons, New York.

Kling, R. (December 1991), 'Cooperation, coordination and control in computer supported work', *Communications of the ACM*, Vol. 34, No. 12, pp. 83–8.

Live Net System—http://livenet-demo.it.uts.edu.au:8080

Neuwirth, C.M., Kaufer, D.S., Chandhok, R. and Morris, J.H. (1990), 'Issues in the design of computer support for co-authoring and commenting', *Proceedings of the CSCW90 Conference*, pp. 183–93.

Sprague, R.H. and McNurin, B.C. (1993), *Information Systems Management and Practice* (3rd edn), Prentice Hall International, Englewood Cliffs, New Jersey.

Suchman, L. (1987), *Plans and Situated Action: The Problem of Human-Machine Communication*, Cambridge University Press, Cambridge.

Suchman, L. (September 1995): 'Making work visible', *Communications of the ACM*.

Winograd, T. and Flores, F. (1986), *Understanding Computers and Cognition: A New Foundation for Design*, Ablex Publishing Corporation, Norwood, New Jersey.

Winograd, T. (1987–88), 'A language/action perspective on the design of cooperative work', *Human-Computer Interaction*, Vol. 3, pp. 3–30.

EVOLVING BUSINESS SYSTEMS

KEY LEARNING OBJECTIVES

The kinds of business units often found in organizations
The relationship between a business unit and a business process
Some common information systems
System integration
The nature of business processes
Trends in business processes
Defining technical strategy
Why there is a need to re-engineer business processes
Why quality is important

INTRODUCTION

Chapter 2 described many ways that people work in organizations and the processes that they follow. Some of this work is carried out in groups, but some is done by individuals. Chapter 2, however, concentrated on work practices and did not describe any specific systems on which people work. This chapter will describe some typical **business units** *found in organizations and the way they can be supported by information technology. The business units described in this chapter are the common units that are needed to keep most organizations going, such as accounting or human resources. There are also many industry-specific units—for example, portfolio management for an investment firm or a reservation system for an airline or bus company.*

You should keep in mind that the way business units operate is continually changing, and this change is often driven by information technology. Thus, although the goals of each individual unit may remain the same, the way these goals are achieved can change, and this change can be significant over relatively short periods of time. The trend toward electronic commerce is adding to the rate of this change. Not only the way business is done is changing, but sometimes the whole nature of business itself changes. The business change is toward greater emphasis on knowledge, which in turn is made possible by information technology. Such emphasis on knowledge applies especially to knowledge-intensive organizations such as consultancies that deal primarily with the marketing of services rather than products. It also applies to businesses faced with increasing competition calling for the rapid development of new products and services using the knowhow that the business has or can bring in.

This chapter describes business systems in the context of the changing environment, as shown in Figure 3.1. The chapter looks at business from the perspective of change rather than the way things are done now—the way analysts must look at systems. It sees things from the perspective of how we do things now, then defines where we want to go and determines the way to get there through a development plan to build the future system.

The chapter first describes some of the more common business units and their processes. It then concentrates on the process of change, especially how change in information systems must support change in the strategic direction. It describes how

Business unit A part of a business responsible for a well-defined business operation.

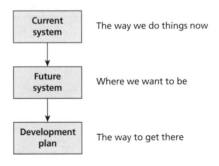

Figure 3.1 *Defining directions*

organizations go about positioning themselves in these volatile environments and how they use information technology to do so. The chapter also identifies a number of potential directions to give the reader an idea of the kinds of systems they will be developing in the future and the skills they will need.

INTRODUCTION TO BUSINESS INFORMATION SYSTEMS

An organization is made up of a number of business units, such as accounting, marketing, and production. Each business unit is described by the unit function and the processes followed by the unit. Although each business unit deals with a specific function, they all nevertheless have one important common requirement— they must work well and accurately and be easy to use. They must help the business grow and achieve its objectives. In addition, they must also be integrated so that they can work together to fulfil the organization's mission rather than working towards their own ends. Working well and accurately is sometimes hard to define. The term quality is often used to describe systems that work well. The term can mean many things to many people. Quality is a general term that is used to define our or our client's expectation of a good system. Quality usually means that there are few errors in the system, the users are happy with the system, and everybody knows what is going on. We will continue to expand on the term quality as we describe systems in more detail.

Organizations are often described by structure charts, which show the organization's business units. A typical structure chart of business units found in many typical organizations is illustrated in Figure 3.2. It shows the major business units, which themselves may be made of lower-level subunits. For example, in Figure 3.2, the human resources unit is made up of personnel and personnel development subunits. Many of the business units in the structure chart will be supported by computer-based information systems. The information system can, therefore, be seen as a number of subsystems, where there is a separate subsystem for each business unit. Each subsystem may be further subdivided if necessary. Some important information subsystems found in most organizations are discussed below.

THE HUMAN RESOURCES SUBSYSTEM

This maintains information about the organization's personnel. It is made up of the **personnel subsystem** and often the **payroll subsystem**, although the trend now is for payroll to become part of the accounting subsystem, while the human resources subsystem concentrates on personnel development. The personnel subsystem keeps track of personal data such as date of birth, address, marital information, employment record, start date and appointments in the organization, personal skills, vacation records, and entitlements. The payroll subsystem stores information about pay entitlements and produces paychecks when required.

Increasing specialization and emphasis on knowledge in the workplace makes it more important to keep track of and develop skills within the organization. Such **skill profiles** are becoming more important in knowledge-intensive organizations,

Personnel subsystem A business system for keeping information about people.

Payroll subsystem A business system for paying the organization's personnel.

Skill profile The skills possessed by a person or by an organization.

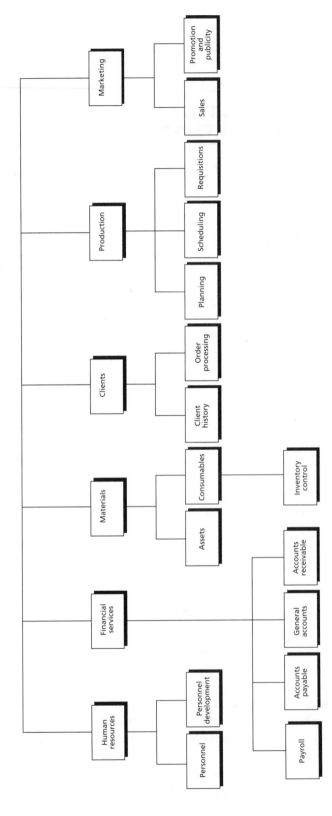

Figure 3.2 *Typical information systems—static structure*

such as consultancies, which require people with different skills to be assembled quickly to provide new products and services. Many knowledge-intensive organizations now maintain databases of personnel skills, which are often stored on an intranet to quickly find the people that can contribute some expertise to a project, even if they are at a distance. Such intranets also provide the collaborative services that enable the experts' knowledge to be quickly obtained. The human resources subsystem is thus beginning to place more importance on education and training, including any courses attended within the organization. In many cases now, personnel subsystems arrange or provide courses for personnel within organizations to develop their skills. System dynamics here include movement of people between organizational units, paying people, and arranging for people to improve their skills.

THE MATERIALS SUBSYSTEM

This keeps track of all the items owned by an organization. It records item quantities, locations, and value. A distinction is often made between the organization's *assets* and *consumable items*. Assets are characterized by the kinds of items needed to carry out an organization's activities. They can be furniture or equipment such as computer laptops. Assets are usually distributed throughout the organization. Organizations therefore need an asset register to keep track of equipment distributed for use in an organization's day-to-day operations. Consumable materials are of two kinds. The *parts inventory* keeps track of material items purchased outside the organization and used by the organization in its activities, such as paper or parts needed to produce the organization's products. The *stock inventory* keeps track of the items produced and sold by the organization. Consumable items are kept in a warehouse or various storage locations within the organization. An organization maintains an inventory of its consumable materials.

A typical parts inventory system includes the components shown in Figure 3.3. Its goal is to purchase parts and issue these parts as required to the other units of the organization. The components of the inventory system include the following:

- The parts issues component is responsible for issuing parts to the rest of the organization.
- The inquiry system allows inquiries to be made about current inventory levels.
- A stocktake component checks for any discrepancies between records and actual store contents.
- The reordering component is responsible for ordering parts from suppliers. It monitors current store levels and determines whether new parts must be ordered. To make this decision, the system keeps statistics about item usage to determine quantities to be ordered. It must also keep information about times needed to deliver items, sometimes called lead times, to make sure that sufficient time is given to suppliers to deliver the needed items.
- The parts delivery system keeps track of parts received from suppliers. It adds the parts to the current stock record and advises accounting that the parts have been received. This advice is needed by accounting to authorize payment to suppliers for the parts.

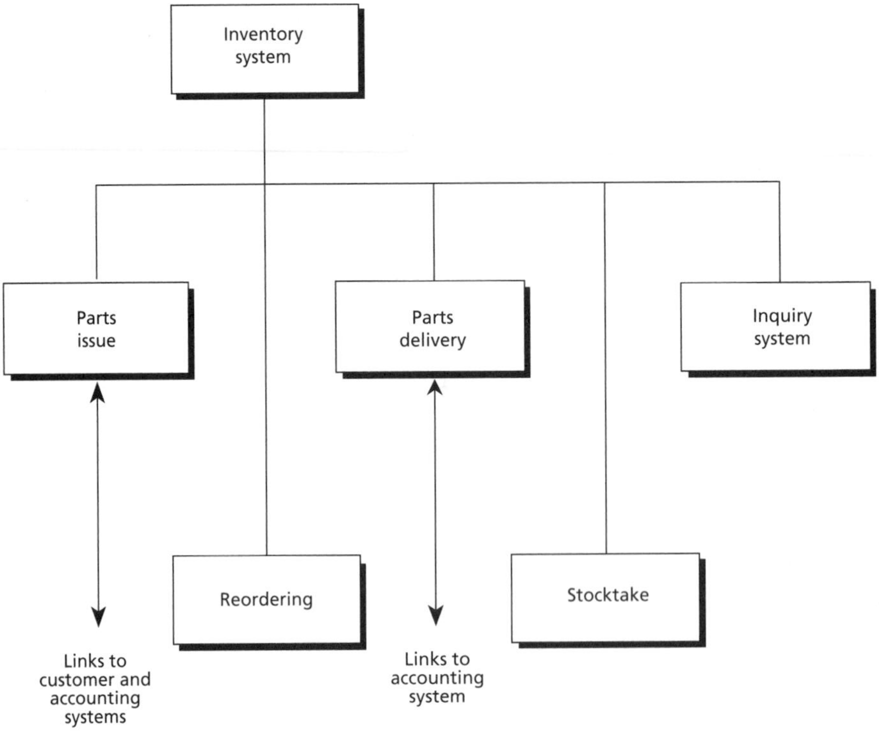

Figure 3.3 *Inventory control system*

Specialized user demands, however, often mean that inventories need to be more sophisticated. Emphasis is being placed on providing parts in a **just-in-time** manner to reduce inventory holdings and costs.

THE FINANCIAL SERVICES SUBSYSTEM

This keeps track of an organization's financial transactions and financial status. The two most common subsystems here are the **accounts receivable** and **accounts payable** subsystems. The accounts receivable subsystem keeps track of moneys owed to the organization. It produces invoices for sales, checks client credit limits, and records payments made against these invoices. It also sends out reminders when payments are overdue. This subsystem has an account for every one of the organization's customers and is responsible for keeping track of customer payments. It checks that the payment for each invoice is made and checks the payment against the invoice. Any differences between the payment and the invoice must be resolved. The system also keeps track of the times taken by customers to make payments. Some systems can identify customers who are continually late in making payments and bar them from receiving further credit.

The accounts payable subsystem is the reverse of the accounts receivable subsystem and keeps track of the money owed by the organization. It keeps track of purchases made by the organization by recording outgoing orders. Invoices received from suppliers are then checked against the orders to ensure that the invoice lists only goods that have actually been received. If the invoices are verified, payments are made to suppliers.

Accounts receivable The subsystem that keeps track of moneys owed to the organization.

Accounts payable The subsystem that keeps track of the moneys owed by the organization.

Most organizations also have a **general accounts** subsystem to keep track of funds used by internal departments and projects. This subsystem produces reports about the organization's assets and its use of resources. It includes setting budgets for the organization's departments and projects, monitoring spending against these budgets, and providing budget reports. It also produces end-of-year reports about the organization's assets and liabilities, and a statement showing the organization's income and expenses for the year.

The **payroll subsystem** is often part of accounting. It uses information about entitlements from the human resources subsystem to produce paychecks on a regular basis.

In addition, the financial services subsystem is often responsible for developing plans and forecasts on the financial status of the organization.

Financial services systems are now increasingly being extended into the Internet. The trends are to support interaction with the customer, where for example clients can observe their account status with an organization, make direct payments, or even submit queries through a WWW interface. Although marketed by banks through automatic teller machines for a number of years, the trends are toward making such interaction more specialized and integrated with other activities.

A MARKETING SUBSYSTEM

This publicizes the organization's products and activities to its current and potential clients. Marketing usually has two functions. One is to sell the services. The other is to determine a strategy for selling the products or even to decide what products to sell and the most likely customers. The WWW is now becoming almost essential in selling services and products. The marketing subsystem is now often responsible for an organization's Web site, which establishes an organization's presence on the Web, describes its products, and lists ways to purchase them. The question that must be addressed by marketing is how to make the Web site interesting and one that people will want to visit over and over again

Determining a strategy is sometimes called **market research**. Market research establishes what customers need and ensures that the organization is actually producing useful products. It will establish the most desirable features of its product and services and build a market strategy around these features. Market research often requires calls to be made to potential clients to determine their likely purchasing preferences. It often includes market surveys to determine potential customer needs. The number of potential clients may be estimated and the services or products needed by them can be determined. This information is used to create a marketing campaign through advertising and other forms of information dissemination, which may include advertising, mailouts, or simply visiting the customer. Again, the Web site is increasingly used for this purpose by capturing knowledge about customers who make purchases through the organization's Web site.

Sales units are often closely related to marketing. They are responsible for selling the marketed products. Each sale results in an order, which is sent to stock inventory to supply the items and accounting to issue an invoice. The sales unit may also predict future sales or keep track of sales by customers.

General accounts
A system that keeps track of funds within an organization.

Payroll subsystem A business system for paying the organization's personnel.

Market research Determining how to make products acceptable to customers.

A CLIENT RELATIONS SUBSYSTEM

Client relationship management Finding ways to get, work with and retain clients.

This maintains contact with the organization's clients. The importance of building a community of clients that do repeat business with an organization is now being recognized. The term **client relationship management** (CRM) is now becoming increasingly used to define relationships with clients. The three major steps are to attract a client, make the relationship profitable, and retain profitable clients for life. The first step is achieved through effective marketing. The next step is to improve client profitability through maximizing sales to the client. To do this, organizations keep personal records about their clients. This includes not only client addresses or their financial and employment status but also their preferences. Organizations can differ in the kinds of clients they have and the kind of information kept about them. Some organizations have customers who order goods from the organization. The organization keeps track of orders made by these customers and deliveries made to them. Other organizations, in particular public agencies, have clients who get help from the organization. In all cases, privacy and security are important issues, as people do not like information about them to be widely distributed.

Loyalty program Providing rewards to clients for continuous support.

Such information helps an organization maximize its services. Knowing a client's preferences enables an organization to keep the client informed of those of the organization's services that are most likely to be of value to the client and thus increase the client's requests for products and services. The next step is to maintain a community of loyal clients. This is usually done through **loyalty programs**. These are usually based on recording clients' activities with the organization and providing bonuses once certain levels are achieved. A common example here is frequent flyer programs offered by most airlines. Points are built up with each trip by a client and can then be recouped for free flights once required point levels are achieved. Loyalty programs often provide the client with easy access to relevant information within the organization. The provision of what are commonly known as 'help desk' facilities is often a first step here. Help desks can quickly give clients any expert assistance they need.

A PRODUCTION SUBSYSTEM

This subsystem is found in organizations that produce physical goods. Such organizations may have one or more factories that use consumable items to produce products, which are later sold to customers. They need to maintain an information system about their production facilities as well as schedules for these facilities and their parts and personnel requirements. A production system includes a planning subsystem to determine what goods are to be produced, and a scheduling subsystem to schedule machines to produce these goods. It may also include a subsystem to requisition consumable items needed to produce these goods and to deliver produced items into inventory.

Each of the business units in Figure 3.2 satisfies some organizational need. In a typical organization, some of these units or parts of them may be manual and some automated. Others are a mixture of manual and automated processes. The trend in most organizations is to increasingly automate their information systems but to do so in ways that meet broad organizational goals.

SYSTEM INTEGRATION

Figure 3.2 is a static representation of information systems. It shows each information subsystem to be self-contained and does not show the information flows between the subsystems. When you look at the dynamics of business operation, you will find that there are many information flows between the subsystems. For example, orders from clients are sent from sales to the accounting subsystem to generate invoices, production subsystems generate requisitions for parts that go to the materials subsystem, and scheduling systems in production usually interact with payroll to establish the amount of time worked by employees during production.

So, to obtain an accurate understanding of what information systems look like, it is also necessary to show how the subsystems fit into the business processes and the information flows between them. Many business processes involve more than one business unit, and as a consequence these units must integrate their activities. Consequently, their information systems cannot often be built independently of one another but must easily exchange information. Such exchange reduces the duplication of data, as business processes can share information rather than building their own databases. It also means that such subsystems are less complex and easier to manage. There are two ways of integrating subsystems.

One way of integration is shown in Figure 3.4. Here there is a layer of software between the business process and the information subsystems to assist integration. This layer of software takes service requests from the business process and sends them to the business unit, which then transmits the response back to the business process. The term **client–server** is often used to describe this kind of operation. Here the business process is the client, and the information systems for the business unit provide services to that process. The client–server approach is now supported by the client–server architectures described in Chapter 1. Again, the pressure is on the business units to provide a quality service to the process.

Client–server process A process that describes how a server provides a service to a client.

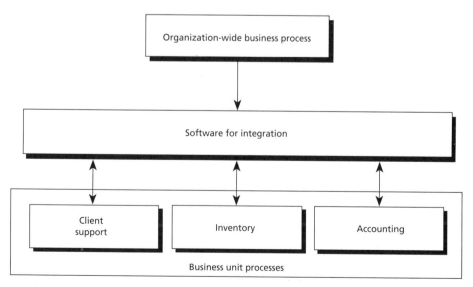

Figure 3.4 *Structure of a business process*

Another approach to integration is to use a data warehouse. This does not mean that all the data would be in the one location, or that one section would be responsible for it. Each subsystem can still be responsible for its data in the data warehouse. Business processes would easily get information from any subsystem through the integrated data warehouse management software. Data warehousing often results following the integration of business units within an enterprise. Often units in an organization developed their own systems, perhaps because of initial physical separation. In this case, similar functions must now be integrated to introduce consistent operating standards throughout the organization and to present a unified enterprise interface.

Methodologies and tools are now being openly marketed to help organizations in system integration. One way is to take existing systems and change them so that they can easily exchange information. Another way is to replace existing processes by standard industry-accepted processes. Such standard processes are now increasingly available though vendors such as SAP, a German company that has developed a set of widely used reference processes. Curran and Ladd (2000) describe many of these processes in the references. We will return to such standard processes later in this chapter after we describe some business processes.

There is a further requirement of such integration. It concerns changes in how organizations deal with their clients' continually changing demands. It is now common to describe systems as front-office and back-office systems. The front-office systems interact directly with clients or business partners. As an example, these may include the Worldwide Web site or help desk services. The back-office systems manage the organization's internal data. These may include accounts, inventory records, and so on. The goal of system integration is to organize the back-office systems in a flexible way to enable them to react quickly to changes in the front-office systems needed to satisfy new client demands.

BUSINESS PROCESSES

Business processes define the way the business or organization does its business. They define the way the organization achieves its goals, interacts with its clients, and carries out its internal operations. Information technology has had and will continue to have considerable impact on the way business is carried out. One goal of using information technology is to improve the quality of business processes. The word quality is widely used but has no universal definition. It simply means doing the best one can. Generally, in the context of a process, quality means that the number of process errors and unpredictable events are reduced, the process goals are achieved at minimum cost, and both the customers and people within the organization who participate in the process are satisfied with the process. Satisfaction can also mean that everyone is aware at all times of what is happening to objects in the process so that information about any activity can be readily obtained. How this is done is specific to a system. What we often do is define our own quality requirements, or a quality policy, and ensure that the processes meet these requirements. A word of caution: simply installing a computer to automate a process does not necessarily improve the process. We should first define what the process must do and then install a computer to support the process. We now describe some characteristics of modern-day processes and some processes found in business units.

There are a variety of business processes in any organization. The best known processes supported by computer systems are the large corporate transaction systems, such as banking systems, insurance policies, and reservation systems. These latter in most cases use data warehousing to store the large volumes of data and record transactions to this data. However, information technology is now providing new ways of working rather than automating predefined transaction processes. It is also providing new ways of doing business with the outside world. It is now common to identify **critical business processes**, which are those organization-wide processes that are crucial to the survival of the business. They are usually processes concerned with providing services to the business clients. They may be the process used in a telephone company to install a telephone, the process used to provide the items ordered by a customer, or the way a hospital treats its patients. Such processes often use information from more than one business unit.

Critical business process A business process that is crucial to the survival of a business and supports the core business of the organization.

HUMAN RESOURCE UNIT PROCESSES

The human resource unit contains a variety of processes, ranging from those that are predefined to those that depend on the current situation. An example of predefined systems is automated payroll processes. These are transaction-oriented, with transactions processed in batches. A batch run is made on each payday to generate checks or payslips. All the data needed for the payroll run are collected before the run. This data may be obtained from other systems or input directly for payroll use. Once the information is available, a payroll run is made. The payroll run will look at the employee's current level and determine the salary for that level. It will then deduct the employee's commitments, such as medical, insurance, or union deductions, and do any computations on overtime payments or special duties undertaken by the employee. All these adjustments will generate the employee's net pay. Finally, the system will refer to tax tables and subtract the tax payable to obtain the employee's net income.

Another process is maintenance of personnel records. Changes to personnel status, such as a promotion, are recorded for each person. Most personnel systems are now interactive: personal records can be displayed on the workstation for either enquiry purposes or for amendment and updating. Security and privacy are very important in personnel systems, and care must be taken to ensure that personal details are not freely available throughout the organization but are restricted to those staff directly concerned with personal records maintenance.

Most processes in human resources are internal, although there are times when a process includes external participants. One example of the latter is the way organizations hire people. The process may include steps such as:

- advertising a position
- creating a team
- selecting applicants for interview
- interviewing applicants

- making an offer
- placing a successful applicant.

The process must be complete, and responsibility for each step should be assigned to someone in the organization. Confidentiality must also be maintained throughout the process.

It is also increasingly common to have an interface between the human resources system and other business units. Such interfaces usually center on providing support to business unit managers in managing their employees, including providing training support and defining employment conditions and standards. The human resources function also interfaces with the financial services unit to provide information used to develop the personnel budget within a business unit and to provide advice on personnel levels for future planning.

PROCESSES IN CLIENT RELATIONSHIPS MANAGEMENT

Relationships with clients are almost always defined to be critical processes. They provide the interface with the organization's clients and must ensure that clients continue to do business with the organization. Most clients nowadays demand better and better services. A consequent emphasis in organizations is to provide what is now commonly known as a **quality service**. Service is now a generic term and can include selling an item to a customer, providing entertainment, or providing assistance or social security payments.

Quality service A service that meets all client needs in a mutually satisfactory way.

Figure 3.5 illustrates a broad and generic view of the customer service process together with two cases. One case is selling products (Case A), the other is making government benefits available to qualified recipients (Case B). Both of these usually begin with publicizing a product or benefit. Clients then approach the organization with a specific request—for example, placing an order or making a request for a benefit payment. This request is recorded and checked to see whether the client is eligible for the benefit in Case B, or whether the order specifies its requirements completely and accurately in Case A. If so, then arrangements are made within the organization to provide the service. This may be arranging for goods to be delivered in Case A or payments to be made to clients in case B. Any problems found during

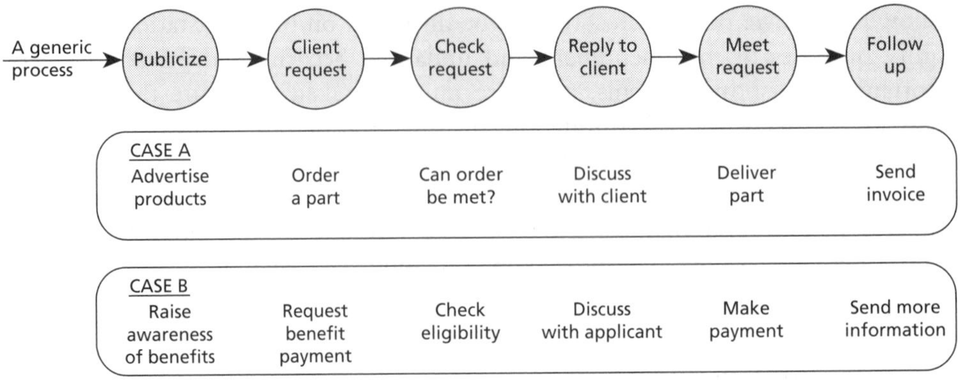

Figure 3.5 *Providing a client service*

the client request are clarified with the customer. A process must also deal with exceptional conditions, such as changes to an order or the order's cancellation. The information system must cater for all these possibilities.

There is usually a follow up on the service. It may involve a payment, in which case the accounting system will become involved in the process, or finding out whether the customer was satisfied with the service. Usually, the client system should follow up such requests to ensure that the service is carried out and also answer any customer queries about the service. This is now usually the function of a help desk system. Emphasis on quality usually means that a client can always be kept informed of what the organization can do for them, and that the organization can quickly answer questions on the progress of any requested service.

Many such services now include additional features, such as maintaining profiles and personalized marketing. Thus, follow-ups may now be more sophisticated. They will probably include an automated help desk to allow client queries to be easily resolved. Follow-ups may also include sending clients additional information that may increase the use of the organization's services. Long-term clients may also be given additional access, such as allowing them to keep track of their business with an organization and the progress of any open-ended service. Typical examples are banks, which allow clients themselves to initiate transactions and to move funds between their accounts.

Providing better client support often requires the front-office systems to be integrated with the back office. Checking a client request and delivering a service may involve the inventory system, whereas the accounting unit would be involved in any financial transactions associated with the service.

SUPPLY CHAIN MANAGEMENT

Many businesses now receive goods from other businesses, process these goods to produce new goods, and then supply their produced goods to a further business. Thus, company A might receive, say, fruit from a supplier, can the food, and then deliver the cans to a supermarket. The term **supply chain** is often used to describe this kind of process between businesses. Part of this process is that received goods must often be stored before the manufacturing process is ready for them. Secondly, manufactured goods must be stored before a client is ready to receive them. The goal of many businesses is to reduce the time that received and manufactured goods are stored in the system. The term 'just-in-time delivery' is one important criterion for the transfer of goods between organizations. Thus, required inputs are received from the supplier just in time to process them, and manufactured goods are ready just in time for the buyer. Other goals often quoted for improving the supply chain are to remove the need for middle people and to add as much value as possible at each point of the supply chain.

Supply chain The movement of parts from the input through to the output of a business process.

WWW technologies often play a significant role here in enabling the information systems of networking partners to interface with each other across the Internet. The Internet can lead to radical changes in the supply chains, such as the evolution of electronic sales that can eliminate not only steps in the process but entire functions. Perhaps the most well-known example here is that of Amazon.com books, which began by selling books through the Internet and is now extending to sell other kinds of items such as CDs. Amazon.com can make a sale on the Internet and deliver books to the customer without the need of a shop distribution network.

TOWARDS 1:1 CLIENT RELATIONSHIPS

Many writers now suggest that one of the more important aspects of electronic commerce will be the provision of specialized services to clients. In fact, rather than including all client transactions within the one function, what should be happening is that each client order should be treated as an individual workspace with all transactions related to that order or client linked to that workspace.

KNOWLEDGE-INTENSIVE PROCESSES

Many organizations are now realizing the importance of using information in effective ways. Such information is used to identify new markets, design new products for these markets, and generally improve internal operations. Such new products must be designed within increasingly shorter time frames within highly volatile environments. Riggins and Rhee (1998) describe the importance of these processes in identifying new products and services to maintain competitiveness in the market place.

Although knowledge-intensive processes are sometimes seen as similar to workflows, they differ in two major respects. One is that tasks are now carried out by groups or communities rather than by individuals. Group membership itself is flexible, and members with the expertise needed for the particular task must be quickly found and brought together. Another is that the processes themselves cannot be predefined as workflows but can change as new opportunities are identified. Thus, typical work processes here are often described as a number of closely coordinated communities each working continually towards the organization's goals.

BUSINESS ALLIANCES

The trend toward electronic commerce is leading to a growth in business-to-business alliances. Information systems in this case must be able to interface with the internal system of the alliance partners while preserving their internal confidentiality. Supply chains are one form of **business alliance**. Supply chains can often lead to closer business networking. A study by Williams (1999) describes how business-to-business networks evolve. Often they start as simple transaction exchanges in a supply chain such as clothing items from a retailer to the distributor, but then grow toward more strategic processes such as suggestions on designs that are likely to be acceptable in the retailer's area.

There are other ways for alliances to grow. One common way to network is when an organization outsources a business unit but then still uses the unit's services. Another is where an organization finds that it can generate additional business by combining resources with another organization to offer a new product or service, combining the resources and expertise of both organizations.

In the field of electronic commerce, business alliances are leading to what are known as **B2B** (business-to-business) processes. The trend for these is go beyond exchange of transactions as now occurs in EDI systems. On the other hand it is to develop systems that keep records about activities, such as relationship with a client,

Business alliance One or more businesses working together for mutual benefit.

B2B Carrying out business electronically between businesses.

rather than about transactions, such as record a payment. Thus the alliances will be transparent to clients who deal with only one of the alliance partners although other alliance members may provide services. The best example now is the alliances between airline companies where a particular travel plan may include a number of carriers. The transaction, however, can be concluded through one carrier. Such trends will eventually penetrate more and more business alliances.

TOWARD VIRTUAL ORGANIZATIONS

Most people assume that business alliances will be between a set of fixed partners and will persist over time. However, once familiarity with working in alliances grows, the need to maintain fixed relationships may change. In such cases one might look at more transient relationships with partners chosen because they can provide a needed service at a particular point in time. Who the partner is may not matter so much as long as the service is what is needed. Ways of doing business with increasingly transient relationships leads to what is commonly known as the virtual organization. Here the business goal and services needed to reach the goal are defined. Who provides the services is not important as long as they are of high quality and satisfy the business process need.

There is now increasing discussion about such virtual organizations. Virtual universities now offer educational services at a distance, provided by experts who may not belong to the one institution.

The book now introduces a new case to illustrate some of these processes.

TEXT CASE C:
Universal Securities

Universal Securities installs security services at a variety of sites ranging from individual homes to large buildings or complexes. Installations can often involve large building complexes, requiring careful planning and specialized installation. Universal Securities operates in a number of regions, and each region has a regional manager. Each region has its own marketing operation, and carries out deliveries and installation at customer sites within the region. Some regions include a production operation and often specialize in the production of specialized components—thus there may be one region that produces video equipment, another that produces alarms, and so on. A region also has facilities and the experts needed to design and advise on installation to cater for various specialized requirements. In addition, each region has marketing personnel and a local marketing manager.

MARKETING

The regional marketing manager is responsible for sales and for the completion and delivery of systems within their region. The following is a typical marketing process:

- A regional market representative negotiates an order with a client in that region. Negotiation includes face-to-face meetings and visits where documents may be exchanged, and exchange of further detailed information through faxes, phones, and so on. During the negotiation, advice is often sought from experts on various aspects of security. Such experts may be located in different regions. Before a design is complete, it must be approved by the local marketing manager and agreed upon by the customer, and an order must be signed.

- The regional marketing manager then develops a plan for the installation of the system. This includes ordering the necessary components, and arranging for delivery and installation.
- The task of assembling the parts and team needed for an installation falls on the marketing group, which usually begins by seeking interest from other regions by telephone or sending faxes to selected regions about the *task requirement*.

ARRANGING PRODUCTION

Regions that have a production facility have a regional production manager. Each such region maintains a production plan. This plan is created in light of expected orders for the coming year, by consultation between the site manager, the marketing manager, and the production manager. The plan is then implemented and monitored by the production manager.

As the year proceeds, incoming orders are committed to the production plan. The production plan defines the items to be produced and the quantities needed by particular dates. The items needed are usually specialized and need some customization to meet special customer requirements.

ORDER PROCESSING

A client order must be integrated into the production plan. Once an order is received, the steps are as follows:

- The local marketing manager must identify the components needed to complete the order and construct an initial order schedule.
- The managers must find the production facilities that can produce these components. To do this, the local marketing manager must negotiate with the production manager at other sites to obtain components that meet time commitments.
- The production managers must accommodate the component requirement in their production plan. To do this, the marketing group may request formal *quotes* from the production sites, then components on the production line are specifically committed to the order. Often a current production schedule must be altered to record the commitment to an order and any special customization features.
- The initial order schedule is amended to describe when particular order components will be ready and when final delivery is expected.
- The marketing manager plans the delivery of the components and installation of the system.

Central office is continually informed of these task negotiations and may help if necessary (for example, through suggestions or reassigning staff if needed).

An initial analysis describes the activities in Universal Securities using the **rich picture** shown in Figure 3.6. It uses ideas from a modeling method known as Soft-Systems Methodologies (SSM), introduced in Chapter 2. Here each of the ellipses is an activity. The rectangular boxes show the main documents exchanged between the activities. The rich picture also shows the people involved in each activity by their role, as for example clients or experts. The 'Sales and design'

Rich picture A pictorial representation of a system.

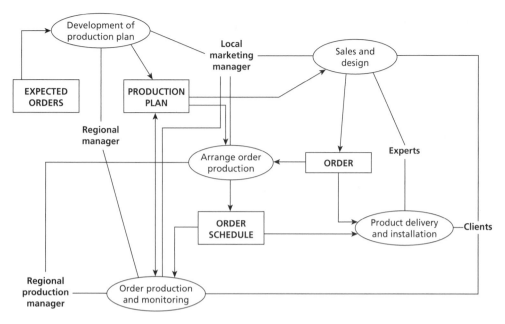

Figure 3.6 *Major activities in Universal Securities*

activity, for example, is the activity that produces orders through interactions between the client, experts and local marketing managers. Orders are then used in the 'Arrange order production' activity to finalize the order schedule. There are also the development of a production plan based on expected sales and the monitoring of the order completion process.

Problems have been found in Universal Securities:

- There are production delays, especially of specialized parts.
- The negotiation to develop parts can sometimes be protracted, leading to delays.
- Often the expertise required to identify what is needed is difficult to find.
- Too much time is spent acquiring new production facilities at the expense of time spent with clients.

OTHER TYPICAL SCENARIOS IN UNIVERSAL SECURITIES

- The regional managers have a number of representatives to help with regional work.
- A customer can change some part of their requirement, which must then be renegotiated with the different sites.
- Monthly reports of project progress and sales results are sent to central office, which may then raise queries about the report.
- The progress of orders is continually monitored by the sales site to keep track of order progress and to answer customer enquiries.
- The progress of orders can also be monitored at central office.

It is important that the latest documents be available to all parties to avoid any confusion in the negotiations, something that occurs frequently at the moment.

SETTING NEW DIRECTIONS

One important question facing many organizations is how to use information technology to give them competitive advantage. Organizations are increasingly realizing the value of information technology. Thus, not only must they use information technology to support individual tasks, but they must also decide how to change the way they work to get the best value from this technology. It is important to remember that an information systems plan cannot be considered in isolation from its other activities. It is necessary to show how the information systems plan fits into an organization's strategic plan. In fact Tolis (1999), after many observations in Sweden, suggests three levels of change: strategic, organizational, and systems. Here, strategic development examines the directions of the organization and defines any new directions. Organizational development looks at how to change the structure and ways the organization can work to fit in with any new directions. Systems development must then support the new organizational structures and its new ways of working.

Changes in strategy are needed when an organization finds it difficult to maintain its strategic position in the marketplace. Changes in strategy often require the existing organizational structure to change, which in turn requires change to existing information systems. Thus, suppose a strategic decision is made to specialize client support so that clients do not seek services elsewhere. This may mean that the field sales force may itself have to develop specialist skills. In turn, better information systems are needed to provide access to specialized knowledge.

The changes in all three levels must be integrated in the sense that organizational development is consistent with strategic objectives, and information systems development helps to realize the strategy while adapting to organizational development. Thus, Figure 3.1 now begins to look more like Figure 3.7. The future

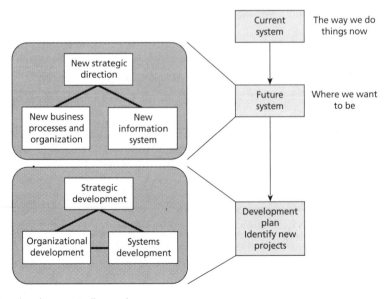

Figure 3.7 *The development dimensions*

system defines the new strategic direction, the new business processes and organizational structure to support them, and the new information system. The development plan defines how each of these is to be achieved and, more important, how they will be integrated.

The system development plan will define the information systems projects to create the new information system. We can also make a distinction between two kinds of projects for information systems, namely:

- projects to support new strategic directions and business innovations—that is, *strategic IS development*
- projects to improve existing systems and processes, sometimes known as *re-engineering the business process.*

The important aspect of planning is that it must develop a vision that is both practical and acceptable throughout the organization. What are the outcomes of planning?—a mission statement and objective statements. These are then elaborated into plans.

WHAT IS INVOLVED IN DEFINING A DEVELOPMENT PLAN?

One way to view planning is shown in Figure 3.8, which shows the external environment that creates opportunities for the organization to make some gains. It requires planning to reach a balance between the pressures of the external environment and the use of total resources to meet these pressures. The plan must decide how the organization must deploy its resources to make these gains. Planning must also make the trade-offs between the various organizational constraints. Given limited resources, a system planner is faced with the problem of deciding what to begin now and what to leave until later. Thus, priorities must be set during planning that will ensure the maximum return for funds invested in information systems development. Such priorities determine the resource to be allocated to information technology and how much to other resources, such as buildings or people.

How is this choice made and how are priorities set? The questions that must be answered are what new demands the business can easily satisfy, or what new equipment can improve the organization's operations. Most of these answers are

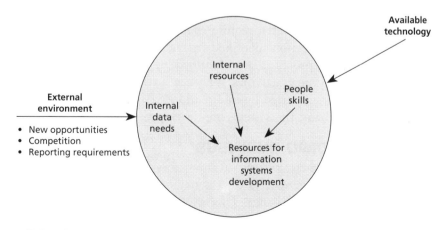

Figure 3.8   *The environment for strategic planning*

usually speculative and based on assumptions. There are now many analysis techniques used in planning. SWOT (Strength, Weakness, Opportunities, and Threats) analysis is one such method; another is Porter's value chain analysis. At an organizational level, many such questions are about the role of information systems. Re-engineering existing information systems is only one possibility for improving an organization's standing. This must be evaluated against other possibilities, such as increasing research into new products, spending more on advertising, or replacing production machines. All these possibilities must compete for the organization's funds.

It is perhaps fair to mention that it is no longer valid to make plans on the assumption that the environment remains constant. Customer preferences change, new laws are enacted, and production methods can change quickly. A lot of organizations cater for this by including contingencies in their planning. These are actions to take if the strategic plan cannot be met. However, if the contingency plan tends to dominate strategic plans, then strategic planning is no longer valid. Alternatively, the strategic plan must continually evolve and not simply consider fixed periods. Another suggestion is that strategic planning should change its emphasis and cover new planning concepts related to change. These include:

- building up human resource capabilities, especially in knowledge-intensive organizations
- greater emphasis on improving links to the external world
- facilitating the learning process.

NEW STRATEGIC DIRECTIONS

Strategic design defines new business opportunities and identifies new ways of working, or defines ways to change existing processes to become more efficient and effective. Many books have been written on defining strategic directions for business in the broad sense. For example, acquisitions of new businesses and entering new markets are a field of their own. This book does not cover them in detail. However, Robson (1997), for example, describes in detail how strategic directions are set. Her book includes ways to carry out strategic analysis and what to consider when choosing between alternatives.

This book concentrates more on how information systems are built to realize the strategic directions, which often require change to existing activities and work practices. It follows the direction identified in Figure 3.9. Here a variety of new business innovations and processes are identified to realize a strategic direction. These are used to specify new activities. The new activities identify new ways of working, which must be supported by information systems. Projects to build these systems are then defined and the systems are developed.

NEW WAYS OF DOING BUSINESS

What are the choices when looking at new ways of doing business? Many developments will suggest themselves to a strategic planner. Businesses now work in environments of continuous change. This is forcing more and more organizations to look at new ways of doing business, especially within the electronic commerce environment. Typical innovations in this environment include:

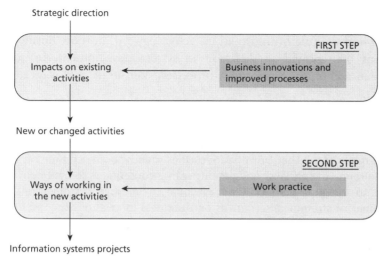

Figure 3.9 *Linking design dimensions*

- *extension to clients*, where clients are given limited access to an organization's database for things like monitoring progress of their orders, getting specialized advice, or looking up the organization's products
- setting up *business-to-business links*, either by outsourcing existing functions or by developing new liaisons to expand opportunities
- improving the supply chain to minimize inventory costs
- introducing virtual work, where organizations make extensive use of external and outsourced services in their operations
- improved access to knowledge through better access to the organization's databases, especially using intranets
- improving communication within and between organizations and relationships between suppliers and distributors.

Improving client relationships through client relationship management (CRM) programs is also becoming an important consideration in business operations. This often includes a variety of loyalty programs, such as airline frequent flyer points. Others include special discounts once a volume of sales is achieved or access to special services. CRM usually requires the building of client profiles so that clients can be kept aware of new developments of particular interest to them or of new services that may suit their particular needs.

All of these business innovations are possible only if they are supported by effective information systems.

THE BUSINESS PLAN

The outcome from these initial stages is often known as the business plan. This defines directions such as participating in electronic commerce, and affects the way the core business process itself takes place. This in turn can significantly affect the information system. The business plan also defines new ways of working within the organization and how information systems will support these ways of working. The business plan must clearly identify the way in which business processes will be carried out and the role of information technology within those business processes. These plans then provide a direction to system development. In addition, the business plan identifies new business processes, skill needs, and ways of interacting with clients.

TEXT CASE C:
Universal Securities—Identifying New Ways of Working

You might recall that Universal Securities spends too much time on developing a wide range of specialist equipment, with a consequent commitment of resources and to the detriment of time available to improve customer relationships. A number of strategic alternatives and business innovations suggest themselves in this case.

One obvious direction is to outsource some of the specialist production and then form alliances with the outsourced and other businesses that produce needed parts. As a result, less time will be spent on developing expertise in the production of a large range of specialist devices, and company personnel will have more time to concentrate on developing good customer relationships and providing specialist installation services. A second strategic change is to have more emphasis on developing client relationships and extending the system to its customers by letting customers trace the progress of their orders and possibly suggest changes as installation identifies new possibilities. In general terms, Universal Securities is looking to outsource production, develop a supply chain for receiving parts from the outsourced suppliers, and install them at clients' premises.

Figure 3.10 is a high-level summary showing how the new business will work. It shows a strategic change with reduced production of parts and greater emphasis on supporting client services. Each of the business processes would, of course, be elaborated in greater detail. Figure 3.10 provides the broad picture that identifies new business processes. One is to allow customers to track their orders. The other

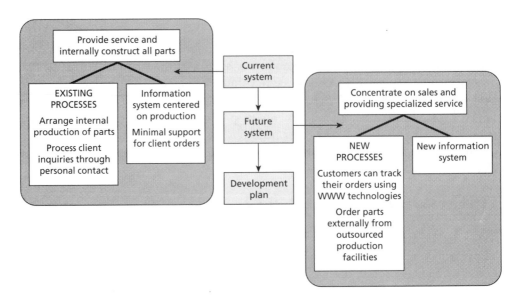

Figure 3.10 *Starting on the business plan*

is to arrange orders with outsourced and other external organizations. The business plan would include a more detailed description of these business processes. It would describe the actual organizational units to be outsourced, the kinds of alliances to be set up with them, contractual arrangements, and so on. It would define the business processes in detail, possibly using broad scenarios, identify any new positions needed in these processes, and define how people in these positions would interface with the external production organizations and clients.

In the strategic sense, some of these changes can have quite dramatic effects on the information system. Any of these will in themselves require innovations to the way business is carried out and the kinds of information systems that will be needed. Information systems will now become externally oriented.

THE INFORMATION SYSTEM PLAN

Information technology strategy must match the business plan. A business can change in both what it does and how it does it. Now that we know what the new business will look like, the next step is to define the supporting information system. There is no such thing as a standard description of an organization's information system. The components of the information systems must be defined. It is also necessary to describe precisely how the information systems will support the new ways of doing business. This in turn requires the choice of an IS organizational structure that fits the business culture and the choice of technical system components. The choice here ranges from the traditional structure of a development-oriented and centralized project team environment to user-oriented organizations where information system units support organizational divisions. The role of centralized units here is to coordinate activities and standardize technical resources to facilitate system integration. A more detailed description of these alternatives is given by Robson (1997).

Some of the technical components of an information systems plan are described below.

DATABASE STRATEGY

One important component of any plan is deciding where to store the organization's data. Databases are now considered a valuable organizational resource. They include client records, inventory, and in some organizations important business transactions or arrangements. Loss of such data could lead to severe repercussions within the organization. Important considerations include how to ensure data security while maintaining its distribution across the organization's computer network. Many strategic plans now include integrating data into data warehouses and giving personnel access to this data through intranets. Providing tools to access this data and use it effectively in the organization's operating procedures is perhaps the next challenge. This includes data mining tools and interpreting the data in the work context.

NETWORK STRATEGY

The equipment strategy includes both the infrastructure and specific items needed for specific systems. The infrastructure defines features that are common to all applications—for example, whether online access is to be supported, whether equipment will be centralized or distributed, how the database is to be distributed, the kinds of workstations to be supported, the operating system needed, and integration of local networks. The equipment strategic plan will define any proposed changes to the infrastructure, how they will be supported, and their timing and effect on existing applications. An increasingly important component of networking is how to provide people with the kinds of networking services described in Chapter 2—in particular, electronic mail and access to WWW sites. Conversely, it should show how the organization will enable external clients and potential clients to access its service through networks.

An important part of planning is to coordinate all these component plans. This is usually done through graphs.

THE INFORMATION TECHNOLOGY INFRASTRUCTURE

Infrastructure A basic structure for supporting a system.

An information technology **infrastructure** requires a major commitment of funds to provide the kind of support needed to develop planned applications. There are several broad considerations in developing an infrastructure strategy. First, we must determine in broad terms how the network will be used in the context of the business plan. Are we to allow individual departments to purchase their own microprocessors and develop minor applications on them, or are all such purchases to be coordinated? Do we have centralized or distributed systems? What information is to be kept in a data warehouse? Should a client–server approach be used? Once we decide on this broad approach, we must ensure that the equipment has sufficient capacity to support our applications. An important decision here is the approach to be used with the WWW, including its integration into the information network.

To satisfy capacity needs, it is necessary to ensure that adequate storage capacity exists to store all the needed data and that there is sufficient computer power to maintain and process this data. We must also have enough workstations to allow access to the data.

Choosing an infrastructure with sufficient capacity places considerable reliance on statements in the strategic plan. We know broadly what we must do from the strategic plan; we now need to get the details. How many users will the inventory system have? Where are the users? What is the expected size of our inventory file? All this information is then put together to estimate the size of equipment.

Once agreement on an infrastructure is reached and the necessary funds are allocated, a search for the new equipment or software to construct the infrastructure begins. Depending on the size of the acquisition, tenders may need to be called and alternative proposals may need to be evaluated. Major tasks in realizing the strategy are the selection of equipment or software, its acquisition, and final installation. The installation may require new accommodation and a new management structure.

TEXT CASE C:
Defining the Information System for Universal Securities

The information technology plan for Universal Securities must fit in with the business plan. This of course can have a major impact on the kind of information systems support that may be needed. A possible future information system structure is shown in Figure 3.11. It shows a number of major interfaces to customers, marketing, and suppliers, each of whom has an interface to the corporate database through the WWW. The major impact on the corporate databases is on customer orders and supplier orders. It is assumed that impact on financial and human resources systems will be minimal.

The link of the information systems plan to the business plan is broadly illustrated in Figure 3.12. The link is often described in more detail, describing how particular technical services will be used in the business processes. The difference from Figure 3.10 is that Figure 3.12 describes the major new components of the information system. This broadly states that customer orders will be warehoused and proposes a Web-based process for interacting with clients and suppliers. The actual plan would include more detail elaborating on the structure shown in Figure 3.11.

Detailed scenarios are also defined to explain how the business will work in the new environment.

DEVELOPMENT PLAN

The next step is to propose a plan to get to the new system. The plan must show how the information system and business plan will be integrated. Often a distinction is made between developing a totally new system and changing an existing system. The latter is often referred to as re-engineering the business process.

RE-ENGINEERING THE BUSINESS PROCESS

Changes to information systems are often seen as a way to improve business operations by reducing costs. A typical business information system may, for example, consist of many programs that were written in the past and still use old methods and technologies. They often support only one business unit. These

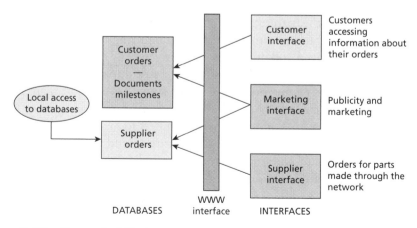

Figure 3.11 *Suggested IS plan*

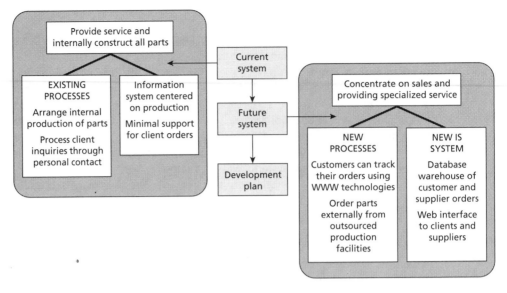

Figure 3.12   *Linking IT and the business plan*

systems are generally called **legacy systems**. On the other hand, most organizations are now demanding support for organization-wide processes that support new ways of working. Organization-wide business processes have tended to depend on the type of facilities provided by individual business units. Systems built for business units did not usually have the business processes as an important priority. There is now considerable pressure on businesses to improve their processes and thus achieve a better way of doing business. The term **business process re-engineering** often appears in this context. It refers to the identification of organization-wide business processes and changing existing systems to support the business process.

Re-engineering applies to both the business process and the information system. The business process may be changed to reduce the cost of running the business or to provide new services to clients. In either case, this often requires a change to be made to the information system.

Re-engineering the business process begins by identifying what the process must do. It must often start with existing systems. We then identify the business process needs and choose the best way to support the process, changing the legacy systems if needed. The important thing to remember in re-engineering the business process is that it must be process-driven and not technology-driven. The criteria here are factors such as improving client service, removing unnecessary duplication, ensuring that everyone is aware of what is going on, and reducing system errors. The main purpose of the business unit, then, is to set the standard that satisfies quality service requirements. After that we look at how to use computers to make it work in this way.

There are many aspects to re-engineering. One is the trend to newer technologies to support the new ways of working. Here are a number of reasons for changing the technical part of the information system:

- Changing a batch mode to a transaction-and-enquiry mode. Users who wish to have a system with better characteristics, though they have no new functional needs, often demand this.

- Improvement to the supply chain by using information systems to ensure just-in-time delivery and eliminating unnecessary steps from the source to the eventual user.
- Changing the technology to improve flexibility; this is often used as a justification for changing to object orientation.
- Changing to a new architecture—for example, going to a client–server system.
- Changing to support task-oriented teams through better coordination and workflow technologies.
- Placing a WWW interface on an existing system to extend it to clients.
- Integration of applications of different business units.
- Moving to a data warehouse system.

The simplest way to change the information system is to change a program, or programs, and the database structure. However, eventually a stage is reached when it is no longer possible to simply change the system by adding a module or changing a data record structure. This may occur because a business application has radically changed, or because of changes in technology.

Figure 3.13 shows the process used to develop a technical re-engineering strategy. Here we start with the organization's legacy systems, or those systems developed in earlier days. We then develop a model for the new systems using a process known as **reverse engineering**, where components of the existing system are included in the model. This is often necessary because the legacy systems were not properly documented, or the incremental changes have made the original documentation obsolete. The model is then further refined to show the changes needed to meet the new requirements of the redesigned business systems. The model can use any of the techniques described in earlier chapters. Then a process known as **forward engineering** is used to propose how the existing system will be changed to meet the new requirements using the existing systems as much as possible.

Reverse engineering
Developing a model for an existing computer-based information system.

Forward engineering
Re-engineering an existing system based on a model.

REFERENCE MODELS

There is also a more radical approach to re-engineering. It is simply to replace existing systems by totally new systems based on reference models or what one company, SAP, calls blueprints. Reference models or blueprints are derived from

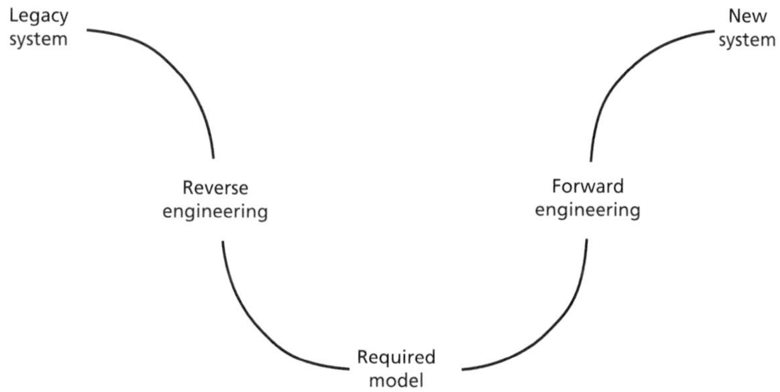

Figure 3.13   *A re-engineering strategy*

best practices found across many organizations. The reference model thus gives designers good checklists or guidelines to follow in the redesign process. There is one further advantage in adapting a reference model. This is the availability of standard software that supports the reference model. This can reduce the time and cost of re-engineering by using or customizing existing programs to build the system that supports the reference model.

One widely used reference model is known as SAP R/3. Curran and Ladd (2000) describe the standard processes that make up SAP R/3. These include sales processes, production logistics, organization and human resource management, and external accounting. The idea is that businesses define their way of operation using a special modeling method, EPC. Once the process is defined, standard modules are chosen and customized to support the process.

IDENTIFYING PROJECTS—SETTING PRIORITIES

An important part of planning is to identify the period over which the systems will be developed. It will define the application, the time when development will commence and finish, and the resources needed for the development. The relationship of applications to each other and to the organization's business plan must be clearly defined.

TEXT CASE C:
Defining the Development Plan

The exact things and their broad timing are now defined, and resources are identified. The broad link between the information systems plan and the business plan is part of the planning process, and is illustrated in Figure 3.14. The difference from Figure 3.12 is that now the development plan is included for all three dimensions. Thus, strategically, the organization will now concentrate on improving the marketing force, raising its level of specialization. Organizationally it includes the training of people in the new practices, distribution of new material to clients, and working with key clients to introduce the new systems. The

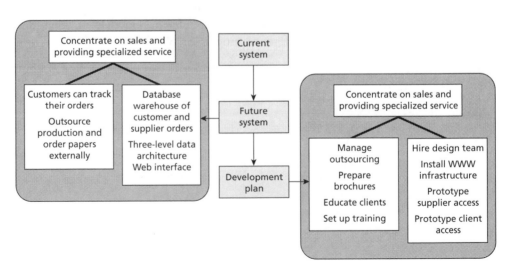

Figure 3.14 *The development plan*

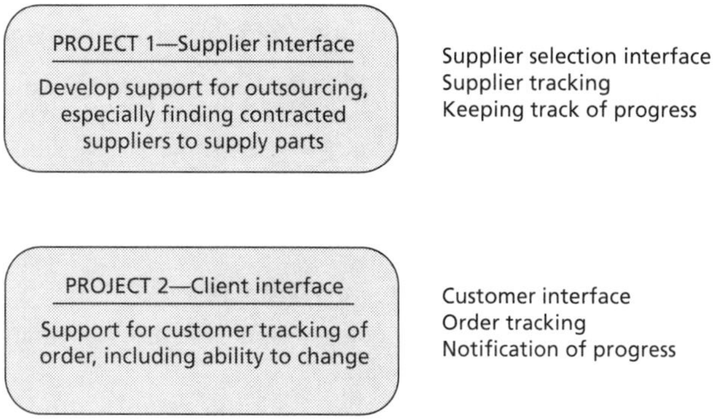

PROJECT 1—Supplier interface

Develop support for outsourcing, especially finding contracted suppliers to supply parts

Supplier selection interface
Supplier tracking
Keeping track of progress

PROJECT 2—Client interface

Support for customer tracking of order, including ability to change

Customer interface
Order tracking
Notification of progress

Figure 3.15 *Selecting projects*

information systems plan calls for the hiring of new staff to prototype the WWW systems. Again, Figure 3.14 illustrates only the broad directions, whereas in reality the plan would include considerable detail.

PROJECT SELECTION

A number of projects are now identified as part of the development plan. The first project as shown in Figure 3.15 is to develop an infrastructure for extending the current system to the WWW and developing the supplier interface. It involves selecting the technology for building Web interfaces, which often includes some experimentation and installation prototyping. The second project is to develop the monitoring interface that is to be extended to the client. These project proposals now require more detailed evaluation during the project development process, which is described in Chapters 6 and 7. Once the projects are defined they will be developed and evaluated in more detail. Such detailed evaluations are described in Chapters 7 and 13.

SUMMARY

This chapter outlines the major business units found in most organizations and how they are evolving. It stresses that changes made to information systems must be consistent with the changes in strategic directions and with new organizational structures. The chapter provides an overview of how information development plans are created and how projects to change information systems are identified.

BIBLIOGRAPHY

Boar, P. (1997), *Strategic Thinking for Information Technology*, John Wiley and Sons, New York.

Chang, L.J. and Powell, P. (1998), 'Towards a framework for business process re-engineering in small and medium sized enterprises', *Information Systems Journal*, Vol. 8, No. 3, pp. 199–216.

Curran, T.A. and Ladd, A. (2000), *SAP R/3: Business Blueprint* (2nd edn), Prentice Hall PTR, Upper Saddle River, New Jersey.

Davenport, R.A. (1993), *Innovation Processes in Business*, Harvard University Press, Boston.

Edwards, C., Ward, J. and Bytheway, A. (1991), *The Essence of Information Systems*, Prentice-Hall, New York.

Espejo, R. (January 1993), 'Strategy, structure and information management', *Journal of Information Systems*, Vol. 3, No. 1, pp. 17–31.

Hsiao, R.L. and Ormerod, R.J. (January 1998), 'A new perspective on the dynamics of IT-enabled strategic change', *Information Systems Journal*, Vol. 8, No. 1, pp. 21–52.

Kalakota, R. and Whinston, A. (1997), *Electronic Commerce*, Addison-Wesley, Reading, Massachusetts.

Lucas, H.C. Jr. (1999), *Information Technology and the Productivity Paradox: Assessing the Value of Investing in IT*, Oxford University Press, New York.

Markus, M. Lynne (June 1983), 'Power, politics, and MIS implementation', *Communications of the ACM*, Vol. 26, No. 6, pp. 430–44.

McFarlan, F.W. (September–October 1981), 'Portfolio approach to information systems', *Harvard Business Review*, pp. 192–250.

McKeen, J.D. and Guinares, T. (1985), 'Selecting MIS projects by steering committee', *Communications of the ACM*, 28(12), pp. 1344–62.

Mowshowitz, A. (ed) (September 1996), 'Virtual organizations', *Special Issue, Communications of the ACM*.

Nolan, R.L. and Gibson. C.F. (January–February 1974) 'Managing the four stages of EDP growth', *Harvard Business Review*, pp. 76–88.

O'Brien, J.A. (1994), *Introduction to Information Systems in Business Management* (7th edn), Irwin, Homewood, Illinois.

O'Connor, D. (April 1993), 'Successful strategic information systems planning', *Journal of Information Systems*, Vol. 3, No. 2, pp. 71–83.

Porter, M.E. (1985), *Competitive Advantage: Creating and Sustaining Superior Performance*, Free Press, New York.

Pyle, R. (June 1996), 'Electronic commerce and the Internet', *Communications of the ACM, Special Issue on Electronic Commerce*.

Rebstock, M. and Hildebrand, K. (eds) (1999), *SAP R/3 Management*, Corioles.

Riggins, F.J. and Rhee, H.-K. (January 1998), 'Developing the learning network using extranets', *Proceedings of the Thirty-First Hawaiian Conference on Systems Sciences*.

Robson, W. (1997), *Strategic Management and Information Systems* (2nd edn), Prentice Hall, Harlow, 1997.

Sprague, R.H. and McNurlin, B.C. (1993), *Information Systems Management in Practice*, Prentice Hall International, Englewood Cliffs, New Jersey.

Tibbets, J. and Bernstein, B. (December 1996), 'Legacy applications on the Web', *American Programmer*, Vol. 9, No. 12, pp. 18–24.

Tolis, C. (1999), 'Facilitating understanding and change: the role of business models in development work' in Nilson, A.G., Tolis, C. and Nellborn, C. (eds), *Perspectives in Business Modeling: Understanding and Changing Organizations*, Springer, Berlin.

Ward, J. and Elvin, R. (July 1999): 'A new framework for managing IT-enabled business change', *Information Systems Journal*, Vol. 9, No. 2, pp. 197–222.

Ward, P. and Griffiths, R. (1996), *Strategy Planning for Information Systems*, John Wiley and Sons, Chichester.

Williams, H. (April 1999), 'Interfirm collaboration through interfirm networks', *Information Systems Journal*, Vol. 8, No. 2, pp. 103–16.

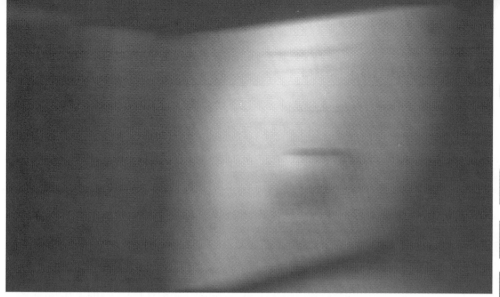

REQUIREMENTS ANALYSIS

CONTENTS

KEY LEARNING OBJECTIVES

Importance of defining requirements
Solving problems in system development
Analysis and requirements models
Translation from usage to system terms
Understanding the system through observation
Prototyping
Storyboard prototyping
Joint application development

INTRODUCTION

Before we continue with the ideas of Chapter 3 and begin to describe project development, we will first describe the kind of things that go on in actually defining what we want the future system to do. Much has been said about building correct information systems that work according to user needs. However, correct systems can be built only if it is known exactly what the user needs and what the system must do. One of the most important factors in building correct systems, therefore, is to first clearly define what the system must do. Identifying detailed user requirements has always been important and is becoming even more so in the complex systems that are being developed now. It has also been extremely important in what are known as critical systems, where malfunctions can lead to severe system failures or even catastrophes.

There are many ways to develop system requirements. One of these, however, is not simply sitting down and drawing a model of the system or setting user requirements in the privacy of an office. It can only be done by going out and discussing with users to find out what they require of the system and then building systems that satisfy these requirements. The requirements gathering methods may themselves depend on the kind of system being studied. There may be differences in searching for information in highly structured systems, or in systems that contain highly volatile group environments, or in unstructured systems such as those often found in decision support. Thus, in summary, it is necessary to determine the best way to identify requirements and then to spend some time studying the system, observing it, talking to its users, and obtaining information in many other ways about how the system works and what is needed of the new system.

In the engineering sense, described in Chapter 1, requirements are used to produce a specification—as happens in work such as building a bridge or a house, where requirements must also be identified. For example, how much traffic should the bridge carry, how long will it be, or how many rooms will there be in a house? It is only when these specifications are clearly defined that we can say whether the finished product is correct—that is, what we wanted. The goal, which is still proving to be elusive, is to develop a precise specification for computer system requirements so that we can say precisely what we want and then say whether the produced system is correct or not.

THE IMPORTANCE OF COMMUNICATION

Requirements analysis must ultimately result in a specification, which unambiguously describes what has to be built. The engineering discipline has now developed many ways of specifying requirements. There are plans, models, prototypes, and simulations. Ways to define computer system requirements are also emerging. A common computer view of requirement analysis is shown in Figure 4.1. Here there is the business process, like running a bank or making a decision about share purchases, which the new system will support. Then there is the final implementation of the system. In the middle is the system specification that defines how

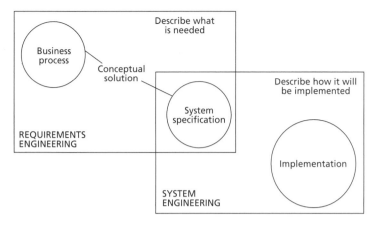

Figure 4.1 *The role of requirements analysis*

the new system is to look. A system specification is preceded by defining a broad conceptual solution, which was described in Chapter 3. Ultimately, however, it is necessary to develop the system specification, which clearly defines what the implemented system will look like, what it must do, and how it fits into the business process. The question, then, is what are the best tools to use to develop the specification. The answer is that precision is achieved by building models, which leads to the next question—what form should these models take? To answer this question, it is perhaps best to look at what it is that people actually do during system development.

WHAT DO WE DO IN SYSTEM DEVELOPMENT?

Figure 4.2 shows the main activities in system development and in the way strategy is defined. To begin with, analysts have discussions with users to familiarize themselves with the system and to get a better idea of what the new system will be required to do. New ideas are then discussed and evaluated, with arguments and positions about the new system developed. Previous knowledge or experience with similar systems is used when developing new ideas. There may also be some experimentation to find out whether some of various proposed ideas can be put into practice; opinions are formed and often used in design. During this time we maintain a record of what was discussed and what conclusions were reached about the system. This record becomes part of our general experience or group memory, which can be used in this or other projects. Many of the activities in Figure 4.2, such as design and decision making or simple explanations, are carried out in ways described in Chapter 2.

There are two other important aspects here. One is communication. In all of these processes it is often necessary to reach agreement between many people, to specify what we are to do and to represent it in an unambiguous way. The second is organization of this work into a process that must eventually finish by producing a new system. This chapter describes the ways we can communicate or negotiate to reach agreements and develop system specifications. Chapter 6 describes organizing the activities into a development process. Perhaps the most important

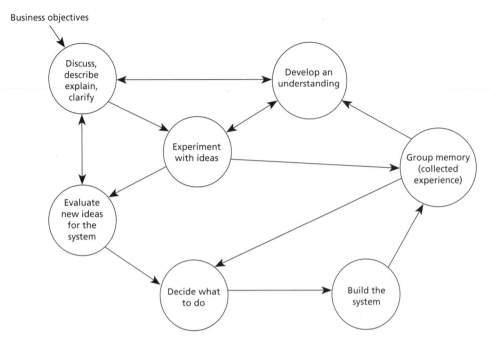

Figure 4.2 *Some activities in problem solving*

aspect of all these activities is to ensure that no ambiguities arise in discussions between the various people involved in analysis. Such ambiguities can easily arise, however, because of the different jargons that are commonly used by different people—in particular, the users, who often speak in terms common to their domain, and computer analysts, whose language uses computer terms. The goal of any analysis must be to ensure that all of these people 'eventually speak the same language' so that the correct requirements are identified. This requires the clear definition of terms so that everyone attaches the same meaning to any statements made.

THE DIFFERENT 'LANGUAGES'

Another way to look at communication is that it involves a number of people each using a language that suits their 'world'. Some users talk about their problem domain—for example, how to manage bank transactions and make decisions. This is the usage world. They like to use **terms** common to their work and describe their work by scenarios—that is, giving examples of particular things that happened in their world. At the other end are the computer system developers, who like to talk about computer systems, using terms particular to this development world—for example, operating systems and databases. What we are doing is finding a way to bring these two worlds together. To do this it is necessary to translate from one language to another in unambiguous ways. This must often go through a number of steps or other languages, as shown in Figure 4.3.

Figure 4.3 identifies some of the other worlds that one finds in building systems. One is the usage world, where the language commonly describes the way the user works, usually in terms of examples. At the other end is the development world,

Terms Words with specific meaning used to describe systems.

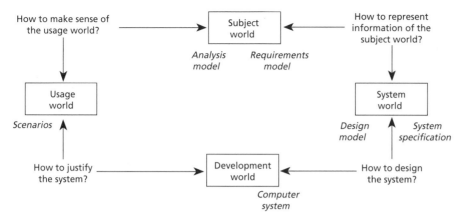

Figure 4.3 *The worlds in system development*

which describes how the computer system works. The other two worlds are those needed to establish a communication path between these two diverse groups.

The first task here is to define the subject world. Here we try to describe the usage world in clear, unambiguous but now well-defined terms rather than specific scenarios. Here we try to make sense of the scenarios and examples and try to find out what the essence of the system is. Some of the user terms will now be unambiguously defined to eliminate any jargon that is common in the usage world. Thus, the precise meaning of terms like quote or premium in a given organization is now defined, not what the analyst thinks they mean using their common knowledge. In fact, now there is increasing demand for subject specialists who can deal with the increasing complexities found in specific subject worlds, such as insurance or banking. A clear picture begins to emerge about how the usage world works in terms that describe general processes rather than individual scenarios. The most essential characteristics also begin to emerge at this stage. These terms are used in the statement of requirements, which as one of its components should include a precise glossary of terms.

The next step is to represent the subject terms in abstract system terms that are later useful to the developers but still can be understood by the users. Such system terms are used to build the analysis models that become part of the system specification and are easily converted to computer systems. These system models are described in detail in Chapters 8–12.

REPRESENTING SYSTEMS

An important activity in system development is to build representations or models in the different worlds. These models are built using the terms identified for each world and thus become precise system representations in that world. The models are used to keep track of what has been done to date and what else needs to be done to complete information gathering.

THE IMPORTANCE OF MODELS

Most practitioners agree that good design occurs only once the designers develop a good understanding of the existing system and what users expect from it. To do this, it is necessary to establish good ways of communication so that everyone in system development clearly understands what anyone else is saying. Models are used to develop such an understanding and to provide such precision in communication. They work on the premise that communication is possible only when everyone uses a consistent terminology and precise concepts. Models achieve this by providing a consistent set of concepts and use well-defined terms. In addition, most models include visual components, which often make it easier to understand proposals. Other important advantages come out of using a model and a consistent set of terms. A model provides a framework for decision making and for generating alternatives by providing ways to precisely evaluate such alternatives.

Models are usually made up of objects that can be easily related to the problem domain. They are there to make it easier for people to understand what systems are all about and to discuss proposals in a meaningful way. A model looks like the thing you are analyzing and designing. For example, the house plan described in Chapter 1 looks like a house and can be easily used to describe various design options.

Modeling also requires analysts to follow a process. They ask questions, like how do you start? Do you build a high-level model and then look at detail or go the other way around? How do you make sure that you did not miss something? Many modeling methods include guidelines and checklists to help people choose alternatives and make sure that nothing is forgotten. An obvious checklist item for a house may be that it must have a bathroom. However, many other requirements come from the building codes relevant to the type of building under construction.

WHAT KINDS OF MODELS?

The next question is what kind of modeling methods are commonly used. Although there are claims that some modeling methods can be used for everything, it is now commonly agreed that different models are better for different things. Thus, you should not learn just one modeling method but learn a variety of methods. A system designer, just like any craftsman, has many tools. So a system designer should view models as tools that help them design a system. They must choose the best tool for a given problem. Just imagine if a furniture maker only had one tool, the hammer. There is not much that this person could do. Similarly a system designer is highly constrained if that designer tries to use the one modeling method for everything.

There are many different kinds of models. They often correspond to the worlds shown in Figure 4.3. Thus, different models are needed for strategic evaluation, organizational design, and information system design from the perspective of the different worlds in Figure 4.3. Thus, we need different models to develop an understanding at different levels. In the overall process described in Chapter 3, models can be seen as method chains and alliances, as shown in Figure 4.4. Here we start with a business model that describes the overall organization. We can use

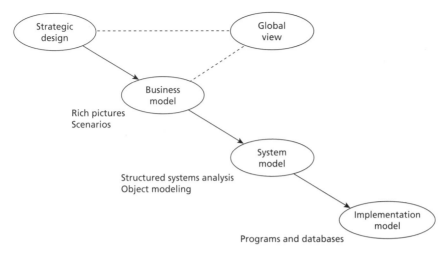

Figure 4.4 *Model chains*

rich pictures to do this. Then system models use representations that describe the way the system works. A number of alternative modeling methods are available for this purpose. The two main modeling methods are structured systems analysis and object modeling. Finally we have an actual implementation as a set of programs and databases.

DEVELOPING AN ORGANIZATIONAL PERSPECTIVE

Perhaps the first step is to develop an overall perspective of the organization. Some writers suggest that methods such as Soft-Systems Methodologies are useful for understanding the problem situation. We have already used ideas from this method in Chapter 3 to describe Universal Securities. Strategic design often uses financial models. Organizational design often uses things like organization charts.

TEXT CASE A:
Interactive Marketing—An Overall View
Figure 4.5 thus describes the overall business model of the interactive marketing situation in terms of rich pictures. It includes a number of activities shown as clouded shapes. The roles associated with each activity are linked to the activity. The artifacts produced in the activities are shown as rectangular boxes. Thus, BUYERS are the main kinds of people involved in purchasing. They examine PRODUCT OFFERS and initiate a SALE.

Strategically, we can then look at each activity and see how it is to change.

WHAT ARE THE MODELS IN SYSTEM DEVELOPMENT?

Most of this book concentrates on modeling of information systems, which includes a variety of modeling techniques such as object modeling, structured analysis, and ways to model data structures. All of these are described in later chapters.

A number of different models may be created as one proceeds through a development process. The development process centers on starting with an initial

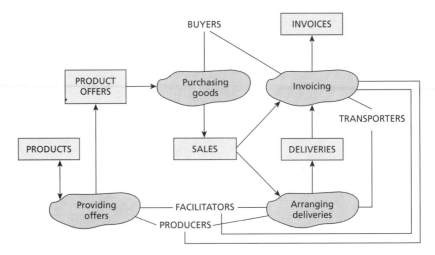

Figure 4.5 *Describing the overall system*

usage world model and converting it, through subject and system models, to a computer system. In fact, development processes are often characterized by the models they support at each of their steps. Examples of models include:

- the analysis model, which describes how the system works now in terms of the subject world
- the requirements model, which describes what users need in terms of the subject world
- the design model, which describes the required computer system, in terms of the system world.

Models and conversion play an important role in quality assurance. In the first instance, models themselves must satisfy some criteria related to good practice. Thus, there are accepted ways of carrying on business, best practices for the usage and subject world, criteria for good systems models, and measures for good computer system performance. The second aspect of quality assurance is that the conversion process must ensure that all characteristics captured in one world model are correctly translated into the next world model.

Quality is an important aspect in the conversion. It is necessary to ensure that requirements are correctly translated from the usage world to the subject world and the system world. Equally important is to judge the quality of the models created in each of the two worlds. System design methods thus often include steps and guidelines to be used in the conversion, and standards to judge models. Thus, for example, best-practice models in particular subject worlds, such as banking or insurance, can be used to guide designers to good models. Equally, standards are being created for system world models. The rising complexity of the subject world is also leading to subject world experts, people who are familiar with both the intricacies of the subject world and their mapping to the system world.

Many development processes are supported by computer tools, known as CASE (Computer Assisted Software Engineering) tools, that help designers develop and maintain their models.

IDENTIFYING REQUIREMENTS

An important aspect of modeling is its role as a communications tool in determining requirements. Usually the following approach is used. An analyst will develop a model following an initial discussion with a user. Then a repeat visit may validate the model with the user. Once agreement is reached on the model, further detailed data may be gathered to elaborate the model. This iterative approach serves a number of purposes. First, it ensures that there is always a record of the information gathered to date. Second, it serves to ensure the correctness of the information by continually verifying with the user the results obtained to date. This ensures that an analyst does not get too far ahead using erroneous assumptions.

Requirements must in the first instance be identified from information collected in the usage world. Analysts must concentrate on the ways the new system can improve this world and express their improvements in usage world terms, by explaining proposed scenarios for the new system. However, the requirements must also be defined in general terms, more common to the subject world, that can then be converted to specifications in the system world. One important aspect here is to use general terms so that systems can support a large range of scenarios, rather than one. Thus, in the subject world, we try to identify general terms, whereas in the system world, we define general system terms. These terms must be chosen in ways that make it easy to translate from one world to another.

DESCRIBING THE USAGE WORLD

One way to describe the usage world is to use rich pictures. This is the premise of what are known as **Soft-Systems Methodologies**, where the usage world is described as closely as possible to match what people see happening in their world, by pictures and scenarios. Examples from the usage world are then used to develop a subject world model by defining the terms that will be used to describe this world. The choice of concepts is left open to the designers.

Soft-Systems Methodology
A development process centering on the user and subject worlds.

> ### TEXT CASE B:
> ### Managing an Agency—Analyzing Broadlevel Interactions
> Chapter 2 briefly introduced the problems in managing the agency. The manager of the agency decided that one of the first steps in supporting the committee would be to define the activities between the people involved and how the activities work with each other. These activities are illustrated in Figure 4.6, which is a simple diagram identifying the roles involved in the project and joining those roles that are likely to interact. The activities are shown in the clouded shapes. The rich picture also shows the main artifacts used in the system. These are shown in rectangular boxes. Thus, draft sections are exchanged between the different groups, whereas a draft report is considered at the committee meeting. The committee also works on a work plan. The next step is to write a scenario for each activity. For example, the script for the Committee Chairperson – Committee Member activity may include headings such as 'arrange meeting', 'discussion', and 'minute verification'.

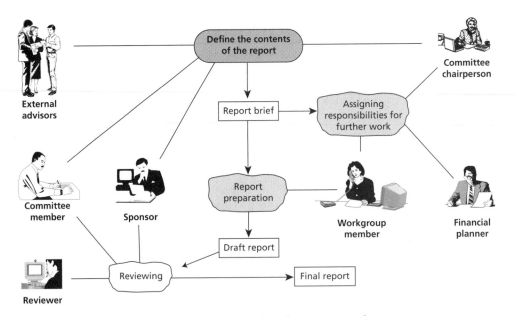

Figure 4.6 *A rich picture showing interactions between people*

SCENARIOS

Scenarios support models such as the rich picture by illustrating system dynamics using typical sequences of actions in the user world. For example, one scenario for the system in Figure 4.6 may be the following:

- The committee chairperson writes a project brief and assigns it to different workgroups.
- Workgroup leaders coordinate their workgroups by discussing the brief and then creating an initial draft, which is passed between the workgroup members.
- Workgroup leaders meet occasionally with the chairperson, external advisors, and other nominated committee members to resolve any conflicts.

Scenarios may take different forms. There may be a specific instance that describes a particular time when a group leader handed a brief, say 'develop a marketing proposal for a new perfume', to individual people. This scenario may describe not only the process followed but the problems found and how they were overcome. Then many such instances may be used to develop a more general scenario that looks at a number of specific instances and defines a general scenario that can cover these instances. It may also identify a variety of exception cases that may arise in the scenario. Ways of defining scenarios are now becoming more formalized. It has, for example, become a very popular way for describing requirements in object-oriented design, where general scenarios are known by the word **use cases**. We will discuss use cases in the chapter on object-oriented analysis.

Use case A
description of work in
usage world terms.

   Scenarios can be expanded to indicate the kind of support to be provided for the interactions. Expansion includes identifying interaction characteristics and choosing appropriate software for them. Thus, for example, a teleconferencing system may be proposed for meetings where all participants are at different locations. This teleconferencing may be integrated with a decision support system if the meeting is required to reach a decision.

THE SUBJECT WORLD

The subject world focuses on business issues. Its goal is to identify the crucial measures of business success—for example, cost or production delays—and propose ways to achieve them. The important factor in subject world modeling is to eliminate jargon that is often found in the usage world and clearly define terms that describe the user world. The definition of such terms is also recognized as an important step in Soft-Systems Methodologies. This approach is also followed in object-oriented design by generalizing individual scenarios into use cases, usually accompanied by a clear definition of subject terms.

With many systems becoming increasingly complex, subject world models are one of the least known in the conversion process. There are few criteria that can be used to evaluate subject world models, although there is a tendency to define best practices in many industries. Eventually these will become the guidelines used to evaluate the subject world. There is also a trend for people to become subject world specialists, such as the bank analysis specialist, whose main activity may be to develop models of the subject world and express them as conceptual schema in the system world.

Soft-Systems Methodologies approach the subject world by defining standard subject terms once the initial analysis is completed. These terms are then used to create a subject world model.

One important activity in the subject world is to define a glossary of terms. These define clearly what is meant by terms such as delayed-back-order or customer-response. Once these are defined, it is easier to find ways to represent them using system models.

THE SYSTEM WORLD

In the system world we begin to talk more in general system ways, ways that can be applied to any problem. This area is often termed conceptual modeling in computer system design. It describes the system in terms that are useful to computer systems designers—terms that can be used to develop a computer system specification. There are a large number of proposals and ways for doing this. One set of terms comes under the general heading of *structured systems analysis*. These terms include models such as data flow diagrams, entities and relationships, and process descriptions. Another set of terms comes from object-oriented design. The book describes these modeling methods in later chapters. This chapter concentrates on collecting data needed to identify requirements.

COLLECTION METHODS

Remember that to define user requirements requires an understanding of how the system works and what its problems are. We now look at methods for collecting information to develop such an understanding. There are many important issues to look at to get a clear picture of a system. One is to look at the current business processes in the system and identify the tasks in these processes. Or one can begin by examining particular system functions and their tasks. The tasks can then be examined in detail. Such examination can identify the users who carry out the tasks, the interactions between the users, the tools they use, and the artifacts on which they operate. Thus, an analyst must always consider the users and what they do. How do they use artifacts? With whom do they interact? Where do they find information? This has an important bearing on the way the system works. Remember also that there is no such thing as a standard system. It is not useful, therefore, to have preconceived ideas about a system, and analysts must approach any study with an open mind. There are three major ways of doing this:

- **Asking questions** by interviewing people in the system, through surveys and questionnaires, or by electronic means using e-mail or a discussion database.
- **Observational studies**, including ethnography or by participating within the user environment.
- **Prototyping** either the requirements or the interface.
- **Formal sessions**, including structured workshops, group discussion, or facilitated teams.

Another important consideration is the methods used to maintain records about information found and for documenting and analyzing this information; that is, translating from the subject to the usage world. The kinds of activities carried out by analysts in this sequence range from simply keeping notes to building models of the system. Computer systems themselves may often support an analyst in keeping the information. Some of these productivity tools are described in a later chapter.

GATHERING INFORMATION BY ASKING QUESTIONS

Interviewing is perhaps the most commonly used technique in analysis. In a way, there is no way to avoid interviews, as they must precede any other method for gathering requirements. It is always necessary to first approach someone and ask them what their problems and priorities are, and later discuss with them the results of your analysis. For this reason we have left interviewing to a chapter of its own, Chapter 19, for those who want a more detailed understanding of interviewing.

QUESTIONNAIRES

Questionnaires provide an alternative to interviews for finding out information about a system. Questionnaires are made up of questions about information sought by the analyst. The questionnaire is then sent to the user, and replies are analyzed by the analyst.

There are a number of disadvantages in using questionnaires. First, they are suited mostly to closed questions and are not effective for open questions such as 'describe your general duties' or 'what are the most important system components?' Secondly, questionnaire questions are usually not answered completely and will often express some current concern rather than long-term concerns. Thus, the response to 'describe your general duties' will often include the things the user did in the last day or so. A set of follow-up questions is usually necessary to establish activities over a longer period of time. This is best done in an interview rather than a long questionnaire.

Questionnaires, however, are useful when the same kind of information is sought from a number of users. This is especially so if that information is of a quantitative nature—for example, finding out how many calls salespeople make each day. A questionnaire that asks that question is simply sent out to all the salespeople in the organization.

It is also possible to use questionnaires to get an idea about the quality and performance of systems. This can be done by using structured questions that include a range of replies and require the responder to tick one of them. For example, a question to customers on the quality of service may read as follows:

Do you feel that our deliveries to you are:

Always prompt?
As good as can be expected?
Sometimes late?
Often late?

Questionnaires, of course, often include many questions to find information about some system component. For example, we might have a questionnaire like that shown in Figure 4.7. Here we are trying to establish the usefulness of a program

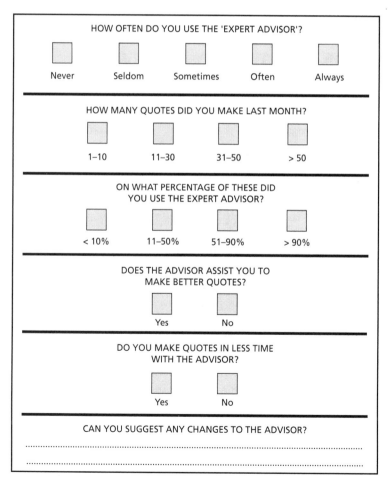

Figure 4.7   *A questionnaire*

developed to support a large number of insurance brokers in making insurance quotes. The first question establishes the user's familiarity with the program. This is followed by a question that indicates the user's general experience. Then questions are asked about whether the user finds the program useful, and finally looks for suggestions for any improvement. A collection of responses can then be used to establish the general acceptability of the program and determine whether it should continue to be used and whether any changes are needed.

In summary, questionnaires are used to supplement other techniques. They are useful for gathering numerical data or getting relatively simple opinions from a number of people, but they are not very effective for in-depth searches or for identifying system problems or solutions. Interviews tend to be more successful for this purpose.

ELECTRONIC DATA GATHERING

Electronic communication systems are now also being increasingly used to gather information through electronic questionnaires. Thus, it is possible to broadcast a question to a number of users in an organization to obtain their viewpoint on a particular issue. Alternatively, a bulletin board can be used to allow potential users to record their requirements.

GATHERING INFORMATION BY OBSERVATION

Interviewing and other ways of asking questions are characterized by having analysts learn about the system mainly by interviewing individuals. Interviews often emphasize what an individual does and how to support them rather than looking at the individual's relationship to their group. The analyst then has to correlate findings about individuals and determine how the whole group works. It is also the interviewer's responsibility to find out the objectives of each individual and try to make them cohesive.

Some other characteristics of interviewing can lead to incorrect assumptions about a system. One is the danger that because the interview is usually carried out outside the working environment, the interviewee may distort replies to questions. There is the possibility of exaggeration or emphasizing the less important aspects of activities. There is also the danger that the interviewer may have preconceived ideas about the system and its needs and try to impose these ideas into the study or interpret replies in terms of these preconceived notions.

A further disadvantage of data gathering based on individual interviews is that it assumes that an individual's work is relatively routine in nature and does not change over time. This, however, may not be the case in some team situations where tasks are dynamically and spontaneously distributed between users. We thus have a situation where there is a dynamic, or time-varying, division of labor, where the rules for this division themselves change. Interviewees often find it difficult to describe these situations in exact terms, thus leading to possible misunderstandings about the system. Ethnography aims to overcome some of these problems.

USING ETHNOGRAPHY

Ethnography is not a new field but it is a new approach to analyzing computer system requirements. One of its most important goals is not to superimpose the interviewer's or analyst's viewpoint on the system but to use the viewpoint of the people within the system. The terms **emic** and **etic** are sometimes used to stress this distinction. An etic view of the system is the 'outside view' or what the analyst sees, whereas the emic view is the 'inside view' or what the system users see. The goal is not to superimpose the outside view onto a system and consequently identify the wrong problems but to actually see what goes on from the insider's viewpoint. The main characteristics of ethnographic studies are listed here:

- Analysts observe or possibly even participate in such activities.
- Any interviews are conducted *in situ*, possibly as informal discussion rather than a formal interview.
- There is emphasis on examining the interaction between users.
- There is emphasis on tracing communication links.
- There is detailed analysis of artifacts.

Ethnographic studies therefore place a different orientation on information gathering. The emphasis is on observation of system activities, and perhaps even active participation in these activities by analysts, and on observing the system from within rather than from the outside. There are a number of advantages in the ethnographic approach. First, because interviewing is carried out *in situ*, the system is directly observed as it actually works. The users are not disturbed in their activities, and information is gathered directly and not from an informal description obtained through interviews.

ANALYSIS BY PARTICIPATION

Ethnographic studies, because of their emphasis on interaction, are particularly important in studying how groups of people work. Their goal is to study the dynamic social situations that occur in such environments. It is usual here to identify communities or workers and analyze their interactions. One way to gather information is by actually participating in group activities. The analyst becomes a member of a team, perhaps in an indirect capacity assisting other team members. Successful studies have been undertaken on the way air traffic controllers work (Bentley et al., 1992).

ANALYSIS BY OBSERVATION

The goal here is to observe what people do in an unobtrusive way. The best way to do this is by video recording. It is important in video recording to ensure that the presence of the camera itself does not alter behavior while at the same time collecting sufficient in-depth information to make useful observations. This requires considerable skill in both the placement of the recording equipment and the setting of the video camera itself. Once a video is complete, analysis commences. Analysis is usually carried out in conjunction with the participants of the work setting. The result of the review session is a script of user activities and their interactions with each other.

Ethnography
Gathering information by observation.

Emic An inside view of a system.

Etic An outside view of a system.

SOME ANALYSIS TECHNIQUES

Ethnography itself includes a number of techniques (Jordan, 1993):

Analyzing people's roles—This identifies how a particular person feels about their work and the kinds of problems they encounter. The goal is to gather information on the practices followed by people carrying out a particular task.

Analyzing interaction—Interaction analysis defines how users work together in groups. A simple diagram may be produced initially, showing the various system roles and their relationships. Then scripts may be developed to describe such relationships in more detail. This often leads to the identification of communities of practice that are spontaneously formed around a particular task.

Analyzing location—A study is made of what happens at a particular place over a period of time. Often the study produces a set of snapshots of activities during that period.

Analyzing artifacts—The emphasis is on how artifact analysis fits into the flow of work rather than on artifact structure in its own right. Important aspects are:

- how an artifact flows through the system
- use of artifacts by teams rather than individuals
- considering the artifact as a workspace in its own right.

Task analysis—The analyst studies the processes within a system and the role of individual users. Emphasis is on the information needed by the user and what the user does with the information and where it is obtained.

Although ethnography has been applied mainly in the study of detailed interactions within dynamic group environments, we can learn from it for more general application analysis. For example, we may identify the interactions at a broad rather than a detailed level and then use them as a guideline in our further analysis.

FROM OBSERVATION TO DESIGN

Another characteristic of the ethnographic approach is that objectives are developed with user participation. Considerable time, for example, may be spent on the design of workspaces through experimentation. Such workspaces not only include the computer screen but also the physical space used for any manual work on associated documentation or paperwork.

Ethnography can, of course, be made to fit in with other methods of information gathering. It can, for example, be used to bring the social issues into analysis because it emphasizes interaction. There is also a clear link between ethnography and prototyping. It should be possible to take parts of information found in analysis and gradually use it to build systems.

GATHERING INFORMATION BY PROTOTYPING

Prototyping is one of those terms that are often used to mean different things. We can build prototypes for many reasons: to test some new idea, as a first step in development, or as a tool for requirements analysis.

Historically, most prototypes were used to test a new idea or design. You often hear of prototypes used in engineering. These were based on models that were used to illustrate some system. For example, we could have a model of a plane in a wind tunnel to test out a design. The model would be 'flown' in the wind tunnel, then measurements would be made on performance and then used, if necessary, to modify the design. Once the designers are satisfied with the results of the model, the real plane would be built. In this case, the prototype is built to illustrate the feasibility of the new system and then virtually thrown away.

The main role of prototyping in information systems design is to improve requirements definition by involving potential system users. Prototyping is primarily an experimental method. A rough system is built, and users can experiment with it and make comments in usage model terms about its suitability for the workplace. Their reactions are obtained and used to define requirements in an iterative way.

Prototyping has the added advantage in that it serves to create a culture of democracy by involving users in the development, rather than a culture where only the analyst and designer play the leading role. It can thus ensure user commitment to the developed system by involving users early in the decision-making process. It removes rigid boundaries that exist between analysts and users and can lead to better elicitation of user requirements and eventually better systems. It is now commonly used in design, especially where systems cannot be easily prespecified. We describe how prototyping is used in system development in Chapter 5.

In many cases, prototypes can be used in the final system, and hence not all parts of a system prototype are discarded. There are, of course, different kinds of prototyping. One way may be to actually build the system while designing it. Another just illustrates some components of the system, which are eventually discarded. A third and very common approach is to simulate interactive processes to see whether they satisfy user needs. We will return to prototyping in Chapter 5. The most common method is to develop interface prototypes.

INTERFACE PROTOTYPING

Perhaps the most common way of prototyping is to develop screens that illustrate what users will have to do in the new system. This can then be used to describe to users the kind of information that would be made available to them from the system and how they can work with this information. For example, we may construct an interface to illustrate to a user the kind of inputs and possible outputs that a proposed system would require. However, the actual system would not be written itself. All we would have is stored example screens that could be shown to a user. Thus, in our interactive marketing system, we could build an interface that illustrates availability of product to potential customers and then get their reaction. We could also propose input screens to gauge their acceptability. Of course, one could argue that this could be done simply by drawing the screens on paper. However, it is found that the actual illustration of screens is a better approach. This is particularly so if the potential user can experiment with some sample inputs and outputs and suggest changes to them.

PROTOTYPING PROCESSES

Prototypes can also be used to describe a process that involves a number of users. The term **storyboarding** is now often used to describe this approach. What we do

Storyboarding
Using a sequence of computer screens to describe how a system will be used.

now is to have a series of small prototypes and tie them together, so that a user can see how the whole system works.

One often develops a **storyboard** as a sequence of screens. It is almost like developing a film script. Thus, there may be a screen that displays what a requisition may look like. The user can experiment with this screen and change its layout to adapt it to a particular method of working. Then we may develop an approval screen. The storyboard then becomes two steps. We can show the kind of input on the requisition screen and illustrate how it results in the approve screen. Then we can develop an order screen, and so on. Thus a storyboard evolves, and users as well as designers can begin to see how the system will work. It should also be noted that initially screen inputs need not result in any actual computations. All that the underlying program does is move data from one screen to another without storing it in a database or checking input for correctness. As the prototype develops, we may add to these programs a database or some checking. Thus, the prototype may gradually evolve to a working system.

Storyboard A sequence of screens that illustrate how a system will work.

Text Case A:
Interactive Marketing—A Storyboard

ISSUES
Producers must accept our interactive marketing system if it is to be successfully implemented. Most of the producers have little familiarity with computers and consequently cannot easily express their requirements. However, the requirements, particularly the user interface, must be known before the system is built.

SOLUTION
The proposed solution here is to use prototyping to elicit user requirements. It is proposed to set up a series of screens to explain to producers all the steps that they would have to follow to use the system. We start with the three screens shown in Figure 4.8, one to show the initial input, the second an offer received, and the third a delivery advice. We can now use the screens as a basis for discussion with the producers.

The screens are developed to follow the sequence of the process used by the producers. In that sense we can use a discussion framework: for example, we could say: 'The first thing that you do is enter the following on the screen.' At that stage you may inquire if something else is needed from the producer, or discuss screen layout in more detail. Then you can discuss the offer screen and how to respond to it. The menu here allows the initial offer to be retrieved or buyer details to be found. The screens for these have not yet been devised, but it is possible to discuss with producers what they would like to have on these screens. Other questions can be asked on issues such as screen layout or, indeed, whether this screen is needed at all or whether offers should be automatically accepted. In this way you can find out what offers can be automatically accepted and what else is needed to make a decision. You may then discuss reporting procedures for automatically accepted offers. The discussion will lead on to deliveries and so on.

One particular issue for discussion is the numbering scheme used for offers and purchases.

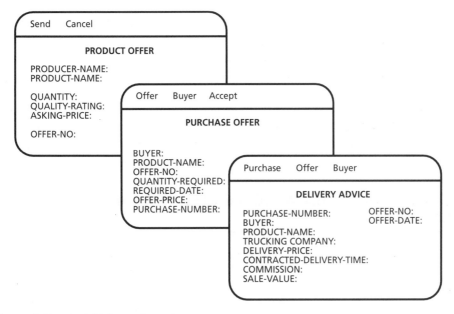

Figure 4.8 *An initial storyboard*

TOOLS FOR PROTOTYPING

Prototyping needs tools that can be used to quickly set up storyboards. Such tools include screen generators and form generators. These tools allow users to quickly define what a screen will look like.

PROTOTYPING ON THE WEB

Web design offers perhaps the easiest way to prototype. Web pages that are not interactive can be quite easily set up using HTML to set up pages that illustrate process steps. These pages can then be linked into a storyboard like that shown in Figure 4.8. Users can then follow the storyboard to provide additional requirements to the designer.

PROTOTYPING TO TEST A NEW IDEA

Here, prototyping is used when a totally novel system is proposed. No early experience exists with a similar system, and some model is needed to gain experience with the kind of problems that can be expected when developing the full system.

A typical example of prototyping is to test out a new algorithm. We may have heard of an algorithm that may be useful for a given problem. We could then test the algorithm on some sample data to determine whether it does what we want and to measure its performance. If it turns out to be satisfactory, then the algorithm will be used to develop the system.

AN EXAMPLE—WINNING AT HORSE RACES

Suppose someone has an idea about how to predict the winners of horse races. This requires considerable amounts of information about horse characteristics and history, as well as track and weather conditions during races. The algorithm proposed is quite experimental. Obviously, most people would be sceptical about any such proposal, but suppose you want to try it anyway. It would be unwise to start by collecting information about all horses and races. This could be quite expensive and nothing may come of it.

A better way to proceed would be to gather information about some small subset of horses and races and perhaps try a modified algorithm on this data. Once such a system is implemented, we can experiment with it to see if we can indeed predict winners. If so, the system is then extended to include all races and horses.

JOINT APPLICATION DEVELOPMENT (JAD)

So far, all the methods outlined in this chapter can take place sequentially. With ever greater emphasis on reduced development times, there is more pressure for requirements determination and design to be expedited. In fact, as early as 1970, a method known as Joint Application Design (JAD) was started by IBM and is increasingly used now to speed up requirements determination. The idea here is that, rather than gathering information from people one at a time and resolving conflicts by repeated discussion, we get all the people together to identify in one session what needs to be done. To do this, all the key personnel are gathered together to propose what is required.

A JAD meeting or workshop is not simply informal meetings with discussion that hopefully leads to a useful outcome. Instead, JAD sessions are organized into structured discussions often supported by tools. Usually JAD sessions are facilitated by people with special roles. They get all the right people to the session and assign them their appropriate roles. A JAD session often involves:

- a sponsor who is supporting and underwriting the development
- a session leader to organize and run the JAD
- the key users of the new system
- managers of the groups that will use the systems
- information systems personnel, including system analysts and designers.

A scribe is also included in the team to record the meeting outcomes.

The meeting is often conducted outside the normal working environment to minimize disruptions and movements of people leaving the meeting to attend to their normal work. The goal is to ensure that all members of the JAD workshop concentrate on the problem at hand and arrive at an acceptable solution.

The leader organizes the meeting by both setting an agenda and facilitating the use of models in the discussions. The session may start with an initial model, like the rich picture shown in Figure 4.5. Then there may be some discussion to verify it and explain whenever necessary. The facilitator may then initiate a brainstorming session to identify activities seen as needing the most change. Strategic goals may be used as guidelines to identify the affected activities. The workshop leader must thus have all this information to remind people of strategic goals through business innovations and then identify which activities need to be changed to bring about these innovations. Comments may be gathered and eventually collated for a decision on setting priorities.

The more modern approaches now use additional tools, like those described in Chapter 2 to assist decision making and collation of information. A good document management system can ensure that everyone gets prior access to the latest documents. Tools such as discussion databases can be used to collect information before the meeting. During the meeting, participants may be provided with information relevant to the meeting and with the facilities to raise comments through the use of computer tools rather than discussion, thus providing ways for all to contribute to the discussion.

SUMMARY

This chapter begins by describing the importance of defining the requirements for the new system. It stresses that communication is one of the most important issues in eliciting requirements, and describes the different levels, or worlds, at which such information can be gathered. Whatever the level, it is important to define unambiguous terms that are used in communication. The advantages of building models and converting from one model to another during the development process are also outlined.

The chapter then describes ways of collecting information about the system. These include gathering information through interviewing, participation in user teams, and prototyping.

DISCUSSION QUESTIONS

4.1   Why is it necessary to go through are a number of 'worlds' in modeling systems?

4.2   Why are unambiguous terms needed in the different worlds?

4.3   Why is modeling important?

4.4   What do you understand by the *subject world*?

4.5   What is the difference between the subject and usage worlds?

4.6   Why are scenarios important?

4.7   Where would participation be the best way of gathering requirements?

4.8   Describe how the use of modeling can improve 'system quality'.

4.9   Are you familiar with any systems where a prototype problem-solving approach would be useful?

4.10   What do you understand by the term *storyboard*?

EXERCISE

Develop a rich picture for the interactive marketing case.

BIBLIOGRAPHY

Anriole, S.J. (1992), *Storyboard Prototyping: A New Approach to User Requirements Analysis* (2nd edn), QED Information Sciences, Inc. Wellesley, Massachusetts.

Bentley, R., Hughes, J.A., Randall, D., Rodden, T., Sawyer, P., Shapiro, D., and Somerville, I. (1992), Ethnographically-informed systems design for air traffic control, Proceedings of the CSCW 92 Conference, Toronto, pp. 123–29.

Brittan, J. N. G. (February 1980), 'Design for a changing environment', *The Computer Journal*, 23(1), pp. 13–19.

Carmel, E.R., Whitaker, R. and George, J.F (June 1993), 'Participatory design and joint application design: a transatlantic comparison', *Communications of the ACM*, Vol. 36, No. 6, pp. 40–8.

Jarke, M. (ed) (December 1998), 'Requirements tracing', *Special Issue of the Communications of the ACM*.

Jeffrey, H.J. and Putman, A.O. (March 1994), 'Relationship definition and management: tools for requirements analysis', *Journal of Systems Software*, Vol. 24, No. 3, pp. 277–94.

Jordan, B. (June 1993): *Ethnographic workplace studies and CSCW*, Proceedings of the 12th Interdisciplinary Workshop on Information Processing, North-Holland Publishing Co. Amsterdam.

Keil, M. and Carmel, E. (May 1995), 'Customer–developer links in software development', *Communications of the ACM*, Vol. 38, No. 5, pp. 33–44.

King, D. (1984), *Current Practices in Software Development*, Yourdon Press, New York.

Madsen, K.H. and Aiken, P.H. (June 1993), 'Experiences using cooperative interactive storyboard prototyping', *Communications of the ACM*, Vol. 36, No. 4, pp. 57–66.

Mason, R.E.A. and Carey, T.T. (May 1983), 'Prototyping interactive information systems', *Communications of the ACM*, 26(5), pp. 347–54.

Mylopoulus, J., Cheung, L. and Yu, E. (January 1999), 'From object-oriented to goal oriented requirements analysis', *Communications of the ACM*, pp. 31–7.

Nilsson, A.G., Tolis, C. and Nellborn, C. (eds) (1999), *Perspectives on Business Modelling*, Springer, Berlin.

Tamai, T. (June 1993), 'Current practices in software processes for system planning and requirements analysis', *Information and Software Technology*, Vol. 35, No. 6, pp. 339–44.

Westup, C. (January 1999), 'Knowledge, legitimacy and progress? Requirements as inscriptions in information systems development', *Information Systems Journal*, Vol. 9, No. 1, pp. 35–54.

THE DEVELOPMENT PROCESS

CONTENTS

KEY LEARNING OBJECTIVES

Work practices in development
Describing development processes
The importance of quality in system development
The linear cycle
Rapid application development
Evolutionary design and prototyping

INTRODUCTION

Chapter 1 outlined the three major processes used to build systems—the development, support, and management processes. This chapter concentrates on the development process, while describing its relationship to the other processes. It sees the development process in the way shown in Figure 5.1, which follows the ideas of earlier chapters by dividing the development process into four activities. These activities are defining a feasible conceptual solution, developing a system specification, design, and the actual development and implementation of the system. In addition, support must be provided to organize the people into effective teams and provide them with documentation to keep track of project progress. Such documentation must ensure that team members always have the latest correct information to work with. The term **configuration management** *is often used to describe the way documentation is managed. Support is also provided through CASE tools, which allow designers to maintain the models that are built during the development process.*

When one looks at the activities in detail, one sees that they may be organized in different ways to suit different kinds of problems. The way the activities are organized is called the development process. Thus, the development process used to design an organization-wide transaction system is different from the process used to design a Web site or a decision support system to decide what assets to buy. Transaction-based systems are usually built in a predefined number of steps and center around planned work, whereas Web-based systems tend to be more experimental and center around situated work. Furthermore, different models may be used to represent systems at different development steps.

Guidelines are now becoming available through standards that suggest both what must be done in the development process and the process steps themselves. The Institute of Electrical and Electronic Engineers (IEEE) has proposed a number of standards for this purpose. This chapter describes a number of development processes that closely follow these standards. It begins by describing different team

Configuration management
Managing a configuration.

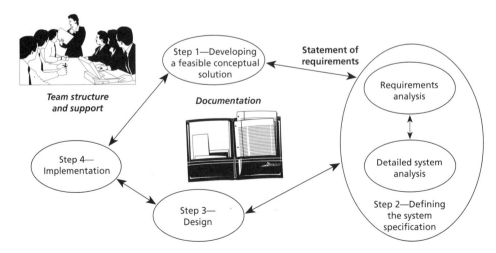

Figure 5.1 *The development environment*

structures and support tools and then describes the kind of team structure needed by the different processes.

DESCRIBING THE DEVELOPMENT PROCESSES

Development processes are also known as the **system life cycle** or the **problem-solving cycle**, or sometimes the **system development cycle**. The term **software process** is now also being used to define the process used in system development.

PROBLEM SOLVING AND THE DEVELOPMENT PROCESS

A development process must include all the problem-solving activities described in Chapter 4 and summarized in Figure 4.2. They include developing an understanding of the system, creating models, making decisions on what is to be done, and planning the work. The problem-solving process is to some extent defined in an abstract way. The development process organizes problem solving into a practical environment to ensure that a product is delivered. This can be done in a variety of ways. One is to follow a defined set of steps, much the same as a business process. The alternative is to allow developers free rein to bring out utmost creativity. But if designers have free rein, how do we ensure that the system is built on time and with minimum cost? Good development processes must reach a balance between structure and freedom.

The question is how the development process accomplishes the practical goal. In the first instance it must ensure that the system captures user requirements. Practically, this means that users are formally required to specify their requirements. In many processes, users must participate throughout the development process. There used to be a time when requirements were defined, the system was built, and then the whole system was tested to see whether it met requirements without error. Any errors found were corrected. The approach of leaving error detection to the end, however, is no longer used. It was usually found that a large number of errors appeared in systems built in this way, and considerable time and effort was needed to clear them up. For this reason, the development process includes various checks during the process to ensure that the number of errors on completion is minimized.

Another goal of the development process is to ensure that the system works correctly. The word *quality* is sometimes used to define the goal of producing error-free systems that meet user requirements with minimum effort. Consequently, **quality assurance** mechanisms are introduced into the development process to ensure that there are no deviations from requirements as development proceeds. Quality assurance mechanisms check the process at its various stages to ensure that requirements are not lost or changed during the process and that errors are minimized. Quality assurance mechanisms include **validation** of outputs from various activities against original requirements. They also include **verification** of individual activities to ensure that each activity correctly converts its output to input. Finally, they include the **testing** of working components. Validation and verification are applied progressively during system development to ensure that the final system meets its requirements.

System life cycle Another term for the problem-solving cycle.

Problem-solving cycle A set of steps that start with a set of user requirements and produce a system that satisfies these requirements.

System development cycle Another term for the problem-solving cycle.

Software process Another term for system life cycle but concentrating on software development.

Quality assurance A process to ensure the development of quality products.

Validation Checking whether a particular product satisfies user requirements.

Verification Checks to ensure that a particular input has been converted correctly to an output.

Testing Checking to see whether a system does what it is supposed to do.

DEFINING THE DEVELOPMENT STEPS

The development process must also ensure completion, while supporting the freedom needed to ensure creativity. System designers should therefore not just continue to create new ideas without ever putting them into practice. At the same time they should not simply implement the first design we think of, using a set of strict process rules. How we reach this balance between structure and freedom often depends on the kind of system being designed.

Perhaps the simplest view of the development process is as a sequence of tasks, an approach that was used in the early days of developing computer-based information systems. Usually, a large number of detailed activities are needed to build an information system. Typical activities include writing a program, designing a form, finding out what a user needs, or selecting a piece of equipment. The first step in creating a development process is to list these activities. In addition to listing activities, we should also specify the sequence for these activities. This can be shown on a chart similar to that in Figure 5.2. This chart shows an initial set of activities, in this case 'Analyze accounts', 'Analyze sales', and 'Examine existing computer system'. The arrows in the diagram show the activities that can be completed before another activity can start. For example, in Figure 5.2, the existing computer system must be examined before the computer data model can be drawn. Similarly, both 'Analyze accounts' and 'Analyze sales' must be completed before 'Redesign forms' and 'Construct data structure for sales accounts' can start.

Charts like that shown in Figure 5.2 were used in the early days of system development, but they are seldom used now, because most projects now consist of a large number of activities linked by complex interdependencies. Plotting all these activities and dependencies on one chart becomes far too complex. In addition, later project activities often depend on the outcomes of earlier project activities, so the

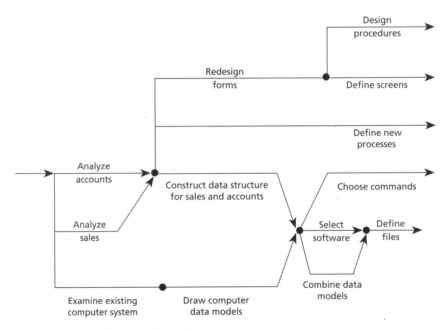

Figure 5.2   *Listing all detailed activities*

planner is required to predict some of these outcomes. If the predictions are wrong, as is often the case, the chart must be changed during the course of the project— sometimes frequently. These changes can become such a problem that they detract from the project work. The last thing we want to do is to introduce another problem into the design of a new system.

For this reason, it is now more common to define development processes in a more abstract way, usually as a set of steps that define the step goal rather than its actual work activity. One widely accepted set of steps is shown in Figure 5.3. This shows development starting with project selection and proceeding through concept formation, system specification, system design, and system development. It also includes installation and post-installation steps that usually follow the completion of development. Each step itself is made up of more detailed activities, and each has a specific goal and produces a well-defined output, usually including a model. However, the detailed activities for each phase are not defined at the start of the project.

It is therefore unnecessary to make a detailed plan for the whole project before it starts; only one plan for the first step is needed at the start. Furthermore, there are now development processes where the project is completed through a number of sequential problem-solving step sequences or where some of the steps proceed concurrently. These kinds of development processes are described in this chapter.

ALTERNATIVE PROCESSES

Figure 5.3 suggests one approach to a development process—to organize the development into a linear process in a sequence that corresponds to the problem-solving sequence. The analyst identifies the problem, defines a solution, and implements it. This is not the only possible process, however. Thus, we might develop a system part at one time, or analyze part of the system, implement some of it, return to analyze the rest, and so on. However, whatever the development process, developers must carry out all the activities shown in Figure 5.3. The activities must be applied in the sequence shown in Figure 5.3, although there may be a number of such problem-solving strands proceeding concurrently in the same development process. It is therefore important to keep in mind the distinction between the problem-solving process and the way this is organized within the development process. This chapter first describes a linear or waterfall process. The

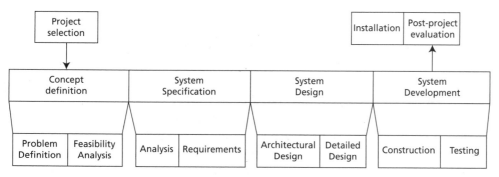

Figure 5.3 *Development steps—main steps*

linear cycle applies the problem-solving sequence sequentially to solve the entire problem. In that sense, it describes what is done in each activity. In practice, these activities are often applied in different ways. They can be applied a number of times, each solving one part of the problem. They can also be combined in a variety of ways to speed up the development process. This chapter describes a number of such different processes.

THE LINEAR OR WATERFALL CYCLE

Linear cycle A set of predefined steps for building a system.

Waterfall cycle The same as a linear cycle.

The **linear** or **waterfall** cycle is a development process that centers around planned work and is best suited to projects where requirements can be clearly defined. The whole project is completed as one sequence of steps, which are called phases. Testing proceeds in parallel with the major phases. A broad test strategy is defined during requirements, a detailed test design takes place during system design, and testing is part of the development phase. A phase in the sequence can commence only after the previous phase has been completed.

Each phase usually produces one or more models, which are shown in more detail in Figure 5.4. Here the models are shown as the rounded boxes with underlined headings. The models are part of a phase report, which describes what has been achieved in one phase and outlining a plan for the next phase. Concept formation produces a statement of requirements predominantly in user world terms. The system specification definition produces a system specification that is usually made up of three models, the requirements model in user terms, an analysis model in system terms, and a broad system specification. Design produces a design specification in detailed system world terms, and development produces the implemented system modules.

The phase reports that include the models serve as important communication tools in the way described in Chapter 4, with communication progressing from the usage to the implementation worlds. The report produced in one phase is used in the next phase. Phase reports are also used to keep management informed of project progress, so that management can use the reports to change project direction and to allocate resources to the project.

Linear cycle phases are chosen to encourage top-down problem solving. Designers must first define the problem to be solved and then use an ordered set of steps to reach a solution. The linear cycle gives the project direction and guidance about what should be done as the project proceeds. It is integrated into the management process through reports on project status and keeping track of resource needs. This integration is described in Chapter 6.

TEAM AND DOCUMENTATION SUPPORT

The linear cycle is usually associated with a structured team and documentation system. Team members are assigned to specific phases and tasks and are required to produce specific documents. Document flows are also highly structured. Documents produced at the end of one phase must be available as inputs to the next phase.

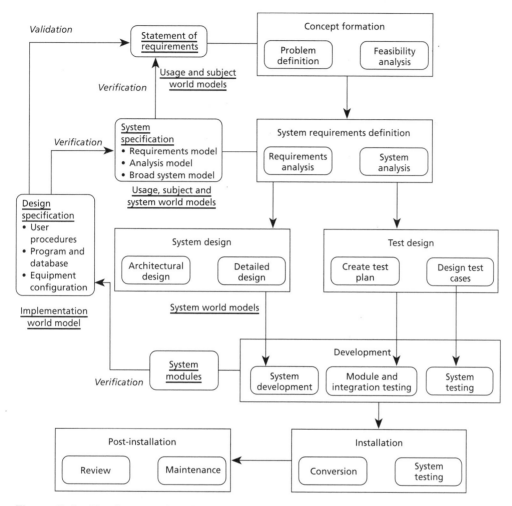

Figure 5.4 *The linear cycle—detailed model*

ASSURING QUALITY

It is also important to take formal steps to ensure that all user requirements are satisfied at the completion of the project; that is, to include the necessary quality assurance mechanisms in the process. Figure 5.4 shows one set of such mechanisms. **Validation** checks phase outputs against the statement of requirements. Validation takes place at the conclusion of the system specification, during design, and then finally following implementation. Verification takes place continuously throughout the process, with the models at one phase verified against those of the previous phase. Finally, any designed software modules must be thoroughly tested during development.

Validation Checking whether a particular product satisfies user requirements.

SEPARATING TESTING AND DEVELOPMENT

Another important issue is the separation of testing from development, where testing is shown as a separate activity in Figure 5.4. The test plan is developed independently from the development team, often starting from the statement of requirements. Each requirement produces one or more test cases, which are then used to test the finished modules. Quality of produced systems is improved if people that test systems are not those that build the system.

LINEAR CYCLE PHASES

The linear cycle begins with concept definition. Concept definition is made up of two parts. The first is problem definition and leads to the project goal. The second, feasibility analysis, determines a feasible way for achieving this goal. Feasibility analysis usually considers a number of project alternatives, one of which is chosen as the most satisfactory solution. It is advisable to investigate as many alternatives as possible to ensure that the best project direction is chosen. Just using the first idea that comes to mind is not the best approach. A certain amount of skill is needed to propose good alternatives and choose the right direction. Many development processes provide guidelines for generating alternatives.

This phase is important because it sets the direction for the remainder of the project and must be completed before the project can continue. Furthermore, the project direction in the linear cycle is set without going into detailed design. There is no point in virtually building a new system to see whether it satisfies the goal. Instead, a number of alternatives should be evaluated first to find the best way to satisfy the project goal.

PHASE 1—CONCEPT FORMATION

Statement of user requirements A formal definition of what the new system must do.

Many people consider concept formation to be the most important phase. It provides a broad **statement of user requirements** in user world terms. The requirements state what the users expect the system to do and thus set the direction for the whole project. This phase also sets the project bounds, which define which parts of the system can be changed by the project and which parts are to remain the same. The resources to be made available to the project are also often specified in this phase. These three important factors—the project goal, project bounds, and resource limits—are sometimes called the project's terms of reference. Because of their importance, they are set by the organization's management.

> ### Text Case A:
> ### Interactive Marketing—Defining the Problem
> As an example, let us consider Text Case A, the interactive marketing of perishable goods. The overriding goal here is to ensure that perishable products are sold and distributed as quickly as possible. This goal may be stated quite explicitly, such as *80% of all products must be distributed within 3 days*. When one looks at the whole system, one can see a number of activities that can be directly related to the goal and others that are not directly related. Those that are directly related include:
>
> * the dissemination of information about available products to consumers
> * arrangements for transportation
> * placing of orders for the products.
>
> Activities not directly related to the goal may include:
>
> * invoicing the consumers

- payment for the product.

This subdivision indicates the bounds of the project, or at least the project priorities. The directly related activities would initially be within the bounds, whereas the others would be outside the project bounds. Or we should at least initially give priority to activities directly related to the goal. Now we might consider all the directly related activities. Perhaps, after some discussion with the users, we may decide that the biggest problem in meeting the goal is letting consumers know what products are available. Transport may contribute to the problem. However, it may be felt at this stage that advertising products is the bigger problem. Furthermore, there may not be enough funds to study the entire system, and we may confine our study initially to the communication links between producers and consumers. These become our problem bounds, and our terms of reference become how to improve communications between producers and consumers in order to improve sales. A major requirement in the statement of requirements is to improve communications so that the consumers become aware of product availability and can obtain products in the minimum time. Thus, the concept becomes one of an easy-to-access product availability register or database with facilities to allow products to be easily ordered.

Part of the concept formation phase is the feasibility study, which proposes one or more conceptual solutions to the problem set for the project. Each conceptual solution gives an idea of what the new system will look like. It defines what will be done on the computer and what will remain manual. It also indicates what input will be needed by the systems and what outputs will be produced. These solutions must be proven to be feasible, and a preferred solution must be accepted using the methods that will be described in Chapter 7.

Many beginners find the idea of a conceptual solution hard to understand. All that is needed at this stage is a very broad idea of the solution, enough to give potential users an estimate of whether it can work and how much it is likely to cost. It is usually a good idea to consider a number of possible solutions at this stage. For example, consider Text Case A.

Text Case A:
Interactive Marketing—Feasibility

There are a number of possibilities for meeting the Phase 1 goal. Not all of them must use computers. For example, we may suggest the following solutions:

- Have an advertising campaign.
- Employ a larger sales force to inform customers of available products.
- Keep a list of customers and their preferences and call them whenever a product is available.
- Have a dial-up service with messages about the latest products.

There is also, of course, a computer solution, especially that of using the Worldwide Web. Here we might have a computer with a database that contains details of available products accessible through the Internet. Producers can enter information about their available products into the database as soon as the product is available. Consumers can browse through this database at any time to inquire about products and place their orders directly through the computer. A more sophisticated approach would be to allow consumers to place *agents* into the system stating their requirements. These agents would automatically notify consumers whenever products that meet their interests became available.

Once we have these possibilities, we may consider them in turn. Often the first step is to eliminate alternatives that are obviously impractical or not operational. For example, dial-up services may not be very useful. If 100 producers each leave a 30-second message, consumers must spend 50 minutes to listen to them all. Again, not a very practical solution. Advertising may also prove not very effective because of the lead times in placing advertisements in newspapers, and a visiting sales force may not work because of the infrequency of visits. This leaves us with the computer solution. We may propose that the marketing database be stored on a computer and then consider a number of alternatives for storing such a database.

It is often necessary to gather supporting data for such arguments. For example, the amount of data needed may be used to estimate the usefulness of dial-up services. Volumes of data and transactions are needed to estimate the costs of the computer solution. Often, initial conceptual solutions may be modified in view of the supporting data. This additional information will lead to more detailed knowledge about the system. It will produce more information about the data needed by consumers and the communication patterns between consumers and producers. It will also define, in more detail, any difficulties in distributing product data to consumers.

At the conclusion, a decision is made here to develop a WWW site for the system. The concept is then elaborated into a broad statement of requirements to support on-demand contact between suppliers and consumers. In detail, it requires:

- direct entry into the WWW database by producers of products for sale
- the ability for consumers to browse the database using specified keywords
- direct ordering through the WWW pages
- contact addresses, and links where possible, to potential transport companies.

Phase output

The statement of requirements from Phase 1 defines what the new system will do in user terms. It defines the proposed business solution, the project goal, and the bounds and terms of reference for the project. Furthermore, it may include any restrictions on the project, such as the parts of the existing system that cannot be changed and those that can. Resource limitations are also often specified at this time to indicate the funds and personnel available for the project. This will include a rough idea of the resource requirements of each of the subsequent project phases. It will contain tentative start and completion dates for each phase and the number of persons expected to be involved in each phase.

The statement of what the new system will do becomes a statement of requirements for further work and sets the guidelines for the next phase, which elaborates these requirements in detail.

PHASE 2—DEVELOPING THE SYSTEM SPECIFICATION

During the development of the system specification it is necessary to find out more about the system problems and what users require of any new or changed system. This whole process is usually characterized by the following activities:

- Producing a detailed **analysis model** describing how the current system works and what it does, usually in subject world terms.
- Using the statement of requirements and the more detailed systems analysis to state what is needed from the new system by a **requirements model** in user terms.
- Producing a detailed model in subject terms of what the new system will do and how it will work, here called the **design model**, which is expressed in system terms.
- Producing a high-level description of computer system requirements using system terms.

The analysts must search for information about the system, including interviews with system users, questionnaires, and other data-gathering methods, to produce the analysis model. They must spend considerable time examining components, such as the various forms used in the system and the operation of existing computer systems. Modeling methods used to create the analysis model are described in the next few chapters. They include methods known as structured systems analysis, described in Chapters 8–10, and object modeling, described in Chapters 11 and 12. Requirements analysis, described in Chapter 4, is usually a set of clear statements about the way the new system is to work. It defines how the users use the system now and how they want to use it in the future, preferably in usage world terms. The design model is usually a revised version of the analysis model that now includes the requirements.

> ## Text Case A:
> ## Interactive Marketing — System Specification
>
> In the interactive marketing case, the first step would be to find out what information is actually sent between producers and consumers. Furthermore, we would carry out a precise analysis of the information that is available about products, where this information is available, how the information is presently distributed, and how long the distribution takes. The analysis in this case would be relatively minimal, as no system exists, because products are traded through wholesalers. The analysis would thus concentrate on what is actually needed. Thus, consumers may now be asked to look at the current situation and suggest what information they need and when they need it. A design model and broad architecture would then be developed. The idea behind these latter components of the specification is shown in Figure 5.5(a). The design model shows the processes that would be supported (circles) and how data flows between them. The two main files, AVAILABLE-PRODUCTS and SALES, are also illustrated. The specification would also contain a list of items that appear in the files and a description of the functions.
>
> The other important activity here is to produce the broad system specification, which describes the main system components. This is shown in Figure 5.5(b). The format of inputs and outputs will be defined at this point, and data to be stored is identified.
>
> The requirements model would be defined in terms of scenarios that describe how the system works from the user perspective and performance requirements.

Analysis model A description of the way in which a system works.

Requirements model A description of what users require the system to do.

Design model A description of the required system using system terms.

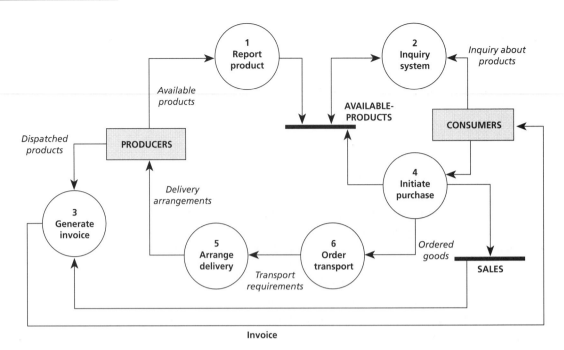

Figure 5.5(a) *The design model*

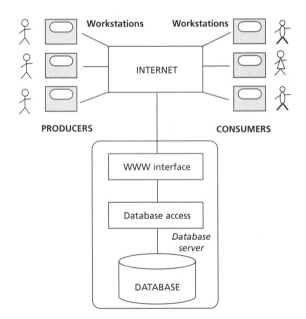

Figure 5.5(b) *The broad system model*

Phase output

A number of outputs are produced at this phase. These can include:

- the analysis models that describe the existing and new systems, preferably in system world terms
- revisions to project goals and cost–benefit estimates
- a requirements model that specifies in usage world terms how the new system will work, including any interfaces presented to the user
- a design model that shows in subject world terms what the system will do and how it will work
- a description of the proposed business process and some scenarios
- a broad system specification in system world terms defining how computers will be integrated into the business process
- a project development plan, defining the expected completion times
- a test plan, in which a number of test cases are created for each detailed requirement.

Part of the management review of this phase is to reach an agreement on detailed requirements and the system specification. Once agreement is reached, design can commence. The system specification is also validated against requirements and the existing system. Validation in this case often incorporates the ideas of structured walkthroughs, which are described in Chapter 20.

PHASE 3—SYSTEM DESIGN

This phase produces a design specification for the new system. There are many things to be done here. Designers must specify the interfaces provided to users and the business procedures. They must specify the databases and system programs and select the equipment needed to implement the system. System design usually proceeds in two steps: broad architectural design—Phase 3A—and detailed design—Phase 3B. One way to see this distinction is that Phase 3A is logical design, or what the system will do, and Phase 3B is physical design, or how the system will do it. The distinction, however, is not always so clear, and often the two phases overlap.

Phase 3A—broad architectural design

The broad architectural design contains three major components—the interface design, the logical database design, and the program structure design. During this phase, the user or subject world models produced in the system specification are converted into computer system terms.

For the interactive marketing stage, a broad architectural design would describe:

- what will be stored in the database
- how interfaces would allow users to access the database
- the programs that will have to be written
- the logical flow of work in the business process and the information needed at each step of the process.

In addition, the directory structure for the WWW pages would be defined, and major pages would be identified.

Architectural design may also suggest the hardware and software that will have to be purchased or built. Thus, broad design may suggest the network configuration, including the size of the computer and the software needed to put the system together. It will also state which software can be purchased off the shelf and which needs to be developed. The design may also suggest whether the computer should be rented or purchased and whether programs are to be developed by internal programmers or external contractors.

Phase 3B—detailed design

The detailed design concentrates on how the architectural design is to be implemented. During the detailed design phase, the database software is chosen, and ways of using it to store the data are proposed. The ways of writing programs and their location, and that of the database, are identified. Detailed user procedures, including their interfaces, are also documented. The interaction between the system users and computers is also defined. These interfaces define exactly what the user will be expected to do to use the system. Methods to do this are described in Chapter 17.

Phase output

The output of the design phase includes an implementation model for the new system. This includes the proposed equipment configuration and specifications for the database and computer programs. Detailed user procedures are also provided. These include any input forms and computer interaction between users and the computer. The user manual is also ready at the end of this phase. This output, particularly the proposed interactions between the computer and users, is validated against user requirements.

PHASE 4—SYSTEM DEVELOPMENT AND CONSTRUCTION

Like the design phase, this phase is also often broken up into two smaller phases: development and implementation. The individual system components are built during the development period. Programs are written and tested, and user interfaces are developed and tried by users. The database is initialized with data.

During implementation, the components built during development are put into operational use. Usually this means that the new and old systems are run in parallel for some time. To complete the changeover, users must be trained in system operation, and any existing procedures must be converted to the new system.

One important part of construction is testing. It is necessary to test all modules to make sure they work without any errors once they are put into operation. Testing itself can be a process on its own. We first test individual modules, then we test their interfaces and see how they work together.

Phase output

At the end of the phase, users are provided with a working system. This includes the set of working programs and an initialized database. In addition, any system documentation describing the programs is also completed. All users have by now been trained and can use the new system.

SYSTEM TESTING

This is recognized as an important part of quality assurance. Testing, as shown in Figure 5.4, proceeds in parallel with system development. This idea is illustrated in Figure 5.6. Here a test plan is developed in parallel with system design. The test plan is then used to develop test cases that are used in system testing. Testing proceeds through a number of steps. First, individual program modules are tested by their developers. Once individual modules are tested, the next step is to test whether they can be combined. This is known as integration testing. During integration testing, groups of modules are combined into test modules and tested together. The goal is to determine whether the interfaces between modules work. Then the entire system is tested. It is important to design test cases that test all the conditions that can arise in system inputs, at the same time ensuring that tests do not take too long.

WHAT HAPPENS AFTER PROJECT COMPLETION?

The system is considered to be working when Phase 5 has been completed. However, a number of activities still take place after a system is completed. The two main activities are the post-installation review and maintenance.

POST-INSTALLATION REVIEW

The post-installation review usually takes place about a year after the system is implemented. It evaluates the new system to see whether it has indeed satisfied the goals set for it. The system is examined to see whether the benefits expected of it have been realized. If they have not, a study is done to see why not. Part of this study is the project life cycle itself. It is important at this stage to go back to the original goals of the project. Thus, in the case of the interactive marketing system, the question would be whether sales of products have increased, not whether the computer system is working. Decisions made during the project are evaluated to see whether they could have been better in the light of experience gained from the project. This evaluation then sets guidelines for decision making in future projects. In addition, post-evaluation may suggest minor changes to be made to the system. In exceptional circumstances, where the system is performing badly, post-evaluation may suggest a total redesign.

MAINTENANCE

Maintenance is necessary to eliminate errors in the system during its working life and to tune the system to any variations in its working environment. There are

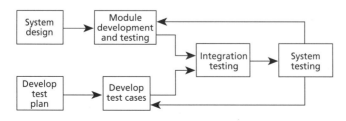

Figure 5.6 *Testing*

always some errors detected that must be corrected. Often, small system deficiencies are found as a system is brought into operation and changes are made to remove these deficiencies. Information system planners must always plan for resource availability to carry out these maintenance functions.

If a major change to a system is needed, a new project may have to be set up to carry out the change. This new project will then proceed through all the above life cycle phases.

SOME PROBLEMS WITH THE LINEAR CYCLE

The linear cycle has one important property: it has definite sign-off points where some activities terminate. All its activities are performed in a strict sequence. One phase must be completed before the next phase starts, and no phase can be repeated.

The cycle assumes that one can develop and precisely specify a system in a top-down manner. Successive phases elaborate the system in increasing detail, with each phase defining a partial solution and then calling for a more detailed evaluation in the next phase.

In practice, however, one often finds that assumptions made in the early phases no longer hold, that some of the early work was incomplete, or (even more likely) that something was overlooked or not completely understood. As a result, it may be necessary to return to an earlier phase and redefine an objective or redo some of the earlier work. The kinds of things that happen here are shown in Figure 5.7. During system design we may discover that some objective defined in the feasibility study cannot be realized as easily as first thought. Hence, a return to the feasibility study may be necessary to redefine the objective. A worse scenario is when problems arise during implementation and these require a system redesign or a return to the feasibility study. Often, implementation detects some previously undiscovered problem in the system, calling for more systems analysis.

Problem-solving cycles like the one shown in Figure 5.7 are often called *loopy*

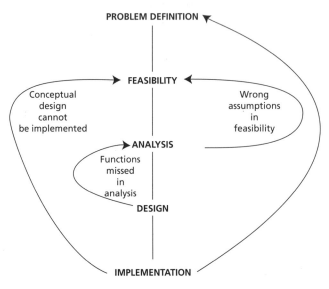

Figure 5.7 *Possible problems in a linear cycle*

linear. They arise by default because something was overlooked in the original plan. Iteration, or 'loopiness', is seen as being costly and is to be avoided. It generally requires wasteful rework of earlier phases.

One reason for project errors and consequent loopiness is that the problem being solved is too large or too uncertain. If it is too large, it is advisable to break it up into smaller stages and build the system a stage at a time.

STAGED DEVELOPMENT

Staged development is illustrated in Figure 5.8. It can be used only where it is possible to break up a problem into a number of distinct subsystems. Thus, the first step may be a global concept phase that finds that the project is very large and suggests that it should be developed in stages. If staged development is approved, the next step will be to break up the whole project into a number of stages. Thus, we could first build a system to disseminate information about our products. Then ordering could be added. We could then add details about transportation, followed by an invoicing system. Each of these can be built up separately and then integrated into a complete working system.

Each stage is developed using the linear cycle, and each subsystem is developed separately. You will note that each stage starts with its own problem definition phase. This is necessary to get a detailed evaluation and plan for each stage, and also to evaluate the feasibility of integrating the system produced in each stage with the systems developed in previous stages. The stage then continues with specification, design, and system implementation. Figure 5.8 shows the stages being carried out sequentially—that is, stage 2 commences when stage 1 is completed. It is also possible in some cases to develop the stages concurrently, leading to a quicker completion time.

USING PROTOTYPING IN SYSTEM DEVELOPMENT

Often designers are faced with the problem of building systems that cannot be precisely specified using a model. This often occurs in organizations that are just starting to use computers or in novel applications where there is no previous experience. Instead, it is more appropriate to develop the system gradually and learn about system capabilities as one goes along.

One example of imprecise systems is the interactive system with a lot of user dialog. Designers do not wish to risk users' rejection by specifying all the dialogs

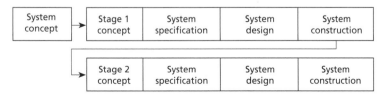

Figure 5.8 *Staged design*

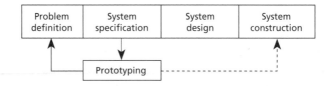

Figure 5.9 *Prototyping in systems development*

and screens. A set of trial screens and dialogs is developed and handed over to users for experimentation, after which changes may be proposed. Then a further experimentation follows until a satisfactory set of screens and dialogs is obtained. Another type of imprecise problem involves workgroup computing. Here again it is not possible to precisely define all the interactions needed to effectively support group work, and it is better to allow the system to evolve in an experimental way. Prototyping is often used in system development to clarify user requirements in imprecise systems. A decision to use prototyping is usually made in the feasibility phase. Within the context of Figure 5.9, prototyping is used to gain a better understanding of possible solutions, and the prototype then becomes the requirements model in the system specification phase. The dotted line indicates that some parts of the prototype may eventually become parts of the new system.

RAPID APPLICATION DEVELOPMENT

Prototyping on its own does not in any way speed up development, as development still proceeds as a sequence of steps. The goal of prototyping is to minimize risk by initially testing the feasibility of conceptual designs.

There is an increasing need to construct systems at an ever faster rate and hence considerable emphasis on providing ways to speed up system development. The most often quoted approach is known as rapid application development (RAD), which was introduced by Martin (1991) in the late 1980s. The idea here is to streamline the development process in a number of ways. There are three steps:

- Reduce the time needed to develop the specification by holding Joint Application Design (JAD) meetings of high-level managers, executives, and those knowledgeable in the applications to determine requirements quickly. Often special workshops like those outlined in Chapter 4 are held to formally agree on a set of requirements. Initial analysis models can be developed at this early stage.
- Speed up development by combining the problem-solving steps so that design is taking place while the analysis model is being developed.
- Provide fast feedback from users. The availability of early prototypes allows users to provide early feedback on how the system is meeting their needs.

One way to combine the problem-solving activities into concurrent processes is shown in Figure 5.10. Here there is considerable overlap of the development phases of Figure 5.4.

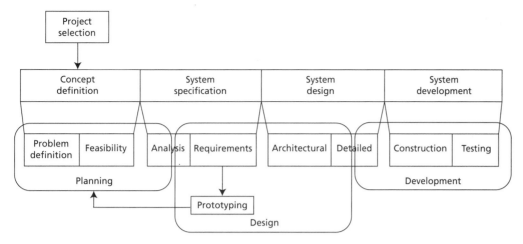

Figure 5.10 *Rapid prototyping*

- The planning phase, whose primary goal is to develop a conceptual model, already includes a substantial analysis component, and part of the analysis model is developed during planning.
- The design commences before the entire analysis model is completed. It finalizes the analysis while the design of earlier analysis models is carried out in parallel.
- Development of some parts of the project commences even before the whole design is complete.

The whole process can be further complicated by the possibility of the development itself being staged, with different stages passing through different phases of the development cycle. There are, of course, variants and refinements of this cycle.

The RAD deliverables are the same as in the linear process but are delivered incrementally in successive phases rather than as a product of one phase. Furthermore, combining the problem-solving steps sometimes suggests modeling tools that combine the outputs of more than one step in the same model.

Such substantial overlap requires RAD to use tools to keep all team members aware of the current state of development. Thus, for example, part of the analysis is carried on in planning and another part in design. It is necessary for people who are working on design to have access to the latest analysis model to avoid repeating what was done before and to ensure that their additional work is consistent with previous work.

The RAD is very applicable in the current climate of Web development. A group made up of client representatives, internal experts, and management may be called for a workshop to propose an initial design for a Web site that is to be extended to clients. This initial workshop may provide an initial solution and suggest a broad design. Then design may build on the initial suggestions and create an initial prototype. This may then be reviewed by a subsequent workshop, with parts assigned to development while other parts are redesigned. Thus, at any time there may be concurrent work proceeding in all three phases.

It has often been suggested that RAD often ignores strategic issues in its haste to produce results satisfying current needs.

DSDM (DYNAMIC SYSTEM DEVELOPMENT METHOD)

Propriety development processes are now coming onto the market to support RAD. One of these methods is known as DSDM, or the Dynamic System Development Method. The method is illustrated in Figure 5.11 and closely approximates the phases described in Figure 5.10.

DSDM starts by defining the feasibility of a proposal and its suitability for a RAD process. It then carries out a business study to define the base-level business functionality to be achieved by the proposed system. The base-level functionality defines the broad objectives of the system but not a precise set of user requirements, as the linear development process does. These steps closely correspond with planning in Figure 5.10.

The next phase is the functional model iteration, which places considerable emphasis on prototyping. The output here is a set of functional prototypes. The goal is to elicit more detailed user requirements while progressing incrementally to an implementation. This closely corresponds with design in Figure 5.10. The design and build phase focuses on improving the prototypes into a form that can be delivered and implemented. The user requirements are now clear and are put into their final form, including all alternatives and error conditions.

Finally the system is put into use.

EVOLUTIONARY DESIGN

Evolutionary design is yet another way of designing systems. It is particularly applicable to systems that cannot be precisely specified. It combines elements of staged design and prototyping. It develops the whole project as a number of stages, with the outcomes of one stage serving to identify the conceptual solutions for the next stage. One way of describing this kind of development is by the spiral model.

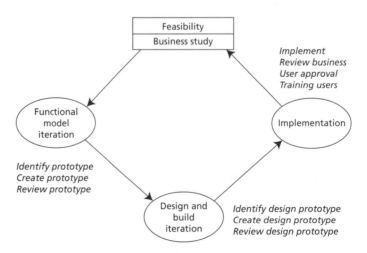

Figure 5.11 *DSDM major phases*

THE SPIRAL MODEL

One formal approach to evolutionary design is the spiral model (Boehm, 1988), which is illustrated in Figure 5.12. It shows system development proceeding as a set of successive prototypes. Each prototype adds functionality to the previous prototype. Each of the cycles in the spiral model can follow the linear development process, where each initial phase defines the goals for the subsequent stages. The goal concentrates on reducing risk at the subsequent stage. The system is developed gradually by successive developments of prototypes, followed by risk analysis to determine the most likely path to successful development.

The spiral model is particularly suitable for designing systems that are experimental in nature. Take a goods ordering system, where ordering procedures differ for different kinds of goods. One way to start would be to develop a system for the subset of goods that are ordered in a straightforward manner. Once this is done and understood, we can go on to the next set of goods.

Such evolutionary design does not assume that we can subdivide the problem into distinct and loosely coupled stages and design the system in one pass through these phases, as is the case with staged development. Instead, the system is developed gradually. We develop a system part and learn more about the problem from the operation of that part. We then use knowledge gained from this operation to define the next part to be developed. This part is then developed, and the process continues.

In evolutionary design, each step adds a new capability to the system, and

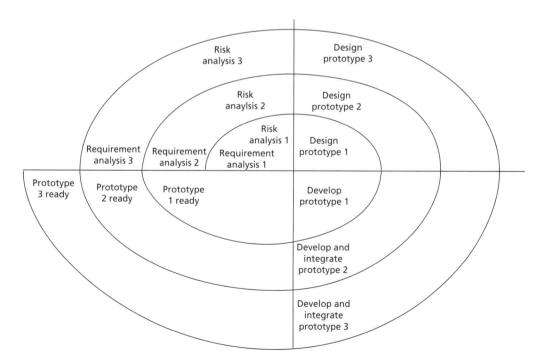

Figure 5.12 *The spiral model*

experience gained with a system is used to define the requirements for the next step. The next step extends system capability a little bit further, and the process continues until no further improvement appears possible or worthwhile.

> ### TEXT CASE C:
> ### Universal Securities—Developing the New System
> The system for Universal Securities will be developed in an evolutionary way. A possible first step is to develop a system that looks at ways of supporting interaction with outsourced and other external producers. It may, for example, experiment with posting requirements on an intranet extended to clients.
>
> The next step may be to provide extensions to clients. This may include presentation of their orders and enable them to make comments and suggestions.

IMPLEMENTING EVOLUTIONARY PROBLEM SOLVING

Another kind of problem that is particularly susceptible to the evolutionary approach is the decision support system, where the interactions used to support the decision making cannot be accurately predicted and an experimental approach is needed. Decision support systems have a further degree of uncertainty because it is not clear whether a computer can actually be used to solve the problem. Alter (1984), for example, describes a taxonomy of decision systems and suggests that any decision system evolves through a number of stages. Alter lists four stages:

- A data analysis system that provides data useful to decision making. The decision maker must still decide on what is relevant in the data.
- Presentation of analysis information that automatically extracts information needed for a decision but still leaves it to the decision maker to compare the data against expectations.
- A decision support system based on a representation model that defines the expected system behavior and compares this with the actual behavior. Accounting models are frequently used here.
- A decision support system using an optimization model that uses the variations to decide on actions that can optimize future behavior.

Decision support systems are transferred into use in an evolutionary way. A method for such gradual transfer has been suggested by Lucas (1978) and is shown in Figure 5.13. Development commences with a pilot system, which may be part

INCEPTION	—user suggestion
INITIAL PROTOTYPING	—initial pilot prototype
MUTUAL PROGRESS	—a number of design/ program/test steps
CONVERSION	—transfer of ownership
MATURITY	—operation of final product
MAINTENANCE	—suggested changes

Figure 5.13   *An evolutionary design method*

of our goods ordering process, a computer interface, or the first part of a decision support system. The pilot system will include some key reports and provide users with an initial appreciation of the proposed system capability. These will be discussed with the users to gauge their satisfaction. A goal of the process is to start by developing some initial prototypes to ensure that the user has sufficient knowledge of the project to constructively contribute to its development. Another goal is to ensure that the user is actively involved in the review of system specifications. As agreement is reached, the system is developed and parts of the system are gradually transferred to the user.

Major functions are defined in conjunction with the user during the initial prototyping, as a result of which the user slowly assumes the responsibility for the functions and puts them to use. Eventually the system evolves to a stage where all the major functions have been transferred to the user area and put into use. The operation is then in its mature stage. Changes can still be made to the mature system, although these are less frequent than in the initial stages.

CHOOSING A DEVELOPMENT PROCESS

Different kinds of problems usually call for different problem-solving cycles, and it is important to choose the most suitable cycle for a given problem. The choice is influenced by the nature of the problem, in particular:
* the degree of system structure
* familiarity with technology
* the project size.

Linear cycles are usually best suited to problems that are well understood and highly structured. In these cases it is possible to make accurate predictions about system behavior in the early design stages and proceed toward a solution in a sequence of well-defined steps. Problems that are unstructured and harder to understand require more experimentation in the early stages. The evolutionary or prototype approaches become attractive for these problems.

Familiarity with technology also influences the choice of problem-solving cycle. Again, if one is familiar with the technology, better predictions can be made earlier, and the linear cycle becomes more attractive. New and unfamiliar technology calls for more early experimentation. So does a new class of problem such as developing Web-based applications that include outside clients. The uncertainties of client or customer acceptance and the simultaneous requirement for quick outcomes make rapid application development attractive.

Problem size also affects the choice. The most pressing need for some problems is to produce early results and maintain user commitment to the project. A staged approach is often attractive for large problems (provided that the whole system can be broken up into useful subsystems).

ADOPTING A DEVELOPMENT PROCESS

Many organizations aim at using standard development processes throughout their projects. Using standard methods offers a number of advantages. First, designers do not have to spend time selecting design techniques and developing new documentation methods for every project. Second, a formal method is likely to be used because a supporting documentation method is available. Finally, system designers and users need be trained in only one method. Thus, training costs are reduced and, over time, most users can become familiar with the modeling techniques used by designers. Users are then able to contribute more toward project development.

An organization can choose standard methods in two ways: it can construct its own development process, or it can purchase an off-the-shelf development process. It is also possible to purchase and modify an off-the-shelf development process. Selecting a development process has four major steps:

- Identify the tasks to be carried out at each development process phase.
- Define the models to be used at each phase.
- Find support tools that can support these models.
- Define the configuration management system to store the models.

  Such a choice must ensure that:

- the chosen models are consistent in the sense that the output of a model at one phase becomes the input to the model at the following phase
- the models contain the three main system components—data, processes, and flows
- there are tools that support the development process.

  Usually models follow either the structured systems analysis method or object orientation. Earlier chapters describe the kinds of techniques available for methodologies and how they can be integrated into system development cycles.

SUMMARY

Different kinds of systems are developed in different ways. This chapter describes some general development processes. For well-defined problems, a linear cycle is suggested. The chapter describes the linear cycle, which is composed of five phases used to build the system. Two additional phases take place after the system is built. Project development proceeds in sequence through these phases. Each phase in the linear cycle has a defined objective and uses a number of activities to achieve this objective. Each phase produces an output that includes the phase product, a management report describing any difficulties encountered in the phase, and a plan for the subsequent phase. The chapter concludes by suggesting that the linear cycle may not be appropriate for all projects, particularly for systems that do not have clearly defined objectives. Alternative cycles may be needed to build such systems.

Where the problem is not well defined, an evolutionary or prototype approach may be more appropriate. Prototypes and evolutionary cycles call for design methodologies that allow easy system changes. Rapid application development is now often used, especially when quick completion times are needed. They also require project management systems that support controlled change.

DISCUSSION QUESTIONS

5.1   Why is a life cycle needed for the development of information systems?

5.2   What is the difference between highly structured teams and adaptive teams?

5.3   Define the phases used in the linear cycle.

5.4   Which kinds of problems would not be amenable to the linear problem-solving cycle?

5.5   What is the difference between the linear cycle and staged design? Where would you use staged design?

5.6   Why is the division between broad and detailed design necessary?

5.7   Describe what information you would keep for the post-implementation review.

5.8   What is the difference between system maintenance and the development of a new system?

5.9   What are the advantages of top-down problem solving?

5.10 Define what you understand by *system quality*.

5.11 Why is it difficult to build decision support systems using a linear cycle?

5.12 Explain the difference between the staged approach described in Chapter 3 and the evolutionary approach.

5.13 What is the difference between the evolutionary approach and rapid prototyping?

5.14 Explain how the linear cycle meets top-down problem-solving requirements.

5.15 Describe any systems you are familiar with where a prototype problem-solving approach would be useful.

5.16 How would you choose a development process?

EXERCISES

5.1   Suppose you are responsible for preparing a budget for advertising products that your organization sells. You have the figures for advertising costs in various media at different times and locations. You also have estimates of sales increases that may occur as a result of different volumes of advertising.

You are responsible for allocating the advertising budget to maximize revenue. However, because of the number of products, media types, and locations, you find it difficult to do so without assistance. A computer-based decision support system has been suggested. How would you break up your problem into the stages suggested by Alter and what kind of output would you expect at each stage?

5.2   What kind of problem-solving cycle would you recommend for the project goal suggested for Text Case B?

BIBLIOGRAPHY

Alavi, M. (June 1984), 'An assessment of the prototyping approach to information systems development', *Communications of the ACM*, 27 (6), pp. 556–63.

Alter, S.L. (1984), *Decision Support Systems: Current Practice and Continuing Challenges*, Addison-Wesley, Reading, Mass.

Avison, D.E. and Fitzgerald, G. (1988), *Information Systems Development*, Blackwell Scientific Publications, Oxford.

Bell, S. and Wood-Harper, T. (1998), *Rapid Information Systems Development*, McGraw-Hill, London.

Bennaton, E.M. (1995), *Software Project Management: A Practitioners Approach*, McGraw-Hill, London.

Bersoff, E.H. and Davis, A.M. (April 1991), 'Impact of life cycle models on software configuration management', *Communication of the ACM*, Vol. 34, No. 8, pp. 105–18.

Boehm, B.W. (May 1988), 'A spiral model of software development and enhancement', *The Computer*, pp. 61–71.

Fitzgerald, B. (July 1997), 'The use of system development methodologies in practice', *Information Systems Journal*, Vol. 7, No. 3, pp. 201–12.

Hirschheim, R. and Klein, H.K. (October 1989), 'Four paradigms of information systems development', *Communications of the ACM*, Vol. 32, No. 10, pp. 1199–216.

Hughes, J. (1998), 'Selection and evaluation of information systems methodologies', *IEE Proceedings Software*, 145(4), pp. 69–82.

IEEE (1993), *Software Engineering, IEEE Standards Collection*, The Institute of Electrical and Electronics Engineers, New York.

Jayaratna, N. (1994), *Understanding and Evaluating Methodologies*, McGraw-Hill, London.

King, D. (1984), *Current Practices in Software Development*, Yourdon Press, New York.

Kumar, K. (February 1990), 'Post-implementation evaluation of computer-based information systems: current practices', *Communications of the ACM*, Vol. 33, No. 2, pp. 203–12.

Lucas, H. C. (Winter 1978), 'The evolution on an information system: from key-man to every person', *Sloan Management Review*, pp. 39–52.

Martin, J. (1991), *Rapid Application Development*, Macmillan Publishing Company, New York.

Mumford, E. (1981), 'Participative systems design: structure and method', *Systems, Objectives and Solutions*, 1(1), pp. 5–19.

Mumford, E. (October 1997), 'The reality of participative systems design: contributing to stability in a rocking boat', *Information Systems Journal*, Vol. 7, No. 4, pp. 309–21.

Rettic, M. (October 1993), 'A project planning and development process for small teams', *Communications of the ACM*, Vol. 36, No. 10, pp. 45–55.

SUPPORTING SYSTEM DEVELOPMENT

CONTENTS

KEY LEARNING OBJECTIVES

Project management
Planning and estimating
How to organize resources for a project
Importance of documentation

Document configurations
CASE tools are used in system development
Vertical and horizontal integration of CASE tools
The necessary characteristics of a good CASE tool
How to monitor project progress
Project management tools

INTRODUCTION

In system development many people work on closely related tasks. They produce many artifacts, starting with written documents that outline initial ideas, followed by models produced during the development process, and finally programs and databases. There are a number of reasons why an organized approach is needed to coordinate team activities and keep track of these artifacts. Perhaps the overriding reason is that we want to complete a project as quickly as possible. To do this we must ensure that team members are aware of what they are required to do at particular points in time and to pass documents from one person to another. The other reason is complementary and equally important. It is to keep track of documents and make sure that they are easily available. We do not want to develop models on loose pieces of paper and misplace them, or fail to maintain records about our work. To do this we need tools to help teams carry out their tasks, provide them with timely access to project documents, and maintain awareness of project progress. Having such tools and document repositories improves the goal of completing projects in time, because it keeps all project members aware of what is happening in a project and avoids repetitive work.

Tools are thus needed for all the three processes used to build information systems and described in Chapter 1—management, development, and support processes. Management is generally supported by **project management tools**. *The development process is supported by* **CASE tools**, *which are used to maintain models. They provide graphical interfaces to enable developers to enter their model constructs. Support processes include ways to manage documentation by using* **configuration management** *systems. This chapter describes some of these tools in a general way. Note, however, that support tools are continually evolving. This chapter concentrates on the evolution of such tools, rather than describing them in detail. Such detailed descriptions can be found in the relevant manuals.*

Project management tools Tools to assist project management.

CASE tools Computer Assisted Software Engineering tools used to keep track of system models.

Configuration management Managing a configuration.

PROJECT MANAGEMENT

Tools cannot be applied in isolation, but their use must be visible and integrated into the development process. Generally it is the role of project management to facilitate such integration. The project management process is integrated with the development process and keeps track of and coordinates all activities. The project management process is made up of three main components—planning, which creates a project plan, organizing the project work, and monitoring to determine whether a project is proceeding according to the plan. Project management also provides the resources for carrying out the work, including the tools needed to expedite the project and procedures for quality assurance.

The management and development processes are integrated in the way shown in Figure 6.1. A common practice in project management is to break development down into a number of well-defined phases with time dependence relationships between them. Thus, in Figure 6.1, Phase 2 can start only when Phase 1 completes, and Phase 4 starts when both Phases 2 and 3 are complete. Phase 3 runs in parallel with Phase 2. Such phases were described in Chapter 5 for the different kinds of development processes. The objective of project management is then to ensure that the project meets its goal with minimal effort. To do this, project management sets the objective for each development phase, which is usually a hard deliverable. It then coordinates the phase activities, sets documentation and development standards, and provides the documentation and tools to support these standards. It then monitors progress and adjusts phase objectives and resources when needed.

PROJECT MANAGEMENT ACTIVITIES

Project management activities center on planning-phase activities, organizing resources for these activities, and monitoring progress. The actual ways of planning, allocation, and monitoring depend on the kind of development process.

PLANNING

Planning identifies the project phases and their broad objectives. It determines how a phase fits into the process. This includes stating how long a phase will take and defining its start and completion times. In addition, a plan defines the phase sequence. To make a plan requires estimates of the effort needed to complete each task to be made. These estimates are then used to allocate resources to a task and in turn determine how long it will take.

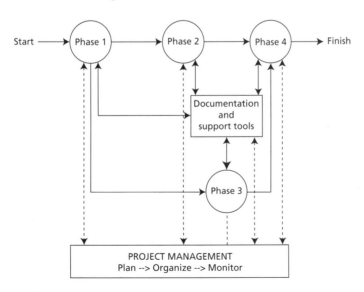

Figure 6.1 *Coordinating tasks*

Planning and estimating differs for the different development processes. The difference arises usually because of the difficulties of defining precise deliverables, especially between the planned processes, like the waterfall process, and the less defined evolutionary processes. In the linear process, the deliverables can be precisely defined. Here project management defines precisely what each phase must achieve, while the time and cost can vary during the process to ensure that these precise requirements are met.

In contrast, in evolutionary development, management is more likely to define a base line, which is primarily a broad objective, not the precise requirements, as is the case with the linear cycle, while the time and cost are kept constant to achieve the objective.

In evolutionary development, resources are allocated for a fixed period of time and the results are closely monitored. If nothing emerges after this time, then the project may be terminated. If a structure gradually evolves, more resources may be added to the project and its time may be gradually expanded. If uncertainties are removed, then we may continue project development using linear controls.

RAD also requires its own management techniques where a phase may deliver a number of related deliverables—for example, some requirements together with an initial analysis model and specification. Usually most RAD processes, for example DSDM, also define a baseline and set a time limit by which the base objective is to be achieved. This is particularly the case in the initial phases, where user requirements are defined. Management can expedite the development of requirements through workshops to ensure that a complete set is captured within a defined time.

ESTIMATING

Estimating involves evaluating the amount and complexity of work to be done in each task. This information is used to determine the resources needed to complete the work. Estimates will depend on the type of work carried out in the task. Most software estimates take into account the organization's experience in the type of task. Four scenarios are likely:

Existing software is used: This involves minimal risk, although often some compromise is needed to ensure a good fit with other system components.
Previous experience exists: Estimates use historical information about resources used in similar tasks to estimate the resource needs for a task in the new project. These estimates are usually quite accurate. However, a check should be made that this experience has been retained—in particular, that people with this experience can participate in the new project.
Similar experience exists: The risk here is higher, although there is always some part that is familiar and can be used as a baseline for an estimate.
It is a totally new development: This is a high-risk task that will call for closer monitoring during development and suggests the need for prototyping.

Details of how estimates are made are out of the scope of this book, and include alternatives for developing cost models for projects. There are methods known as function point analysis, for example, whose goal is to measure software complexity.

They do this by identifying functions such as external input/output, file operations, data, and control operations. The complexity of each function is then estimated and used to compute a function point measure.

IDENTIFYING AND EVALUATING RISK

The idea of risk management has also become important. There are two major risk concerns:

- The system will not meet the needs of the users.
- The project will overrun on costs or timings.

The idea is to determine the most likely factors that could cause projects either to fail to meet requirements or to overrun cost and time estimates. Then the goal is to minimize the probability of their occurring. Proper risk management requires close identification of factors that eliminate risk. Possibilities here include the following:

- Risk of incorrect requirements leading to frequent change. Management must provide for change management, use evolutionary or prototype development cycles or prototyping, and better customer involvement.
- Risk posed by new technologies. Management must hire or train experts in the technology. It may be useful to initially use a new technology in a limited way.
- Risk of resistance by users to adopting the new system. This is particularly important in developing Web-based applications that require changes in work practices. The usual approach here is to implement the systems gradually, allowing users to change their work practices gradually.

One way to minimize risk is to use evolutionary or prototype cycles, where we are dealing with systems that are unique in some way and have no historical precedents to suggest needed resources.

RISK MANAGEMENT

Risk is beginning to be seen as a major element in many projects, and there are suggestions that it should be used as the cornerstone of project management. Thus, the risk at the beginning of each step would be evaluated and decisions would be made on the basis of reducing risk. The spiral cycle described in Chapter 5 has been suggested as ideal for this way of managing projects. Prototyping is also seen as a way of reducing risk, as its goal is to reduce uncertainties, especially in the specification phase.

REDUCING RISK WITH IMPRECISE SYSTEMS

Imprecision can arise for a number of reasons. Example here include:

- dialog design for a given system. Hence, a management entity may be to develop the user dialog, and a certain amount of time will be devoted to it. The dialog at the completion of that time will be the dialog used in the system
- seeking knowledge about the capability of a technical system or procedure
- testing the feasibility of some algorithm.

Prototyping is also often used in imprecise systems. Its goal is to develop a better understanding of the system. The only project control is a time restriction. The experiments are run for a given period of time, and if no satisfactory result is found in that time, the project is terminated.

Decision support systems are another example. In a decision support system, it is never clear how much of the decision-making abilities can be passed on to the system; that is, how far it is possible to satisfy user requirements. Rather than trying to build the whole system, it is advisable to break up the project into parts, so that each part resolves some well-defined problem. The framework suggested by Alter (1984) and briefly described in Chapter 5 can be usefully applied for this purpose. Each of Alter's phases becomes a management entity. Resources are allocated to one management entity at a time.

The suggested phases are:

- a data-analysis phase to collect data useful to decision making
- an information-analysis phase to automatically extract data needed for decision making
- a phase to develop a representation model to define the expected system behavior
- a phase to develop optimization models.

Each phase can be allocated a certain resource level, and a completion time can be specified for the task. The phase is then managed using the linear life cycle approach and often ideas from the spiral model. If we find at one phase the likelihood of getting no further useful outcomes, then the project stops.

ORGANIZING RESOURCES

An important part of the project management process is to organize the resources needed to carry out a project. Project organization includes a variety of resources, perhaps the most important of which is to choose the right people for the project and to organize them into teams that support the chosen methodology. The documentation methods and supporting tools must also be chosen. The tools will make it easier for users to become involved in system design.

Other resources also need to be organized, such as the necessary computers and terminals, building space, or the needed productivity tools. Arrangements must be made for user departments to contribute to the project.

CHOOSING TEAM STRUCTURES

An important part of project management is to select the team structures for a project. Different development processes often need different kinds of work practices to be followed. Thus, in some cases, work can be easily subdivided and allocated to people who work independently; in other cases, people may have to work closely together. The work practices in turn affect the way people exchange information and require different support processes and team structures. A good system will provide the right match of teams to processes, support them with proper tools and documentation, and provide the correct level of management. Users must also be included in each team, thus ensuring the user involvement that is essential to good system design. Users know what the system does and its problems and objectives, so they should contribute to its development.

It is now increasingly recognized that teamwork is important in system development. A good environment will provide a match between team structures and the development process. It is common to distinguish between two ways of doing work—planned work and situated work. In planned work, steps and tasks are predefined and assigned to people. Work is often planned in large projects where different tasks are assigned to different people, who then closely coordinate their activities. Such assignment includes clearly defining the roles and responsibilities of team members, organizing ways for them to coordinate their activities, and providing them with the supporting documentation.

In situated work, the goal is known, but tasks cannot be predefined, as they depend on outcomes of previous tasks. People in teams, however, have the knowledge needed to quickly evaluate the outcome of tasks and decide on subsequent action. Team structures here are different and require each team member to be aware of what the other team members are doing and to make joint decisions on subsequent actions.

Team structures can also change in different project phases. The initial phases require considerable user involvement to set system objectives and develop system specifications. They thus tend to become more situated in nature. Work becomes more planned as specifications become clearer. The role of computer professionals in early phases may be minimal, but their role increases in subsequent phases, where technical system design becomes important. Different projects may also call for team members with different backgrounds. For example, people with database expertise should be available for database design. Communications specialists should be consulted for any computer system distribution problems. Obviously the right mix of people and resources can go a long way toward ensuring that problems are being solved in the most effective manner.

TYPES OF TEAM STRUCTURE

Constantine (1993) has developed a framework for characterizing teams involved in system development, making a distinction between the following four structures:

- *Structured or closed* teams usually working towards well-defined requirements and where work can be subdivided into well-defined tasks. People are assigned specific tasks and are required to carry them out in a predefined sequence. Here communication requires the smooth flow of information from one task to the next.
- *Open teams* working on projects where requirements are well defined but there can be continual negotiation on how to achieve them. Here team members easily move between roles, and new roles are created as new needs are identified.
- *Synchronous or mission-oriented* teams who work toward a relatively stable goal but where specific requirements can change. The team is characterized by continual planning and agreement on requirements changes. Planning in this sense is continuous but there is a common mission. Team members, guided by the mission, are aware of what must be done in particular circumstances.

- *Random* team structures, where the requirements are broad and can change rapidly, often depending on earlier outcomes. This is usually found in research environments or highly innovative projects. Here all team members have wide scope and can make *ad hoc* contributions to work, and communication requires everyone to know what everyone else is doing.

There is no optimum team structure that can be used in all projects. The question is whether to formally structure the teams or let a structure evolve as the project develops. Different system development processes need different team structures. The chosen team structure will also depend on the variety and complexity of a system, the system size, the type of problem, and the development process phase.

STRUCTURED TEAMS

Structured teams are most common in planned work, where everyone is assigned a task and tasks must be carried out in a predefined sequence. It can be effective in projects that include many repetitive and similar tasks, such as an environment where:

- the technology is stable
- the problem is well defined
- the project size is large.

One example may be a system that requires the development of a large number of relatively straightforward screens. Or it may be a batch system made of many program modules, each requiring considerable editing. Resource control is often considered of paramount importance in these environments. Many of the project tasks, although not identical, are similar, and hence many practitioners feel that the tasks can be effectively standardized and coordinated using highly structured project management techniques. Most team members have very similar roles—that is, as programmers developing code for some system part. The major project problem here is to coordinate the activities of these programmers. Structured teams require a structured documentation system, and team members are required to produce specific documents at the completion of their tasks.

CHIEF PROGRAMMER TEAMS

One of the earliest proposals in planned work was the chief programmer team (Baker, 1972). Such teams have well-defined but different roles for all team members. A typical team structure is shown in Figure 6.2. The core of the team consists of the chief programmer, the assistant chief programmer, and the program librarian. Briefly, the chief programmer is responsible for the functional structure of the programming system and writes the most difficult system programs. In small projects, the chief programmer may be the only programmer in the team, the remainder of the team acting in a supportive role. In most cases, the chief programmer also has an assistant to take over if necessary. The program librarian is there to maintain up-to-date program documentation.

Bigger projects require many more members. For example, the team may include additional programmers and, if the project size is large, there may be a manager to

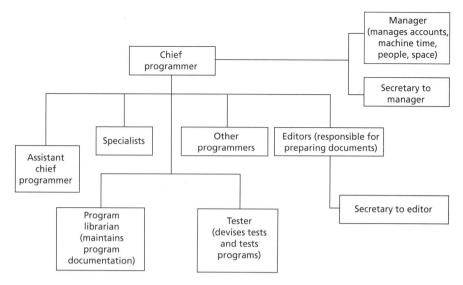

Figure 6.2 *Chief programmer team*

do the administrative tasks and an editor to prepare documents. Each of them may have a secretary, and there may also be someone to devise system tests. The team may have a number of specialists—for example, a language specialist or someone who has a very good understanding of system editors and utilities. Variants of chief programmer structures have been developed over time.

TEAMS IN EVOLUTIONARY ENVIRONMENTS

Evolutionary environments generally require more open and synchronous team structures. Random teams usually create roles as needed and assign people to them. Rettic (1990) describes a number of roles and how they evolve for different projects. Usually, a team begins with a team leader and team members. The team leader in such teams does not usually assign tasks to other members, but is someone who facilitates the work of the rest of the team by arranging meetings, removing obstacles to progress, and recording progress toward a goal. Then, as work evolves, new roles may develop. Particular roles can include an implementer, who develops the programs, a technical reviewer, who validates outputs against requirements, a scribe, who keeps records of meetings, or a specialist consultant. Some of these roles, such as a scribe, may rotate between team members. Teams in such projects have members with a variety of skills, and no person is an expert in all project aspects. Instead, individuals with particular skills are called on to contribute at different times to the project as the need for their skills arises. Furthermore, continuous adjustment of goals also calls for continuous reassignment of tasks. This is done as needed, with team structures changing as people pick up tasks when needed, hence the term random, as the kind of team structure that evolves is basically random.

A typical example is the development of WWW sites, where user acceptance plays a vital role in determining subsequent tasks. Such feedback requires agreement to be reached on subsequent tasks and who will carry them out.

RAPID APPLICATION DEVELOPMENT—DSDM TEAMS

DSDM, for example, places considerable emphasis on including users in the development process. Here there is high degree of end-user development. End-user teams are usually composed of functional specialists whose main concern is the effective operation of their business system. DSDM defines roles such as sponsor, end-user ambassador, and advisor user, who may be someone that uses the interface. Members of such teams often do not have the time to keep up with all computer developments and need specialist support on computing tools and techniques. These include technical coordinator, team leader, developer, and senior developer.

USING TOOLS IN PROJECT MANAGEMENT

Apart from planning the tasks and setting up team structures, project management also provides the necessary tools. One particularly important requirement is for project management to put in place ways to maintain documents and distribute them among team members. Support for documentation becomes more crucial as development becomes more distributed and uses development processes other than linear.

In linear development processes, documents are delivered at the conclusion of each phase. Thus, in theory anyway, design, for example, will not start until a set of specification documents is available. This is no longer the case when processes such as rapid application development are used. Now it is necessary to keep track of documents in different steps as they evolve in parallel. Designers in this case must keep constant track of specifications as they evolve and be sure that they are always using the most up-to-date information.

THE DOCUMENT CONFIGURATION

Documentation support must be integrated with project management. It provides information about the latest work to all project personnel and reduces the chance of work having to be repeated. It is also the only project deliverable, especially in the early project phases, and thus serves to determine project status and progress. We know what documentation must be provided at the end of each project phase, and can measure phase progress by estimating the proportion of phase documentation that has been completed. Finally, documentation becomes part of the phase output. This output includes the system model and plans for subsequent project phases.

Another view of documentation is as a communication tool that is crucial in coordinating the work of team members. It is a communication tool because it contains a repository of all work done to date and makes it available to all persons working on a large project. Such a repository can prevent unnecessary repetition when someone leaves the project team. Proper documentation ensures that all the information developed about the system is always available to new people joining the project. In addition, documentation is needed in project management to manage changes to requirements that inevitably occur in any project. The impact of change can be assessed by using cross-references between documents. This enables team members to be continually aware of what is happening and for new team members to become quickly aware of the current state of development.

Project management must provide a document configuration system for use in system development. The system includes the models constructed during development, any specifications, and program modules themselves. The document configuration can also include the organizational structure and user responsibilities within this structure. Entries in a configuration can be a document or a model, which is produced during development.

A **document configuration** system keeps records of all the documents created during the development process. Different organizations use different terms to refer to the collection of all documents about the system. Terms such as **system directory** and **project dictionary** are still common. The terms **configuration** and **configuration management** are also commonly used to describe documents associated with a project. The documentation is organized in a way that makes it easy for any user to find information about the system. For this reason, it contains many indexes and cross-references between documents. It is also usually maintained in one place. A configuration management system:

- stores all documents created during the development process
- provides access to the latest versions of documents or allows earlier versions to be examined
- organizes documents in ways that allow work to be tracked throughout the development process, usually by linking documents to the development phases
- supports change by maintaining document **versions** and their status. Documents can go through status states such as 'initial', 'accepted', and 'approved'. They may go through a number of versions as the system undergoes change.

The documents stored in a configuration management system depend on the development process and the modeling methods used. Figure 6.3, for example, illustrates a typical set of documents created in a development process that uses structured systems analysis. The document configuration contains the major documents for each phase: the statement of requirements produced during the concept formation phase; the systems specification, analysis, and requirements models produced during the system specification phase; and the design model and system modules. Each of these major documents can contain any number of parts. Figure 6.3 shows only the parts contained in the analysis model for structured systems analysis. These are data flow diagrams, entity–relationship (E–R) diagrams, process descriptions, and detailed descriptions of these components. These latter are often produced with a CASE tool and include descriptions of individual data flows, data stores, or the items that appear in the system. It is also possible to create entries for system users and how they use the system, or descriptions of forms within the system.

Documentation must be mandatory and provide deliverables at the conclusion of project phases. Mandatory documentation requires formal techniques to be used in systems analysis and design, and an *ad hoc* approach is then avoided. It also ensures that the organization obtains the communication and management advantages of good documentation practices.

Document configuration
The structure of documents within a project dictionary.

System directory
Another term for project dictionary.

Project dictionary A record of all documents produced during system design.

Configuration See configuration management in the glossary.

Configuration management
Managing a configuration.

Version A variation of the same document.

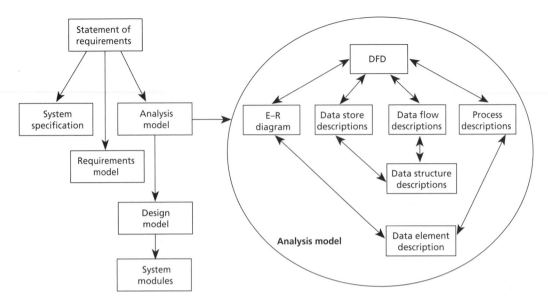

Figure 6.3 *Configuration components*

CASE TOOLS

By now you will have noticed that there is considerable work associated in building and keeping track of models in any development process. First, for any system other than the trivial, there are many diagrams and models. Furthermore, the iterative nature of design means that these diagrams and models are continually changing. As a result, designers must continually redraw their models, validate them against each other, and eventually convert them to an implementation.

CASE (Computer-Assisted Software Engineering) tools are used to help designers develop the models needed during the development process. They help team members build models and convert models developed in one phase to those needed in another phase. CASE tools support the modeling techniques used in the development process. Different development processes can use different models. CASE tools first became prominent in the mid 1980s by supporting modeling techniques used in structured systems analysis and design and by relieving designers of much of the manual work. For instance, in structures, systems analysis entity–relationship methods may be used to describe data, and data flow analysis may be used to describe data flows in analysis models. More CASE tools now support object-oriented methods.

A development process may require more than one CASE tool. Thus, there may be one set of tools for developing the analysis model and another for the design model. Designers must select the right tools for their work and use them during system development. Outputs produced by one tool should serve as inputs to the next. The trend is thus toward developing a suite of CASE tools that are integrated and can support the entire development process. One way to distinguish CASE tools is by the degree of integration and assistance they offer.

CASE TOOL INTEGRATION

Tools can be **integrated** both horizontally and vertically. CASE tool **horizontal integration** is where we connect tools at the same stage of the life-cycle—for example, a connection between the variety of models during analysis. This allows a designer to cross-check the data in the two tools to validate both models. It also allows a designer to use information in one model as an input to the other. CASE tool **vertical integration** applies to different stages of the life cycle, and the output from one tool can become the input to the next tool. Thus, for example, we may use a tool to convert one model to another model. A sequence of such tools can automate the whole design process.

Most support systems provide integrated sets of tools that cover either major parts or the entire development process. An example of integrated tools that cover a part of the development process is shown as the shaded part in Figure 6.4. Three tools cover the analysis part of the cycle and convert the E–R diagram to a database design. These tools cover only some of the development tasks: the shaded tasks in Figure 6.4. They do not, for example, cover project planning or the implementation phases. Some of these may be integrated and others may be standalone tools. The goal is to provide a set of integrated tools that cover the entire development process.

CASE tool integration
The ability of one CASE tool to accept the input from another.

CASE tool horizontal integration
CASE tools integrated from same life-cycle stages.

CASE tool vertical integration
CASE tools integrated at the adjacent life-cycle stage.

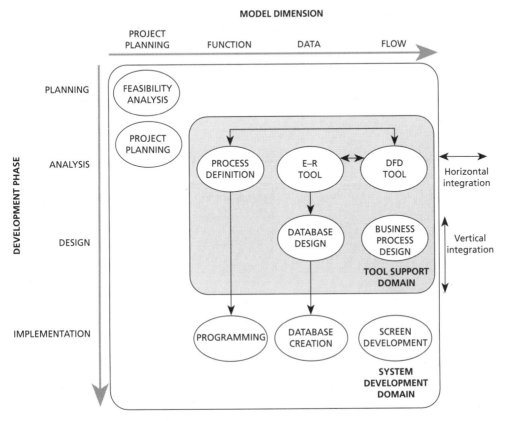

Figure 6.4 *Support domain*

USING CASE TOOLS

Most CASE tools simply act as repositories of models and design decisions. They do not help in the creative part of model development or in the decisions that lead to the model. They simply record the modeling decisions made by analysts or designers. Designers should not make the mistake of thinking that once a model is accepted by a CASE tool, the model is correct. It simply means that the model satisfies the syntactic rules of the CASE tool but may not be the correct representation of the system. A number of CASE tools attempt to provide some assistance, ranging from simple prompts, through syntax checking, to explanations of possible errors and suggestions for their correctness.

Ideally, as more routine work is passed to tools, more time can be spent on the creative aspects of design. However, this holds true only if CASE tools are easy to use. If they are not, then using them may become a problem in itself and detract effort from the real design problem. CASE tools are only one of a set of possible productivity tools. There are also trends to develop other and more powerful tools. The trend is toward tools that can be used to develop a specification in user terms and generate a working system from this specification. This means that we do not have to go through any conversions, and as a result the system can be developed much quicker.

The interface presented by CASE tools is thus important. The trend now is for window support, where different model dimensions are represented by different windows on a screen. Designers can select items in more than one window either to validate the models or to link the items in modeling. However, screen limitations often place constraints on what can be displayed in each window, and tool designers must use ingenuity to ensure that displays can assist designers.

Figure 6.5 illustrates the general structure of displays. Usually in an integrated system there will be one window for each model dimension. Figure 6.5 has one

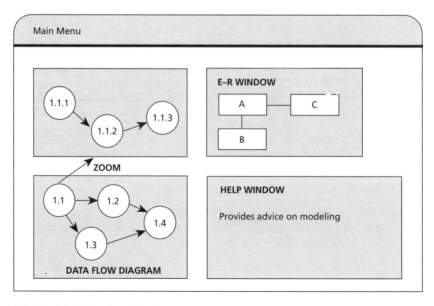

Figure 6.5 *Multi-window displays*

window for E–R diagrams and another for processes. There could also be another one for data flows. The display of all windows simultaneously has the advantage of allowing the designer to correlate the models and verify them. Most CASE tools also allow designers to select one component of the diagram and explode or zoom in on it.

INTEGRATION INTO SYSTEM DEVELOPMENT

CASE tools were introduced in the early 1980s to support the modeling techniques used in structured systems analysis. Further development of CASE tools took place in the early 1990s to support object-oriented design methods. Perhaps the two most common ways of using CASE tools are:

- as independent tools used in some phases of the development process
- as a collection of integrated tools supporting the complete development process, usually following a prescribed set of steps.

Structured design methodologies are likely to be found with highly structured cycles, whereas adaptive teams or prototyping are more likely to use the tools that suit their purpose. Using a tool in one development phase is usually the personal preference of the person responsible for the task, and the tool is often not integrated into the development process. It is simply used to help people in that task. In an uncoordinated approach based on personal preferences for tools used at different phases, it may not be possible to integrate the tools at the different phases, thus losing this potential advantage. Proper use of CASE tools calls for a more integrated approach, where CASE tools are used throughout the entire development process. However, this is not always simple to achieve. First of all there must be more than one tool, and all the tools must be integrated. The tools themselves must be purchased and installed, and people must be trained to use them. What is even more important is that management must be sufficiently committed to their use and require their outputs as part of phase reviews. Studies show that without such commitment, CASE tools are not likely to be used in an organization.

TOOLS FOR STRUCTURED SYSTEMS ANALYSIS METHODOLOGIES

CASE tools are continually evolving in the marketplace. This book gives only a brief overview of the trend in CASE tools, and readers are advised to read manuals of particular tools for the most up-to-date information. Historically, CASE tools began by supporting structured systems analysis. Tools were used either:

- to support a prescribed set of steps, or
- to be more open, allow them to be selected as needed.

The first approach is usually useful in the development of large projects, whereas the second is more commonly used in prototyping or in a system where an adaptive team experiments and builds a system.

The flexible approach is attractive but assumes that both vertical and horizontal tool integration is easy to achieve. This is not the case, and it is often difficult to take a model produced by one tool and use it as an input to another. However, there are now vendors who will produce a set of tools that can be integrated in this way. Apart from providing such tools, the vendor can suggest a structured methodology for using them without requiring this methodology to be followed.

SSADM—A METHODOLOGY COMBINING DATA ANALYSIS AND DATA FLOWS

A number of CASE products support structured systems analysis. Usually vendors provide a number of products integrated to support a number of phases of system development. Oracle Corporation, for example, provides a set of tools.

SSADM (Structured Systems Analysis and Design Method) has become a standard methodology in the British civil service. It has been evolving over a number of years, and is now into its fourth version (Hares, 1994). It is a relatively prescriptive methodology that combines many of the techniques described in the previous chapters of this book and integrates them into a system development life cycle. The tasks are divided into major phases, which in turn are divided into stages. An early version of these phases and stages is shown in Figure 6.6. The techniques used at each stage are also shown in Figure 6.6.

The SSADM stages in Figure 6.6 follow the linear development process described in Chapter 5. SSADM begins with problem definition and then continues through to feasibility analysis, systems analysis, system design, and finally implementation. However, in the main, SSADM concentrates on system analysis and design; that is, Phases 2 and 3. Phase 1 (feasibility analysis) is not mandatory, and Phase 2 can start using either the output from Phase 1 or a project outline from some other source. Methods used in construction and implementation are also left to the implementers, although some guidelines are given. Furthermore, the

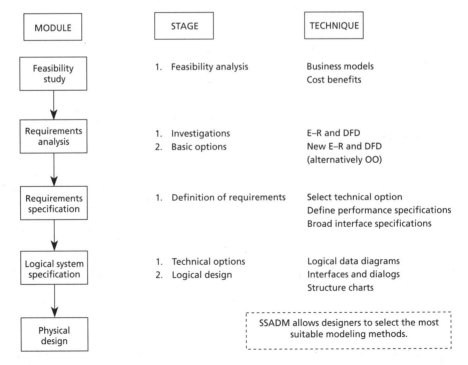

Figure 6.6 *SSADM stages*

SSADM stages do not directly correspond to the phases described in Chapter 3 but include all the tasks described in Chapter 3. These tasks are subdivided between the phases and stages in SSADM, thus illustrating an alternative implementation of the linear cycle. Let us now look at the SSADM phases in more detail.

Phase 1 corresponds closely to the first phases of the development process described in Chapter 5. It has two stages, called 'Problem Definition' and 'Project Identification', which correspond to problem definition and feasibility study. They identify the major problems to be solved by the project.

An analysis model is developed in Stage 1, namely 'Analysis of Current Systems'. A 'Requirements Specification', which also includes some parts of broad design, is produced in Stage 2. The next stage, Stage 3, is part of the broad design and considers alternative physical implementations using ideas like automation boundaries (described in Chapter 16). The remainder of the process covers system design and construction.

CASE TOOL LIMITATIONS AND POSSIBILITIES

In summary, CASE tools do not guarantee a correct or good design, they simply help the designers build their models. There is very little experimental data to claim that CASE tools in fact require fewer resources to build a system. However, it is acknowledged that they improve the correctness of systems and are easier to maintain.

Research in the area of CASE is concentrating on how to provide better assistance in both modeling and supporting a process. It addresses the issue of developing a knowledge base about system design and using it to help designers.

CASE tools help designers build new systems and become used to generating systems from existing software. The original approach was using packages.

TOOLS FOR PROJECT MANAGEMENT

Once the project tasks are determined and their time requirements are known, it becomes necessary to determine their sequence. One choice that needs to be made is which problem-solving cycle to use. Do we use a linear cycle, a staged cycle, or some other cycle? In our particular problem we have opted for a staged cycle. The project plan will define the timing of the tasks, including its start and end times, and the resources, including people needed for each task.

You will find that considerable information is necessary to keep track of project estimates and monitor project progress. It is usually necessary to maintain a database on project status and resources. As shown in Figure 6.7, such a database must allow inputs that indicate any changes to project status and resources and to produce reports that can be used in project reviews.

The database helps management with quantitative controls. Planning and usage data is input to the database, and reports about current resource status can be produced. The project management database also contains project scheduling data. Activity start and end dates can form part of the inputs. These start and end dates can be monitored, and deviations from planned schedules can be output as exception reports.

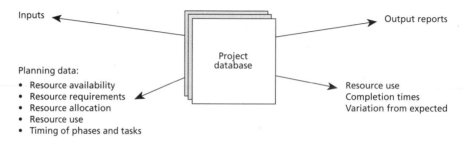

Figure 6.7 *A project database*

Management may request status reports on project progress from the project database. A project database, maintained on a computer, can provide a large variety of such reports.

Typical reports produced by existing systems include:

* personnel availability by skill and by time
* personnel schedules for a project or for a number of related projects
* resource summary reports by project and by tasks within projects
* summaries by project size and type
* summaries of resource usage within regular time periods
* possible manpower shortages over a given period
* budget reports.

Schedule data is often presented in the form of bar charts or by project networks, sometimes better known as project evaluation and review techniques (PERT) charts. These charts or networks can be prominently displayed to make sure everyone is aware of project progress.

BAR CHARTS

A bar chart is illustrated in Figure 6.8. Four project phases are included in the bar chart: analysis, design, development, and implementation. Two bars are plotted for each phase. One bar is plotted with the letter P, and the other bar with the letter A. The bar plotted with the letter P shows the planned time for each phase, and the bar plotted with the letter A shows the actual time for each phase. A bar terminated with a C shows that a phase has been completed. Thus, Figure 6.8 shows that analysis has been completed and we are now in the design phase.

NETWORK CHARTS

As well as showing the scheduled times for each task, a network chart also shows the dependence of one task on other tasks. There is one line for each task on the PERT chart. The line for each task is a continuation of the lines of a previous task. A task cannot start until all its incoming tasks have been completed. Figure 6.9 illustrates a network chart for eight project tasks. A bar chart for the same tasks is also shown for comparison. This figure shows that Tasks 1 and 2 can both start as soon as a project is initiated. It also shows that Task 6 cannot start until Task 4 and 5 are completed. Outputs from PERT networks:

* determine a critical path; these are the activities that must commence on time

Activity	Start-date	End-date	September	October	November	December	January	February	March	April	May	June
Analysis	5 Sept. 1993	11 Oct. 1993	PPPPP									
Design	1 Dec. 1993	3 Mar. 1994				AAAAAAAAC						
Development	2 Feb. 1994	5 May 1995						PPPPPPPP AAA				
Implementation	1 May 1994	6 June 1995									PPPPP	

P = Plan   A = Actual   C = Completion

Figure 6.8 *A bar chart*

- determine the earliest and latest start time for each activity.

To illustrate these outputs, we need the expected time to complete each task. We assume these times to be as shown in Table 6.1.

Table 6.1 *Time to complete tasks*

Task	Time needed
1	3 weeks
2	4 weeks
3	8 weeks
4	5 weeks
5	6 weeks
6	3 weeks
7	4 weeks
8	7 weeks

From these times we can compute the earliest starting time for each task. The earliest starting times for Tasks 1 and 2 will be Week 1, because they can both start as soon as the project is initiated. The earliest starting time for Tasks 3 and 4 will be Week 4. These two tasks only can start when Task 1 is completed. This task requires three weeks and will finish at the end of Week 3. Thus Tasks 3 and 4 cannot commence until Week 4. In this way, you can compute the earliest start time for each task. These earliest start times, shown in Figure 6.9, are the first numbers in parentheses following each task.

Note that the earliest start time for Task 6 is Week 11. The two preceding tasks are Tasks 2 and 4. Task 4 starts in Week 4 and requires five weeks. Thus, the earliest that Task 4 can be completed is the end of Week 8. Similarly, the earliest that Task 5 can be completed is the end of Week 10. Task 5 starts on Week 5 and requires six weeks. Task 6 cannot start until both Tasks 4 and 5 have been completed. Because Task 5 ends later, Task 6 cannot start until Task 5 has been

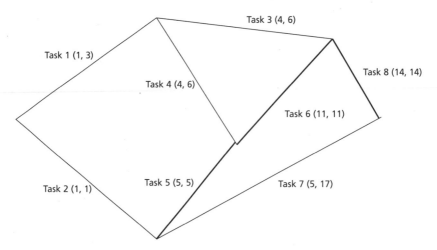

Figure 6.9 *A network chart and a bar chart*

completed at the end of Week 10. Thus, Task 6 cannot start until the beginning of Week 11.

Similarly, Task 8 cannot start until Week 14, because this is the earliest completion time for Task 6. The whole project is completed when both Task 7 and Task 8 are completed. Task 7 can finish in Week 8, but Task 8 cannot end until Week 20. Thus, the earliest completion time for the project is the end of Week 20.

You will find that some of the tasks must always start on time to make sure that the project is completed in Week 20. However, Task 7 need not start on time. It can start in Week 5, but if it actually starts at the beginning of Week 9, it will be complete at the end of Week 12. This will not delay completion of the project. However, if Task 8 starts a week late, the project will end a week later. The sequence of tasks that must always start on time to ensure the earliest project completion time is known as the critical path.

The critical path is found by starting with the earliest project completion time and working backwards to compute the latest start time for each task. For Task 7, the latest start time is the beginning of Week 16. If Task 7 starts in Week 17 it will be complete at the end of Week 20 and project completion will not be delayed. The latest start time for Task 8 is Week 14. The latest start times for each task in Figure 6.9 are shown as the second number in parentheses following each task.

Tasks whose earliest and latest starting times are the same have no slack and must start on time if project completion is not to be delayed. In Figure 6.9, these are Tasks 2, 5, 6, and 8. They lie on the critical path. Other tasks have some slack and their starting time can be delayed. Task 7 can start anywhere between the beginning of Week 5 to the beginning of Week 17. It may, in fact, make sense to delay Task 7. If you look at the bar chart in Figure 6.8, you will see that there are four tasks in progress during Weeks 5 to 8. If we delay the start of Task 7 till after Week 8, there will never be a week that has more than three tasks in progress. This can mean that we reduce peak demands for resources and the project cost.

TEXT CASE D:
Construction Company—A Project Plan

The problem for this case seems to be well defined and not too large, so an evolutionary or prototype approach would not be appropriate. There seem to be two well-defined tasks, namely improving and rearranging current operations and then adding the inventory system. It is proposed that these tasks be done in two stages. First, we will change existing operations and then add inventory. This means that some of the benefits (in particular the replacement of manual checking) can be obtained more quickly with fewer staff.

Once a decision to use a staged approach is made, we specify the times and resources needed for each phase of each stage. Table 6.2 shows one way of representing such an allocation. It includes the start and finish times for each phase and the user and data processing personnel needed. Thus, for example, an average of one person from the information systems (IS) department would work in Stage 1 of the analysis for two months. There would also be an average of one person from the user area working on the analysis for two months, giving a total of four employee-months for Stage 1 of the analysis.

Table 6.2 *A project plan*

(a) Stage 1—Convert project request processing

	Personnel IS	User	Employee-month effort	Start	End
Analysis	1	1	4	1 May 1998	30 June 1998
System design	1	1	4	1 July 1998	30 August 1998
Detailed design	1	1/2	3	1 September l998	30 October 1998
Implementation	2	1	12	1 November 1998	28 February 1998

(b) Stage 2—Inventory function

	Personnel IS	User	Employee-month effort	Start	End
Analysis	1/2	1/2	2	1 November 1998	31 December 1998
System design	1/2	1/2	1.5	15 January 1999	28 February 1999
Detailed design	1	1/2	2.25	1 March 1999	15 April 1999
Implementation	1	1/2	3.75	16 April 1999	30 June 1999
Total employee-month effort			32.5		

Another common way to illustrate project plans is to make a chart that shows project phases graphically over time, and particularly how different projects overlap. Such charts are known as Gantt charts. A Gantt chart for our project is shown in Figure 6.10. The horizontal dimension shows time, and each project or project stage is plotted on the time scale. The overlap of the projects is clearly shown here. For example, the implementation phase of Stage 1 and analysis of Stage 2 quite

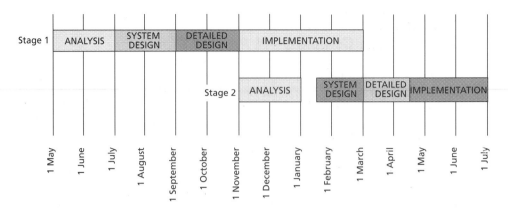

Figure 6.10 *A Gantt chart*

clearly overlap. A chart like the one in Figure 6.10 gives a total picture of what will be going on at each time, when phases finish, and when they start.

PROJECT MONITORING

Monitoring keeps track of project progress. Project schedules contain statements of how long each task should take and how it uses resources. Monitoring collects information to determine whether these goals are being met. Any problems or defects found are usually traced back through the process to find the cause of the problem, and the process is amended to prevent its future occurrence. For example, if it is found that a requirement has not been met, the part of the process where the deviation from requirement first appeared is found, and a further validation requirement is introduced at that point. Successful monitoring requires regular collection of the status of each task.

REVIEWING PROJECT PROGRESS

Project reviews have now been accepted as an essential activity in good project management practice. The project reviews may be quite complex and there may be more than one kind of review. There may be technical reviews to see whether the best technical approach has been adopted. There may also be user reviews that check system models, see whether user requirements are met, and seek agreement on future actions. Finally, there are management reviews to check resource levels and decide whether project objectives have been met. These reviews are also necessary to decide what is to be done in further project phases.

It is usual, if not mandatory, to review a project at the completion of each phase and produce a project phase report that describes the outcome of each task, including:

- the actual starting and completion dates of activities
- the actual use of resources compared with the estimates
- the level of expenditure in each activity
- the quality of output
- any problems found with the supporting process.

Any variations from plans are analyzed, and appropriate corrective action is taken. Management can take action in a number of ways. First, it can vary project scope. Alternatively, project resources can be increased or changes can be made to activity schedules. Finally, the process itself can be improved by changing it in light of experiences with this project.

MEASURING THE PROCESS

A good project management practice is to measure the performance of the development process itself and, if necessary, change and improve the development process from one project to the next. Measurement of the process is different. Its goals, as outlined by Humphrey (1989) of the Software Engineering Institute (SEI), are first to measure the process and then to identify ways of improving it. The process is measured in terms of maturity levels:

- Level 1, Initial, where the process is *ad hoc*.
- Level 2, Repeatable, where there are a basic development process and a project management process to track cost, schedules, and functionality.
- Level 3, Defined, where the project management process is documented and integrated with the development process, with quality reviews in place.
- Level 4, Managed, where detailed measures of the development process are made and correlated with product quality.
- Level 5, Optimized, where measures are used as feedback to improve the process.

Formal assessment procedures are provided by the SEI to measure the maturity level of development processes in an organization. The goal is to reach the optimized level where any problems encountered in development are identified and used to improve the development process, to ensure that such problems do not occur in future projects. In an optimized process, those parts of the process that caused the problem are changed to prevent the problem from recurring. Thus, for example, if some user requirement was not met, a check is made to find where the deviation from user requirements took place, and a validation check is added to that part of the process.

Most processes, in practice, have not reached an optimized level, although there are an increasing number that are reaching Level 3. A Level 3 process must have a development process that is adhered to and monitored by a management process.

SUMMARY

This chapter describes ways of supporting system development. It describes three major supporting tools: project management, document configurations, and CASE tools. Project management is based on defining a set of management entities. These entities are set up to support project phases that have an objective as well as a start and an end. Project management organizes resources to ensure that the project phase objective is achieved in the most effective way.

The chapter describes CASE tools as supporting modeling in analysis and design. It also discusses integration of such tools so that outputs of one can be used as inputs to another.

The third supporting tool is configuration management documentation, which must be integrated with project management. Documentation preserves continuity should personnel change, disseminates any decisions reached, and distributes procedures developed during the phase to appropriate personnel in the organization.

DISCUSSION QUESTIONS

6.1   Why must the management and development processes be integrated?

6.2   Why is project planning different for the waterfall development process and rapid application development?

6.3   What kind of team structure would you expect to be best for evolutionary problem solving or prototyping?

6.4   Describe some roles that you could expect to find in a team.

6.5   Explain what you understand by the term *risk*.

6.6   Why is it difficult to make project estimates?

6.7   . Explain why project reviews are important.

6.8   Explain why documentation is essential for an effective review process.

6.9   What must be included in a configuration management system?

6.10   Describe how you would expect documentation to help analysts and designers.

6.11   What is needed to support an effective documentation system?

6.12   What are the components of a documentation system?

6.13   What is the difference between tools that keep track of documents and those that actively help analysts to develop system models?

6.14   Describe the idea behind CASE tools.

6.15   What are the current trends in the design of CASE tools?

6.16   What are the different ways of using CASE tools?

6.17   How should a case tool help a design methodology?

6.18   What is the difference between prescribed and flexible support tools?

6.19   Why is it preferable to buy an off-the-shelf methodology rather than developing one yourself?

6.20   Describe the SSADM stages and how they relate to the linear life cycle.

6.21   Do you think that the same team should work through the whole project or that project teams should be changed as a project proceeds?

6.22   What would you expect the documentation to include in a feasibility study that uses prototyping?

6.23   Why does time become one of the major controls for imprecise or decision support systems?

6.24   Discuss the importance of making correct estimates in project management.

6.25   What is the difference between project and process measurement?

EXERCISES

6.1   Take the spiral development process described in Figure 5.12. How would you devise a documentation system that fits in with this process? In particular, consider the reusable aspects of documentation. How would you expect documents produced at one cycle to be amended in subsequent cycles and how would you keep track of the changes?

6.2   One goal of successful project management is to break the project up into well-defined tasks and establish methods for coordinating these tasks. PERT and Gantt charts can be used for coordination.

Suppose you have the following set of tasks and task relationships:

Task	Duration	Waits for
T1	6	–
T2	9	T1
T3	7	T2
T4	11	T1
T5	3	–
T6	6	T5
T7	3	T4, T6
T8	6	T5
T9	3	T4, T6
T10	11	T8, T9
T11	4	T3, T7, T10

1. Draw Gantt and PERT charts for these tasks.

2. What are the earliest and latest start times for each task?

3. What is the earliest completion time for the project?

4. What tasks make up the critical path?

BIBLIOGRAPHY

Alter, S.L. (1984), *Decision Support Systems: Current Practice and Continuing Challenges*, Addison-Wesley, Reading, Mass.

Baker, J.M. (1972), 'Project management utilizing an advanced CASE environment', *International Journal on Software Engineering and Knowledge Engineering*, Vol. 2, No. 1, pp. 251–61.

Brooks, F.P. Jr. (1974), *The Mythical Man-Month: Essays on Software Engineering*, Addison-Wesley Publishing Company, Reading, Mass.

Berlack, H.R. (1991), *Software Configuration Management*, John Wiley, New York.

Conradi, R. and Bernhard, B. (June 1998), 'Version models for software configuration management', *ACM Computing Surveys*, Vol. 30, No. 2, pp. 232–82.

Conrow, E.H. and Shishido, P.S. (May–June 1997), 'Implementing risk management on software intensive projects', *IEEE Software*, pp. 83–9.

Constantine, L.L. (October 1993), 'Work organization: paradigms for project management and organization', *Communications of the ACM*, Vol. 36, No. 10, pp. 34–43.

Cotterell, M. and Hughes, R. (1995), *Software Project Management*, International Thomson Computer Press, London.

Curtis, B., Kellner, M.I. and Goldberg, A. (September 1992), 'Process modeling', *Communications of the ACM*, Vol. 35, No. 9, pp. 75–90.

Hares, J.S. (1992), *Information Engineering for the Advanced Practitioner*, John Wiley and Sons, Chichester.

Hares, J.S. (1994), *SSADM: Version 4*, Wiley, Chichester.

Humphrey, W. (1989), *Managing the Software Process*, Addison-Wesley, Reading, Massachusetts.

Iivari, J. (October 1996), 'Why are CASE tools not used?', *Communications of the ACM*, Vol. 39, No. 10, pp. 94–103.

Jarzabek, S. and Huang, R. (August 1998), 'The case for user-centered CASE tools', *Communications of the ACM*, Vol. 41, No. 8, pp. 94–9.

Jones, C. (July 1996): 'How software estimation tools work', *The American Programmer*, Vol. 9, No. 7, pp. 18–27.

King, S.F. (July 1996), 'CASE tools and organizational action', *Information Systems Journal*, pp. 173–94.

Norman, R.J. and Forte, G. (eds) (April 1992), *Special Issue on CASE Tools, Communications of the ACM*. Vol. 35, No. 4.

Perry, D. (March 1991), 'Models of software development environments', *IEEE Transactions on Software Engineering*, Vol. 17, No. 3, pp. 283–95.

Putnam, L.H., Putnam, D.T. and Myers, W. (June 1996), 'Adapting project estimation to advancing technologies', *The American Programmer*, Vol. 9, No. 6, pp. 23–9.

Rettic, M. (October 1990), 'Software teams', *Communications of the ACM*, Vol. 33, No. 10, pp. 23–7.

Rothstein, M., Rosner, B., Senatore, M. and Mulligan, D. (1993), *Structured Analysis and Design for the CASE User*, McGraw-Hill, New York.

Sabherwal, R. (1999), 'The role of trust in outsourcing IS development projects', *Communications of the ACM*, Vol. 42, No. 2, pp. 80–6.

Vessey, I. and Sravanapudi, A.P. (January 1995), 'CASE tools as collaborative support technologies', *Communications of the ACM*, Vol. 38, No. 1, pp. 83–95.

Walz, D.B., Elam, J.J. and Curtis, B. (October 1993), 'Inside a software design team: knowledge acquisition, sharing and integration' *Communications of the ACM*, Vol. 36, No. 10, pp. 62–77.

Whitgift, D. (1991), *Methods and Tools for Software Configuration Management*, Wiley, Chichester.

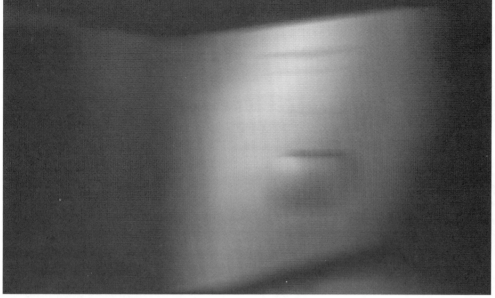

CONCEPTUAL DESIGN

CONTENTS

KEY LEARNING OBJECTIVES

How to identify ways of solving system problems
The importance of finding the right problem to solve

Defining conceptual solutions
How to justify solutions
Defining economic, operational and technical feasibilities
Preparing project proposals

INTRODUCTION

This chapter describes the initial project development phase, concept formation. Concept formation uses the project objectives to produce a broad feasible solution. It does this in the context of continual change in the business environment. Projects eventuate because the way of doing business changes, and the business reason for the project should always be in the background of an analyst's thinking. Hence, any conceptual design must fit within the framework defined in Chapter 3 and illustrated in Figure 7.1. Here, strategic development defines new directions for the whole organization. This leads to organizational development, which changes the organizational structure and ways of working to fit in with new strategic directions. Systems development must then create the information systems to support the new organizational structure and ways of working.

This chapter assumes that projects have been chosen and their objectives have been set as described in Chapter 3. Now the first phase of project development, concept definition, can commence. Concept definition is a continuation of the process of change that follows identification of the project. It proceeds in two parts:

- *First clearly identify the problems that must be solved to meet project objectives.*
- *Follow design procedures to find the most feasible way to meet the project objectives.*

On completion, the phase output will describe how the new system will work in the business environment and set the direction for the remainder of the project.

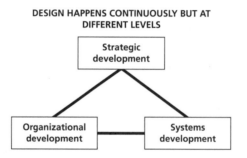

Figure 7.1 *The development dimensions*

PROBLEM DEFINITION

It is perhaps fair to say that that problem definition or identification is not simply a characteristic of information system development. It takes place throughout the entire organization and is a continuous process. It is also important to stress that problem definition is one of the most important issues in development. It is almost imperative to identify the right problems to solve rather than wasting resources on solving problems that have little impact on improving the organization or the operation of its information system. Once the right problems are defined, then good design objectives can be set for the project.

Problem definition is the first activity in the development process. As shown in Figure 7.2, it uses project objectives to identify those problems that stand in the way of reaching the project objective. It then sets design objectives to overcome the problems. Solutions are found both formally and informally. For example, solutions are found informally by simply listening to what people are saying, or there might be something we want to do that someone else is doing, or something might be said at a conference or meeting. We also get ideas about what should be happening both internally and externally.

FINDING PROBLEMS USING INTERNAL CONSIDERATIONS

Internally, it might become obvious that something is not being done the way it should be, simply by examining problem reports or listening to people's viewpoints. This can identify deficiencies such as:

* missing functions and capabilities
* unsatisfactory performance
* excessively costly operations.

One question that often arises here is when to judge an operation as deficient. For example, is a two-day delivery cycle for a product acceptable or not? Unfortunately, there is no such thing as a universal acceptability measure—for example, that a three-day delivery is the universally acceptable delivery time. Obviously delivery times will depend on the type of product and the industry. What one often has to do is look at similar systems or at competitor operations to make judgements on whether a particular operation is deficient.

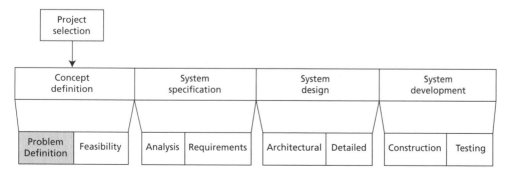

Figure 7.2 *Problem definition*

IDENTIFYING AND SOLVING PROBLEMS USING EXTERNAL CONSIDERATIONS

Some of the ways for finding and solving problems externally are:

- by comparing current operations to normative models, which describe an accepted or conventional way of doing something
- by using historical models of the way organizations develop; this is particularly useful in information systems design because of the development of technology
- by comparing an organization's activities against a competitor's activities or to identify best practices by using some of the guidelines in Chapter 3
- by analyzing changes to government policy, client preferences, and community attitudes.

These external conditions can be used to identify differences between the way things are done in our organization and the accepted way of doing things outside. One obvious observation here is to note the areas where computers are used effectively by others and then look into using them for that purpose as well. This is now a common occurrence in the field of electronic commerce, where more businesses are developing WWW sites to maintain the same presence as their competitors. Another external factor is changes to government rules such as tax policy, for example, which will lead to changes to the organization's systems.

Competitor performance is another important factor in setting goals, and analysts should always examine the methods of similar organizations to see whether some of these can be used in the proposed system. Technical developments outside an organization are also an important source of information, especially for systems that use computers. New external technical developments must be evaluated to see whether they can be used to improve internal system operation.

OUTCOME OF PROBLEM DEFINITION

Problem definition identifies the things that must be addressed by the project and sets the design objectives. The problems are in a way detailed elaborations of the objectives specified during project selection. However, they now define specific problems that must be addressed in the feasibility study.

TEXT CASE C:
Universal Securities—Problem Definition

If you turn back to Figure 3.14, the objectives for Universal Securities were defined. These were to develop support for outsourcing, especially to find contracted suppliers when ordering parts. New systems are required to select suppliers, observe supplier performance, and keep track of order progress.

The problems here are both technical and operational. From the business plan, problems include ways to identify and select suppliers. There are also the initial step of actually outsourcing the operations, and arrangements to be made with existing personnel. Then there are the technical problems. One is simply the lack of expertise in building Web-based systems within the organization. Another problem is the acceptance of suppliers of working electronically and how to phase such work processes into the organization.

The output of this analysis is an actual list of problems and consequent design objectives for feasibility analysis. An initial list may be the following:

- Potential problems to maintain contact with suppliers after outsourcing— develop the most appropriate outsourcing strategy.
- Lack of expertise in Web development leads to the design objective— find a way of arranging Web development.
- Maintaining contact with clients about order progress.
- Suppliers not familiar with ways of working electronically—find the best way to encourage electronic liaison with suppliers.

Each problem identifies an objective that must be considered in the feasibility study.

DESIGN AND FEASIBILITY

Having identified the problems and design objectives, the next step, as shown in Figure 7.3, is to propose solutions, evaluate their feasibility, and select one of them.

Problem definition and feasibility analysis are strongly linked, as shown in Figure 7.4. Feasibility analysis begins once problems are defined and design objectives are set. Feasibility analysis usually includes two steps. The first is to generate broad possible solutions. The second is to evaluate the feasibility of such solutions. Such evaluation often indicates shortcomings in the initial goals—for instance, people may be asking for too much. Alternatively, new insights gained through broad evaluations can identify new opportunities, which can, in turn, lead to goal modifications. This process can go on for a while as project goals are adjusted and alternative solutions are evaluated. The two feasibility steps are now described in more detail.

SOLUTION GENERATION

The important activities in solution generation are to overcome an identified problem, meet the project objective, and show the resulting benefits. Good designers always consider a number of possible alternative solutions. Models often help designers to generate and evaluate alternative solutions. There are also guidelines such as requiring solutions to be developed within the practical bounds of the organization. They should not be ideal models that are impractical. It is advisable

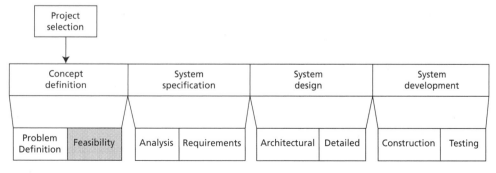

Figure 7.3 *Defining feasible solutions*

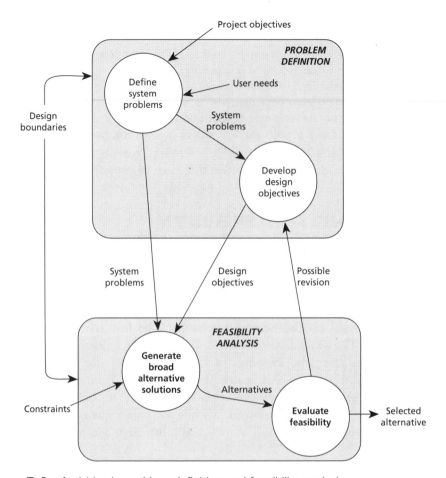

Figure 7.4 *Activities in problem definition and feasibility analysis*

to investigate as many alternatives as possible to ensure that the best solution is chosen. Just using the first idea that comes to mind is often not the best approach. The alternative solutions are then evaluated to determine the most satisfactory solution.

A certain amount of skill is needed to propose good alternatives and choose the right direction. This is where creativity and imagination are used. Analysts must think up new ways of doing things—generate new ideas. There is no need to go into the detailed system operation yet. The solution should provide enough information to make reasonable estimates about project cost and give users an indication of how the new system will fit into the organization. It is important not to exert considerable effort at this stage only to find out that the project is not worthwhile or that there is a need to significantly change the original goal.

TEXT CASE C:
Universal Securities

Universal Securities is an example where we are looking for new systems to meet the strategic objectives set during project identification (summarized in Figure 3.15). One identified project objective was to outsource some of the production

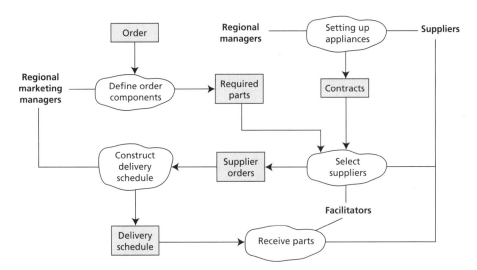

Figure 7.5 *The new 'arrange order production' activity*

facilities to concentrate on sales and service. Problem definition has defined some of the problems and set design objectives.

One way is to look at the broad model of activities and identify those that will be affected. Earlier we defined a model of Universal Securities and described it by the rich picture shown in Figure 3.6. Analysts now consider ways of changing current activities, or adding new ones, to meet the objectives. They may now sketch a broad-level picture of some of the significantly affected activities and their subactivities. One of these is the 'arrange order production' activity. The proposed broad-level picture of 'arrange order production' is shown in Figure 7.5. It shows the more detailed subactivities that must be considered in any feasibility study. One subactivity is to set up alliances with suppliers through contracts that specify the parts to be supplied by each supplier. Another is to define the parts required by a customer order and select suppliers to supply those parts. Another is to construct the delivery schedule and make sure that suppliers adhere to this schedule. A new role of facilitator is proposed to select suppliers and keep track of supplier orders.

A number of alternative implementations of the subactivities are then considered. These concentrate on the major objectives, in particular on ways of getting suppliers to use electronic working methods—an identified design objective. Possible alternatives for selecting suppliers and keeping track of supplier orders are given here:

Alternative 1: Most selection and discussion with suppliers continues on a personal basis, with outcomes recorded on the intranet by Universal Securities personnel.

Alternative 2: Notifications are sent to suppliers through e-mail in the 'select suppliers' subactivity whenever there is a new requirement for parts. Suppliers respond to requirements through the Web interface or by e-mail.

Alternative 3: Orders are posted on the WWW in the 'select suppliers' subactivity and suppliers bid for them electronically. Successful suppliers can then keep track of the delivery schedule to deliver parts just in time.

Because of uncertainties in all of these alternatives, a decision is made to use rapid application development with prototyping to improve the rate of acceptance of electronic trade by suppliers. A number of JAD sessions with suppliers are proposed to identify social issues involved in getting their acceptance of Web-based working and to look at ways of mixing Web-based and traditional processing. Some Web pages will be prototyped before the meetings to raise supplier awareness of the various possibilities.

Alternatives would also be proposed for all the other subactivities in Figure 7.5.

EVALUATION

Once solutions are generated they are evaluated. Three things must be done to establish feasibility. First, it is necessary to check that the project is technically feasible. Does the organization have the technology and skills necessary to carry out the project and, if not, how should the required technology and skills be obtained? Second, operational feasibility must be established. To do this, it is necessary to consult the system users to see whether the proposed solution satisfies user objectives and can be fitted into current system operations. Third, project economic feasibility needs to be checked. The study must determine whether the project's goals can be achieved within the resource limits allocated to it. It must also determine whether it is worthwhile to proceed with the project at all or whether the benefits obtained from the new system are not worth the costs (in which case, the project will be terminated).

Technical feasibility An evaluation to determine whether a system can be technically built.

Technical feasibility. This evaluation determines whether the technology needed for the proposed system is available and how it can be integrated within the organization. Technical evaluation must also assess whether the existing systems can be upgraded to use the new technology and whether the organization has the expertise to use it.

Operational feasibility An evaluation to determine whether a system is operationally acceptable.

Operational feasibility. Operational feasibility covers two aspects. One is a technical performance aspect and the other is acceptance within the organization. Technical performance includes determining whether the system can provide the right information for the organization's personnel and whether the system can be organized so that it always delivers this information at the right place and in time. Acceptance revolves around the current system and its personnel. Operational feasibility must determine how the proposed system will fit in with the current operations and what, if any, job restructuring and retraining may be needed to implement the system. The evaluation must then determine the general attitudes and skills of existing personnel and whether any such restructuring of jobs will be acceptable to the current users.

Economic feasibility An evaluation to determine whether a system is economically acceptable.

Economic feasibility. This evaluation looks at the financial aspects of the project. It determines whether the investment needed to implement the system will be recovered.

RISK ANALYSIS

Risk analysis is another approach used in defining feasibility. The kinds of risk were discussed earlier in Chapter 6. The usual approach is to call for evolutionary development and prototyping where the risk is judged to be high. Analysis centers on identifying those aspects of a project where there is the largest uncertainty about getting a successful outcome. This can be used with new technology, for ill-defined problems, or where there is the possibility of rejection by users. Once such areas are identified they are given special attention. This may be to delay some part of the implementation or to adopt a prototyping approach to more closely evaluate the risk. Thus, a path of least risk may be followed in the development.

ECONOMIC FEASIBILITY

Economic feasibility concerns returns from investments in a project. It determines whether it is worthwhile to invest the money in the proposed project or whether something else should be done with it. Some organizations, especially those with large projects, place great emphasis on economic analysis. It is not worthwhile spending a lot of money on a project for no returns, especially if there are many other things that could be done with that money.

To carry out an economic feasibility study, it is necessary to place actual money values against any purchases or activities needed to implement the project. It is also necessary to place money values against any benefits that will accrue from a new system created by the project. Such calculations are often described as cost–benefit analysis.

COST–BENEFIT ANALYSIS

Cost–benefit analysis usually includes two steps: producing the estimates of costs and benefits, and determining whether the project is worthwhile once these costs are ascertained.

PRODUCING COSTS AND BENEFITS

The goal is to produce a list of what is required to implement the system and a list of the new system's benefits.

Cost–benefit analysis is always clouded by both tangible and intangible items. Tangible items are those to which direct values can be attached (e.g. the purchase of equipment, time spent by people writing programs, insurance costs, or the cost of borrowing money). Tangible costs often associated with computer system development include the following:

- Equipment costs for the new system. Various items of computing equipment as well as items such as accommodation costs and furniture are included here.
- Personnel costs. These include personnel needed to develop the new system and those who will subsequently run the system when it is established. Analysts, designers, and programmers will be needed to build the system. Also included are any costs incurred to train system users.
- Material costs. These include stationery, manual production, and other documentation costs.
- Conversion costs. The costs of designing new forms and procedures and of the possible parallel running of the existing and new systems are included here.
- Training costs. These include both training users of the new system and any training required for developers who may be required to use new technologies.
- Other costs. Sometimes consultants' costs are included here, together with

management overheads, secretarial support and travel budgets. Any training costs for users of the new system can also be included here.

The values of intangible items, on the other hand, cannot be precisely determined and are made by subjective judgement. For example, how much is saved by completing a project earlier or providing new information to decision makers? Considerable argument can take place before agreement is reached on such intangible costs.

The sum value of costs of items needed to implement the system is known as the cost of the system. The sum value of the savings made is known as the benefit of the new system. Once we agree on the costs and benefits, we can evaluate whether the project is economically viable.

The cost estimates are usually used to set the project budget. Often it is convenient to divide these costs into project phases to give management an idea of when funds and personnel will be needed. The cost estimates need to be worked out very carefully. One should avoid any omission from the estimates that necessitates requests for more funds because something was forgotten.

On the other side of the evaluation are the benefits of the project. These can also range from the tangible to the intangible. Tangible benefits include those benefits that can be measured in actual dollar terms. Such benefits can include reduced production costs through the introduction of new technologies, or reduced processing costs through the use of computers. Less tangible benefits include the possibility of increased sales through changes to improve the ordering system, or a wider market through better distribution of marketing data. Measurements of these benefits are not direct but are based on estimates of what can happen when a new ordering or marketing system is introduced. Management must be convinced that the estimates are accurate if they are to accept the evaluation.

Intangible benefits cannot be measured. For example, what is the benefit of better decision making through computer support or the benefit of maintaining a good business image?

DETERMINING WHETHER A PROJECT IS WORTHWHILE

The costs and benefits are used to determine whether a project is economically feasible. There are two ways to do this: the payback method and the present value method.

THE PAYBACK METHOD

The payback method defines the time required to recover the money spent on a project. The concept is quite simple. We know how much a project will cost to start. We also know the costs and benefits for each succeeding year. The difference between the cost and the benefit for each year will be the saving or net benefit for the year. The computation to determine the number of years needed to recover the costs is quite simple. As an example, consider Table 7.1, which shows the cost of implementing a project in the year 0 to be $110,000. Benefits from the project are obtained over the next five years, and the net benefits are also shown in Table 7.1. If you add the values in the benefit column, you will note that the original $110,000 invested is returned some time towards the end of Year 3. Hence the payback period is about 2.9 years.

THE PRESENT VALUE METHOD

The payback method is not always the best way to determine economic feasibility. It does not seem to make much sense to put, say, $10,000 into a project and recover $12,000 in Year 5. You might as well put the $10,000 into an investment account at 10% and get $16,105 in that time. Given an investment rate of 10%, one would have to recover more than $16,105 in Year 5 (assuming no returns in Years 1 to 4) to make the project worthwhile. This is where the present value method comes in.

Table 7.1 *A present value evaluation*

Year	Cost	Benefit	Present value of benefit	Discount factor (10%)	Discount factor (15%)
0	$110,000				
1	–	$20,000	$18,180	0.909	0.866
2	–	$40,000	$33,040	0.826	0.756
3	–	$60,000	$45,060	0.751	0.658
4	–	$30,000	$20,490	0.683	0.571
5	–	$10,000	$6,210	0.621	0.497
	Total amount		$122,980		

The idea of the present value method is to determine how much money it is worthwhile investing now in order to receive a given return in some years' time. The answer obviously depends on the interest rate used in the evaluation. Thus, the present value of $16,105 in Year 5 is $10,000 today at 10% interest. If a project that pays back $16,105 in Year 5 costs $11,000 today, that project is not worthwhile. On the other hand, it is worthwhile if it costs only $9000 today.

To some extent the present value method works backwards. First, the project benefits are estimated for each year from today. Then we compute the present value of these savings. If the project cost exceeds the present value then it is not worthwhile.

Let us look again at Table 7.1 to see whether the project is worthwhile at a discount rate of 10%. To do this we find the present value of the benefit at each year. The formula is:

Present value $\times (1 + r/100)^n$ = benefit at Year n

We call $1/(1 + r/100)^n$ the discount factor in Table 7.1.

Thus, for example, the present value of the $40,000 benefit at Year 2 is computed as:

Present value $= 40,000/ (1 + 10/100)^2$

$= 33,057$

Note that the sum of the present values in Table 7.1 is $122,980. As this exceeds $110,000, the project is worthwhile.

You may wish to compute the present value of the benefits in Table 7.1 at 15% interest. The result is $109,140.

Let us now have a brief look at another class of objectives—business process improvement.

BUSINESS PROCESS IMPROVEMENT

Developing new ways of doing business is one kind of project. Another common kind of project is to improve existing operations to reduce costs and improve throughput. A typical business information system at the present time often consists of many systems that were written in the past and still use old methods and technologies. Each such system often supports only one business unit. These systems are generally called **legacy systems**. It is now common for organizations to integrate such systems to improve their supply chains and internal production processes. The term **business process re-engineering** often appears in this context. It refers to the identification of organization-wide business processes and integrating existing systems to support organization-wide business processes.

There are two alternative ways to re-engineer processes. One is to make adjustments to existing processes. The other is to adopt a strategy to use standard processes or blueprints.

Legacy system An existing working computer system that is to be used in a new business process.

Business process re-engineering Changing an existing business process.

TEXT CASE D:
Construction Company

SITUATION
Two departments in a construction company have independently set up their own computer systems, shown in Figure 7.6, which includes the files kept by the systems and their inputs and outputs. These are the project ordering system (POS) set up by the purchasing department and the goods received system (GRS) developed by the dispatch department.

POS
The POS is used to order parts required by projects by issuing purchase orders (PO) to suppliers to supply these parts. Project teams put in PROJECT-REQUISITIONS for parts. The purchasing department selects suppliers to supply them. Over time, the purchasing department has developed a SUPPLIER-FILE which contains information such as supplier address, items available from suppliers, and item prices. It also contains more subjective information, such as comments about supplier reliability and expected delivery time. The SUPPLIER-FILE is used by personnel in the purchasing department to negotiate purchases with suppliers. Once a purchase is negotiated, a purchase order is issued to the selected supplier. A copy of the purchase order is also stored in the PURCHASE-ORDERS-FILE. It is identified by a unique purchase order number (PO-NO). A project request may be split up between a number of suppliers, although all items with the same ITEM-NO in a PROJECT-REQUISITION are always purchased from the same supplier. A purchase order may also contain items requested by more than one project team. You should note that a purchase order sent to a supplier does not contain any

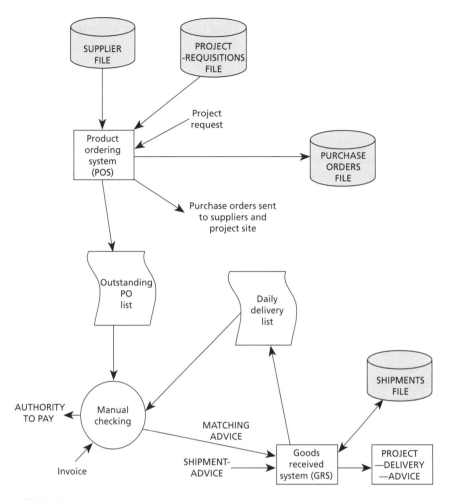

Figure 7.6 *Project ordering and goods received systems*

reference to projects. The reference to purchase orders for items in PROJECT-REQUISITIONS is stored in the REQUISITIONS-FILE, which contains all project requests with references to their purchase orders. These references appear as PO-NO for each ITEM-NO in the REQUISITIONS-FILE.

GRS
The GRS has been developed by the dispatch department to receive parts from suppliers and forward them to project sites. It also uses another small computer system to record shipments received from suppliers and to record the store where these shipments are held. A SHIPMENTS-ADVICE is received with each shipment. The shipment is stored somewhere in the organization, and the information in SHIPMENTS-ADVICE together with its LOCATION is stored in the SHIPMENTS-FILE. You should remember that a shipment advice from suppliers does not include any information about projects but includes only PO-NOs. This means that the PO-NO in the shipment must be matched against PO-NOs in the REQUISITIONS-FILE to locate the project that needs the parts, and to track whether all ordered parts have been received. This is further complicated by the fact that each shipment

advice can contain items ordered from the supplier in more than one purchase order.

MATCHING—THE CURRENT SYSTEM INTEGRATION

Matching, which is currently manual, refers to the POS, to find cross-references in the REQUISITIONS-FILE between project team requests and purchase orders, which appear in both lists. This is done by creating an outstanding purchase order list generated by POS, matching this list against lines on a daily delivery list generated by GRS, and manually finding entries with matching PO-NOs. Once a match is found, the GRS system is informed (using a MATCHING-ADVICE sent from a workstation) of the project to which the items in a shipment are to be dispatched. The GRS uses this advice to generate a PROJECT-DELIVERY-ADVICE, which is sent to the project site with the parts. Once a delivery is made at the site, the corresponding purchase order must be cleared from the POS. Partial deliveries of purchase orders and consequent parts deliveries to project sites also complicate the matching process. The process is not of a high quality, because manual matching often results in delayed deliveries or deliveries made to the wrong places.

ISSUES

There has been some concern about the supply chain in this process with its long elapsed time between project requisitions and project deliveries. This is caused by many factors, including not having an inventory control system, supplier delivery delays, the time taken to place a purchase order, and the time-consuming manual checking of shipments against project requests.

It has been decided to analyze this system to see whether any improvements can be made and how they can be achieved.

The project goals are to remove these deficiencies. For instance, if it takes too long to process an order, the goal would be to speed up order processing. The statement of the goal should, in the first instance, be concise and specific. It should not be the detailed specification that is developed as a next step. If it is a new system, all that is needed is a broad and overriding objective stating what is to be done and its effect on the information system. For example, a broad goal statement may take the following form:

- Project objective: Reduce the time between the project requisitions and goods delivery.
- Project objective: Eliminate errors and delays in order processing.

TEXT CASE D:
Construction Company—Defining the Problems

We can define a number of problems for the construction company. It does not maintain an inventory of frequently used parts, which could satisfy many parts requests by withdrawing parts from the store. The lack of an inventory would probably emerge in the course of interviews. Project personnel would no doubt complain about delays caused by waiting for suppliers to deliver and they might say, 'Why can't some of these parts be held in store?'

Defined problems are often converted into design objectives that ask questions such as: 'How can we reduce the elapsed time between project request and parts delivery?' There are a number of possibilities here. For example, we could:

- create an inventory to hold frequently used parts in local stores
- improve the method for determining project delivery sites for a particular supplier shipment
- improve the procedures for verifying invoices and shipments
- create a help desk environment where requisitions are allocated to one person who pursues it through to completion
- re-engineer the process to integrate the goods received system (GRS) and the project ordering system (POS) and eliminate the time-consuming manual checking. However, the incompatible computers used in the two systems make it difficult to transfer data between computers.

These design objectives are used to propose alternative solutions.

PROPOSING SOLUTIONS FOR IMPROVEMENT

For a novice, it is often difficult to say what a broad solution is and how it differs from a detailed one. The judgement of what to include in a broad solution is often very subjective. However, two things must be kept in mind when proposing a broad solution: it should give people a good idea of what the new system will look like and also convince them that it will work. Another objective of the broad solution is to form an estimate of cost. To do this, a broad solution should include the major parts of a proposed system.

COMPONENTS OF A BROAD SOLUTION

A broad solution should include only the information needed to estimate the cost of projects and how the system will be used. It need not be a detailed description of the system. It is generally agreed that broad solutions should include:

- any additional equipment that will have to be purchased for the project, in order to estimate some of the direct costs of the project
- any computer networking requirements to estimate costs of communications equipment
- any new computer systems that will have to be developed to estimate the development cost
- what is to be done by the computer and what will remain manual
- the information that will be made available by the system
- the services that will be provided to customers and especially any expected improvements
- any computer interfaces provided to computer users
- rough ideas of processes that will be followed using the new system.

From this, we can find the amount of data to be stored on the machine and any transmission costs of getting data to and from the machine.

Only broad descriptions need be given here. As an example, at this stage we may suggest a report, give it a name, and state what its contents are to be. However, detailed report layouts, interfaces, or handling error conditions should not be specified. All that need be specified is the information provided in reports and interfaces. Details of these will be covered in later phases of the development process.

What is needed now is a broad statement of the kind of information that will be made available to users and its effect on user operations.

One should not, however, become extreme here and propose alternatives that obviously cannot be supported in the organization. The organization has certain constraints on the amounts of available funds and personnel skills, and on working and accepted standards. No proposed solutions should obviously exceed such funding limits or ignore some critical system operation or data need.

TEXT CASE D:
Construction Company—Broad Alternative Solutions

Some broad solutions for the construction company are illustrated and a number of alternative solutions are proposed. There are, of course, totally manual solutions that could improve matching and information flows. However, we will concentrate only on computer-based solutions, which are illustrated in Figure 7.7.

Solution 1 is a totally technical solution—to install network connections between the two systems so that they can exchange information between themselves, in fact going to a client–server architecture. However, this may be rejected on technical grounds because there is no vendor software for easily creating a client–server environment that include clients based on the different software systems used in POS and GRS. Development of software in-house to set up a client–server environment may not be considered feasible because of lack of expertise.

Solution 2 is to move one of the systems from one computer to the other. Thus we can move the GRS onto the POS computer and develop the inventory system and the matching program there. Of course, the reverse is also possible, to move the POS onto the GRS computer. Hence, there are two solutions here:

* Solution 2A—move the GRS to the POS computer
* Solution 2B—move the POS to the GRS computer.

Both these solutions will require an entire rewrite of either the GRS or the POS.

Solution 3 is to redesign the entire system based on reference processes. This may have some support, particularly if some of the data generated by the GRS or POS applications is used by other processes.

Solution 4 is to leave the GRS and POS systems as they are and develop yet another system for the checking and inventory functions. This solution may be rejected on operational grounds, as there will still be too much manual work needed to move information between the systems. In fact, the amount of such manual work may be increased because there are now three and not two systems.

Finally, Solution 5 is proposed as a compromise solution. In this solution, a part of the POS is moved to the GRS so that GRS now receives project requests. An inventory subsystem is also proposed for the GRS, and invoice checking is added to the POS. When a project request arrives, a check is made in the GRS to see whether it can be met from inventory. If it cannot, then a purchase requirement with a purchase order number is generated and sent to the POS, which selects a supplier and issues a purchase order. Note, however, that a reference to the purchase order is kept in the new PO-REQUEST in the GRS. This reference is used to match the project request to the shipment when it arrives at the GRS. The shipment record is sent to the POS, where it is checked against the invoice.

Solution 5 was proposed as an alternative to Solution 2. It requires less

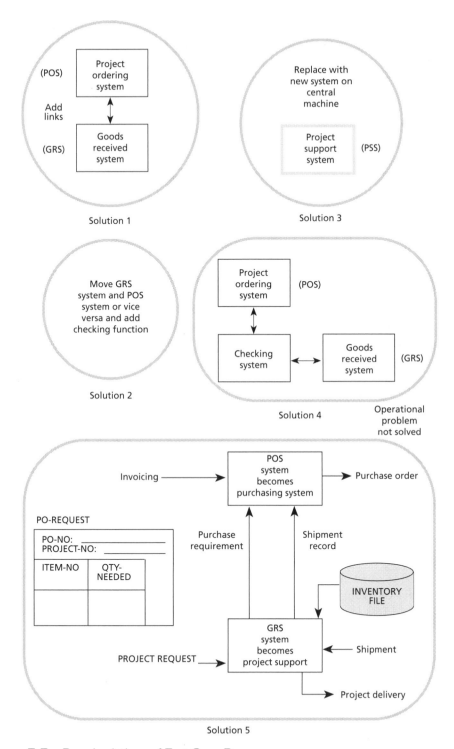

Figure 7.7 *Broad solutions of Text Case D*

redevelopment, because only part of the existing system is redeveloped. However, in comparison with Solution 2, which is totally automated, this proposal still requires some manual work to transfer shipment records and purchase requisitions from the GRS to the POS. Such a transfer can be achieved without manual transcription, either by copying information on disks that are moved between machines, or by using a local area network. In this way, all checking is now automated.

TEXT CASE D:
Construction Company—Evaluating Feasibilities

Of the alternatives proposed in Figure 7.7, Solution 1 can be eliminated on technical feasibility grounds and Solution 4 on operational feasibility grounds. This leaves Solutions 2, 3, and 5 for further consideration, using economic feasibility.

The costs relevant to the broad solutions outlined for the construction company are illustrated in Table 7.2, which shows the costs of converting existing systems to new systems and the costs of developing new systems. You will note that it costs more to convert the POS to the GRS computer than to convert the GRS to the POS computer. This is because of the more complex interface used in supplier selection and the conversion of the supplier database.

Table 7.2 *Initial development cost estimates*

	Tasks	Cost
A	Convert POS to the GRS computer	$90,000
B	Convert GRS to the POS computer	$60,000
C	Develop delivery checking system (DCS) to replace manual checking	$30,000
D	Develop inventory control system (INV)	$50,000
E	Develop invoice input interface (INIT)	$20,000
F	Move project order interface from POS to GRS (Solution 5)	$20,000
G	Redesign to use reference processes	$150,000

Table 7.3 applies the costs in Table 7.2 to the alternative broad solutions. Each broad solution includes more than one activity. Thus, solution 2A converts the POS to the GRS computer (cost of Task A), and develops a delivery checking system (cost of Task C), an inventory control system (cost of Task D), and an invoice input interface (cost of Task E). These four costs make up the total cost of solution 2A. The component costs of all the other solutions are shown in Table 7.3.

Table 7.3 *Initial alternative cost estimates*

Solution	Tasks in solution	Cost
2A	A ($90,000) + C ($30,000) + D ($50,000) + E ($20,000)	$190,000
2B	B ($60 000) + C ($30,000) + D ($50,000) + E ($20,000)	$160,000
3	G ($150,000) + C ($30,000) + D ($50,000) + E ($20,000)	$250,000
4	Hardware cost ($15,000) + C ($30,000) + D ($50,000) + E ($20,000)	$115,000
5	F ($20,000) + C ($30,000) + E ($20,000) + D ($50,000)	$120,000

We have costed Solution 4 here for illustration. The cost includes $15,000 for some additional hardware to implement checking on a separate computer.

SELECTING AN ALTERNATIVE

Ideally, a detailed cost analysis should be made for each alternative, but this is seldom possible. We would not choose one alternative simply because it is slightly cheaper than another. There are other considerations. Some alternatives are quickly ruled out on technical or operational grounds as well as on other internal organizational considerations. These may be things like selecting an alternative that matches skills within the organization. Or perhaps choosing a site for expansion that has been ignored for some time.

TEXT CASE D:
Construction Company—Selecting the Alternative

In Text Case D, Solution 1 has already been ruled out on technical grounds and Solution 4 on operational grounds. Usually only two or three alternatives are evaluated in detail. Once the evaluation is complete, selection begins.

An economic comparison such as that shown in Table 7.3 is usually a good start for selection. Some solutions, such as 2A, can be eliminated straight away. Obviously 2B would be preferable to 2A because it is cheaper. Now we must choose between Solutions 2B, 3, and 5. Because of its cost, Solution 3 could also be quickly eliminated unless it is part of a larger strategy to adopt standard processes in the construction company. Assuming this strategy is not in place, then we are reduced to choosing between 2B and 5. This is resolved by determining the value of doing invoice checking automatically (something done in 2B and not 5). Is the value of such a check worth the extra $40,000?

The final determination of the value of the project is shown by the present value computations in Table 7.4. These computations use a discount factor of 10%. They use only direct benefits, and no intangible benefits are included. The direct benefits are the savings in staff who previously did the checking. The savings per year are $40,000 for Solution 5 and $50,000 for Solution 2B. The difference is the extra manual work to transfer shipment notices between the two systems in Solution 5. The present value for Solution 2B is $155,350 compared with the cost of $160,000. This solution would not be economically acceptable on direct economic costs. The present value for Solution 5 is $124,280 compared with the cost of $120,000. Thus, on direct economic costs, this solution is acceptable but, as you will note, not much is gained by the project on direct costs alone. However, if intangible savings or quicker project completions because of quicker parts deliveries were added, the returns would be more substantial. Thus, in all probability, the project would go ahead using Solution 5, because it requires less initial investment than solution 2B.

Table 7.4 *Present value analysis for Text Case D (using a discount rate of 10%)*

(a) Savings (Solution 5)

Year	Amount	Present Value
1	$40,000	$36,360
2	$40,000	$33,040
3	$40,000	$30,040
4	$40,000	$24,840
Total present value		$124,280

(b) Savings (Solution 2B)

Year	Amount	Present value
1	$50,000	$45,450
2	$50,000	$41,300
3	$50,000	$37,550
4	$50,000	$31,050
Total present value		$155,350

PREPARING A PROJECT PROPOSAL

Once our analysis is complete, it is necessary to prepare a project proposal. The proposal itself may be long and detailed or relatively short, depending on the size and importance of the project. The proposal must be presented in a form that clearly specifies the advantages of the project to the organization and its users. It must be clear and precise and specify the goals. It must also present the arguments, using the kind of information produced in this chapter. Above all, it must stress the advantages it will bring to the organization and the improvements that will be made once the new system is in place.

The proposal must instill in its readers, who are usually the ones who will contribute the funds for the project, confidence that the proponent has clearly thought out the goals, the risks, and the alternatives and that it does provide the best course of action. The proposal should include all the issues covered in this chapter and present them in an order that convinces management and users to proceed with the project. Suggested proposal sections include:

- a statement that defines the business problem being solved
- the chosen solution explaining why it was chosen and briefly indicating the other alternatives
- a description of how the new system will work and its impact on external clients and internal users
- the justification for choosing the preferred alternative and its economic, technical, and operational advantages
- what various people in the organization will have to do to implement the solution
- the effect of the solution on the way people work, including any new skills needed by people and how they will acquire these skills.

The proposal usually goes to management, so it must be written in a language that is understood by them. It must be precise and concentrate on identifying the problem and convincing management that the proposed solution will be beneficial to the organization. Rejected solutions may also be briefly listed and discussed.

SUMMARY

This chapter describes the first phase of the project—developing the project concept and using it to develop the project proposal. The project concept is developed by defining a number of broad alternative solutions and then evaluating the technical, operational, and economic feasibility of these solutions. The most satisfactory of these solutions is then selected and a project proposal is prepared.

DISCUSSION QUESTIONS

7.1 Why is identifying a problem important?

7.2 How would you go about identifying problems?

7.3 'Maximizing benefits for given costs' is one possible criterion for comparing alternative proposals. In what environment would this criterion be most appropriate? Name some other criteria with which you are familiar and the environments to which they apply.

7.4 Why are constraints important when alternative solutions to a problem are proposed?

7.5 Why is a system proposal needed?

7.6 Define what you understand to be the difference between project and design objectives.

EXERCISES

7.1 Propose alternative broad solutions for the truck-scheduling system described in Text Case A.

7.2 Suppose you have done a cost–benefit analysis and found that the estimated cost of a proposed project is $230,000. The benefits of the project over an estimated life of six years are:

- Year 1—$50,000

- Year 2—$80,000

- Year 3—$65,000

- Year 4—$50,000

- Year 5—$40,000

- Year 6—$20,000

Would you proceed with the project, given a 10% discount rate?

7.3 Have another look at the proposed Solution 1, the client–server approach for the construction company, and describe ways of implementing it.

BIBLIOGRAPHY

Goldberg, R. and Lorin, M. (eds) (1982), *The Economics of Information Processing*, Wiley-Interscience, New York.

King, J. L. and Schrems, E. L. (March 1978), 'Cost–benefit analysis in information systems development and operation', *ACM Computing Surveys*, 13(1), pp. 19–34.

Liang, Y., West, D. and Stowell, F. (April 1998), 'An approach to object identification, selection and specification in object-oriented analysis', *Information Systems Journal*, Vol. 8, No. 2, pp. 163–80.

Tolis, C. (1999), 'Facilitating understanding and change: the role of business models in development work', in Nilson, A.G., Tolis, C. and Nellborn, C. (eds), *Perspectives in Business Modeling: Understanding and Changing Organizations*, Springer, Berlin.

DATA FLOW DIAGRAMS

CONTENTS

KEY LEARNING OBJECTIVES

Why it is necessary to model data flows?
Data flow diagrams
Context diagrams
The difference between physical and logical data flow diagrams
How to level data flow diagrams
Good practices in naming and conservation of data

CHAPTER 8

INTRODUCTION

This and the next two chapters describe the modeling methods used in structured systems analysis to create analysis models and, very often, the requirements model during the system specification phase. These methods center around modeling both the system and subject worlds, although they use mainly system world terms. Consequently, they require analysts to have particular skills in expressing subject world requirements in system world terms. Models in structured systems analysis are made up of three components: the process, the data, and the system functions. Data flow diagrams model system processes and are one of the most important modeling tools used by systems analysts. The use of data flow diagrams as modeling tools was popularised by DeMarco (1978) and Gane and Sarson (1979) through their structured systems analysis methodologies. They suggested that a data flow diagram should be the first tool used by systems analysts to model system components. These components are the system processes, the data used by these processes, any external entities that interact with the system, and the information flows in the system.

This chapter describes data flow diagrams and how to use them to model systems. Subsequent chapters describe the other data and function components.

DATA FLOW DIAGRAM SYMBOLS

Data flow diagram (DFD) A method to illustrate how data flows in a system.

Data flow diagrams (DFDs) use a number of symbols to represent systems. Most data flow modeling methods use four kinds of symbols to represent four kinds of system components: processes, data stores, data flows, and external entities. Various writers use different symbols to represent these components. In this book we use the symbols first introduced by DeMarco, but we also indicate some alternative symbols that are often used. The symbols used by DeMarco are shown in Figure 8.1 and described in the text that follows.

PROCESSES

Processes show what systems do. Each process has one or more data inputs and produces one or more data outputs. Processes are represented by circles in a DFD. Each process has a unique name and number, which appear inside the circle that represents the process in a DFD.

FILES OR DATA STORES

Data store A component of a DFD that describes the repositories of data in a system.

A file or **data store** is a repository of data. It contains data that is retained in the system. Processes can enter data into a data store or retrieve data from the data store. Each data store is represented by a thin line in the DFD, and each data store has a unique name.

EXTERNAL ENTITIES

External entity An object outside the scope of the system.

External entities are outside the system, but they either supply input data into the system or use the system output. They are entities over which the designer has no control. They may be an organization's customers or other bodies with which the system interacts. Alternatively, if we are modeling one section in an organization,

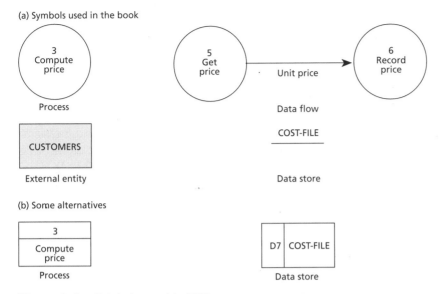

Figure 8.1 *Symbols used in DFDs*

other sections are modeled as external entities. External entities are represented by a square or a rectangle.

External entities that supply data into a system are sometimes called *sources*. External entities that use system data are sometimes called *sinks*.

DATA FLOWS

Data flows model the passage of data in the system and are represented by lines joining system components. The direction of the flow is indicated by an arrow, and the line is labeled with the name of the data flow. Flows of data in the system can take place:

> **Data flow** Data flowing between processes, data stores, and external entities.

- between two processes
- from a data store to a process
- from a process to a data store
- from a source to a process
- from a process to a sink.

We have no control over flows between external entities, so we do not model them. Similarly, stores are passive and there cannot be a data flow between one data store and another. A process is needed to move data between two data stores.

Figure 8.1 also shows some alternative symbols that sometimes appear in DFDs. A process is sometimes represented by a box, with the number appearing on top of the box. Data stores are also sometimes represented by a rectangular box, with a special number and a name inside the box.

DESCRIBING SYSTEMS BY DATA FLOW DIAGRAMS

Context diagram A diagram that shows the inputs and outputs of a system.

We will now illustrate how systems are modeled using these symbols. A common way to begin is to model the whole system by one process. The DFD that does this is known as the **context diagram**. It shows all the external entities that interact with the system and the data flows between these external entities and the system.

Figure 8.2 is an example of a context diagram. It models the 'budget monitoring system'. This system interacts with three external entities: DEPARTMENTS, MANAGEMENT, and SUPPLIERS. In Figure 8.2, the main data flows from DEPARTMENTS are 'spending request' data flows. In response, DEPARTMENTS receive either 'rejected request' data flows or 'delivery advice' data flows. MANAGEMENT receives 'request for special approval' data flows to which it responds. MANAGEMENT also sends 'budget allocation' data flows to the system and gets 'spending summaries' data flows. Suppliers receive 'part order' data flows and return 'supplier delivery advice' data flows.

Of course, a model like that shown in Figure 8.2 does not describe the system in detail. For more detail it is necessary to identify the major system processes and draw a DFD made up of these processes and the data flows between them. The DFD that shows the major system processes is called the top-level DFD (shown in Figure 8.3). The top-level DFD shows the various processes that make up the system. Each process in the top-level DFD has a unique name. Numbering starts from 1 and continues sequentially until all processes are numbered. Note that the context diagram does not have a number. From Figure 8.3 we see that data flow 'spending request' from DEPARTMENTS goes to the 'check funding' process. This process looks up the allocated budget and determines whether special permission is needed from MANAGEMENT to proceed with the request. If special permission is needed, the request is entered in the SUSPENDED-REQUESTS data store to wait for the response from management. It is then referred to when the response is received. The double-sided arrow between SUSPENDED-REQUESTS and the process 'Check funding' indicates that the process stores and retrieves information from this

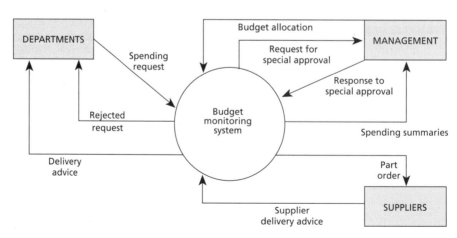

Figure 8.2 *Context diagram*

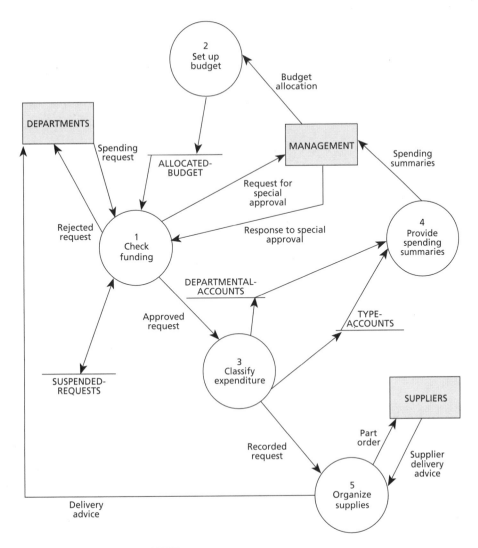

Figure 8.3 *A top-level DFD*

data store. Approved requests and those that do not need special permission go to the 'classify expenditure' process, where they are entered into data stores, DEPARTMENTAL-ACCOUNTS, and TYPE-ACCOUNTS. Once requests are classified they become a 'reorder request' and are passed to 'organize supplies', which places a 'part order' with suppliers. There are also two other processes in Figure 8.3. One is to 'set up budget' and the other to 'provide spending summaries'.

We could continue to expand each process in Figure 8.3 into a more detailed DFD. As an example, Figure 8.4 is a detailed description of the 'classify expenditure' process labeled Process 3 in Figure 8.3. Each process in this detailed DFD is labeled with a 3 and followed by a number to show that each is an expansion of Process 3. The input to this DFD is the data flow 'approved request'. Process 3.1 classifies the entries in each 'approved request' by expense type (such as material and capital overhead), and Process 3.2 uses this classification to update data store TYPE-ACCOUNTS. All the entries for the whole request are summed by Process 3.3,

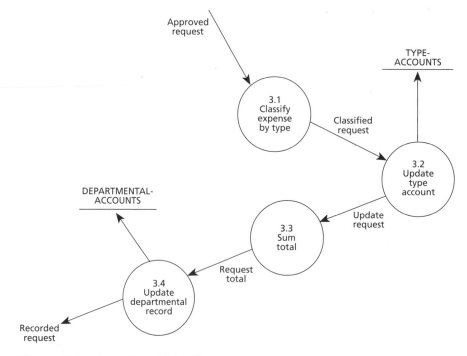

Figure 8.4 *Expansion of 'classify expenditure' process*

and Process 3.4 uses the total to update the data store DEPARTMENTAL-ACCOUNTS.

We could expand each of the processes in Figure 8.4 by even more detailed DFDs. However, eventually we must stop this expansion. If we don't, we will get DFDs containing processes describing simple computations such as the addition of two numbers. DFDs that reached that level would be awkward and unduly complex. However, it is necessary to get down to a level that describes detailed arithmetic computation. This is done by using process specifications rather than DFDs.

Process specifications and detailed data definitions are described in detail in later chapters. A brief description of process and data specifications is given here to show you how all the constructs of a structured system description fit together.

Figure 8.5 illustrates a detailed description of processes, data stores, and data flows. In Figure 8.5, detailed process descriptions are marked by the symbols * and detailed data descriptions by the symbols *1 for data store descriptions and *2 for data flow descriptions. Normally, process and data descriptions do not appear on the DFD but are stored as part of the document configuration, which is described in Chapter 6. Also, at this stage, you should not worry about the details of process and data description syntax. All you really want to appreciate is how all the pieces of the system description fit together. You will find out about the details in later chapters.

Figure 8.5 shows that an 'approved request' is made up of the data items DEPT-NO, REQUEST-NO, and a number (as indicated by the asterisks) of request lines, each made up of the data items AMOUNT and DESCRIPTION. The data changes as it passes through the processes. After Process 3.1, DESCRIPTION in data flow

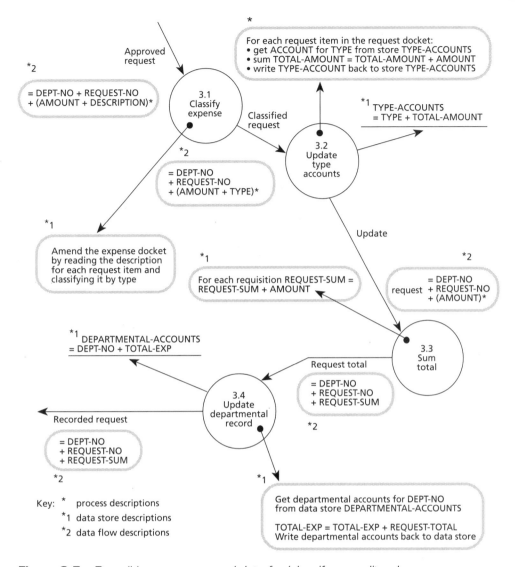

Figure 8.5 *Describing processes and data for 'classify expenditure' process*

'approved request' is replaced by TYPE in data flow 'classified request', and then, after Process 3.3, all the values of AMOUNT are summed to REQUEST-SUM.

The data stores are also defined in Figure 8.5. Thus, the data store DEPARTMENTAL-ACCOUNTS contains any number of entries made up of two items, DEPT-NO and TOTAL-EXP. The TOTAL-EXP means the total expenditure in requests so far made by the department identified by DEPT-NO.

In Figure 8.5, process specifications use formal methods and refer to data item names defined for data structures. For example, the description of Process 3.2 shows that TYPE in the classified request is used to locate an entry in the TYPE-ACCOUNTS data store. The TOTAL-AMOUNT in this entry is updated, and the entry is written back to the data store.

So far we have tried to provide an idea of what structured analysis is all about.

It may look quite simple on the surface, but a number of difficulties can arise once you get into detailed work. First, you should note that there are many ways to draw a DFD for a system. Some of these may be better than others. Ideally, a DFD is self-explanatory, complete, and unambiguous. Nevertheless, a number of conventions and rules have been developed to make your work easier. The rest of this chapter describes some of these conventions, beginning with a description of the properties of a good DFD.

Let us now see how we can apply some of these techniques to the case we introduced in Chapter 4.

TEXT CASE D:
Construction Company—A Data Flow Model
of the Company

We first give some additional information about this case—in particular, the structure of the files and forms of the system. The details are shown in Figure 8.6. This differs from Figure 4.1 in that it now shows the details of the file structure. Thus, for example, it shows that the PO-REQUEST file has three fields that identify the request, the project that made it, and the date by which parts are needed. It also includes a number of lines, one for each needed item.

We also have some more details of the data transferred between the two incompatible systems. The POS generates an outstanding purchase order list daily, whereas the GRS generates a daily delivery list.

The outstanding purchase order list looks like this:

PO-NO	SUPPLIER	ITEM-NO	QTY	PROJECT-NO

It contains the purchase order number, the name of the supplier who will supply the items, the number of the project that needs the items, and item numbers.

The daily delivery list looks like this:

PO-NO	SUPPLIER	ITEM-NO	QTY-RECEIVED

It contains the purchase order number, the quantity of the goods received, the name of the supplier, and item numbers. Manual checking matches the PO-NOs in both lists. Once a match is found, the GRS is informed (using a MATCHING-ADVICE sent via a terminal) of the project to which the items in a shipment are to be dispatched. The GRS uses this advice to generate a PROJECT-DELIVERY-ADVICE, which is sent to the project site with the parts.

The first step of modeling is to draw the context diagram. This is shown in Figure 8.7, which also shows the external entities and the inputs and outputs to the system. Note that the analyst's terms of reference do not include the ability to change project operations. PROJECTS are therefore outside the scope of the system and are treated as external entities.

Before drawing the top-level DFD, we describe one important aspect of data flow diagrams: the difference between logical and physical DFDs.

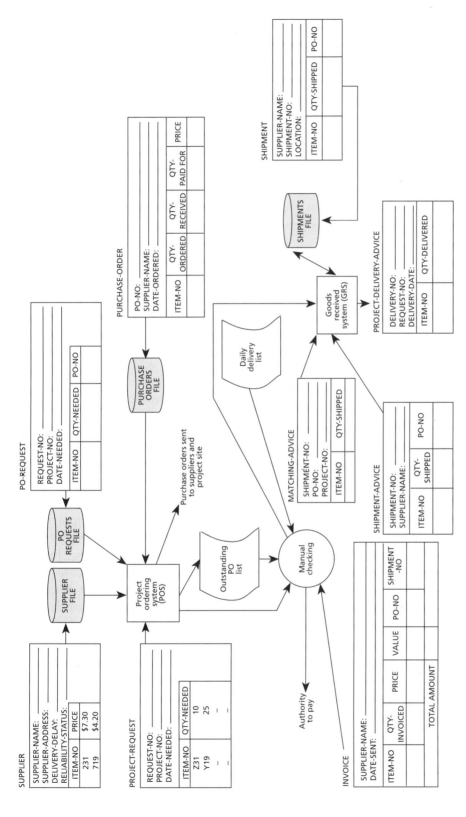

Figure 8.6 *Project ordering and goods received systems*

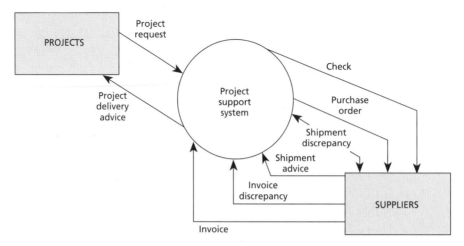

Figure 8.7   *Context diagram*

LOGICAL VERSUS PHYSICAL FUNCTIONS

Many designers make a distinction between logical and physical functions. Another way to consider the difference is to see the physical-level functions as closer to the usage world, whereas the logical diagram more closely mirrors the subject world. Figure 8.8 illustrates the difference between these two kinds of functions. Figure 8.8(a) is a DFD that models what is going on. It says that orders are received, the location of ordered parts is determined, and delivery notes are dispatched. It does not, however, tell us how these things are done or who does them. Are they done by computer or manually, and, if manually, by whom?

Figure 8.8(b) is the opposite. It shows the actual devices that perform the functions. Thus, there are an 'order processing clerk', an 'enter into computer file' process, and a 'run LOCATE program' process. DFDs like Figure 8.8(a), which illustrate only *what* occurs without showing *how* it occurs, are known as **logical DFDs**. DFDs that show *how* things happen, or the physical components, are called **physical DFDs**. Typical processes that appear in physical DFDs are methods of data entry, specific data transfer or processing methods, and processes that depend on the physical arrangement of data. Often DFDs include both physical and logical processes.

What are the advantages of these two kinds of DFDs? It is often argued that logical models help designers gain a clear perception of what the system is to achieve without getting confused by its current implementation details. After all, one of the objectives of systems analysis is to improve system operation. It is easier to do this if one initially concentrates on what is to be done without considering the limitations imposed by physical devices.

However, it is easier to start modeling the physical system, because physical components can be readily identified during analysis. Analysts therefore often begin by building a physical model and then convert this physical model to a logical

Logical DFD Describes the flow of logical data components between logical processes in a system.
Physical DFD Describes the flow of physical data components between physical operations in a system.

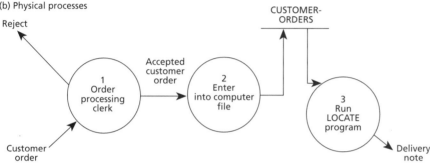

(a) Logical processes

Reject

Accepted customer order

1 Receive customer order

2 Find location of parts

Part location

3 Arrange delivery

Customer order

Delivery note

(b) Physical processes

CUSTOMER-ORDERS

Reject

Accepted customer order

1 Order processing clerk

2 Enter into computer file

3 Run LOCATE program

Customer order

Delivery note

Figure 8.8   *Logical and physical data flow design*

model. In fact, analysts often begin with a DFD that is a mix of physical and logical processes. This is the initial DFD that is then converted to a logical DFD.

CONVERTING INITIAL DFDS TO LOGICAL DFDS

A number of steps can be followed when converting physical DFDs to logical DFDs. These steps are illustrated by using the DFD shown in Figure 8.9, which includes four processes:

1. *The reception clerk* receives an order and checks to determine whether it is of the type made by the organization. If the answer is no, the order is not accepted; if yes, it is accepted and sent to production.
2. Accepted orders are *sorted* into areas.
3. *The production section* checks whether the machines for making the order are available. If not, the order is not accepted; otherwise, resources for order production are committed.
4. *The production section* produces the ordered part.

All of these processes are **physical processes**. During conversion to logical DFDs, we first remove all the processes that refer to physical activities only and do not transform information. The two processes of this kind in Figure 8.9, for example, are 'sort into areas' and 'send to production section'.

The remaining processes are physical because they describe physical components. However, they cannot be removed from the DFD, because they transform data. These two processes are then expanded into their logical functions. To do this, take

Physical process Usually a physical device used to transform data—for example, computer or person.

(a) Top-level physical DFD

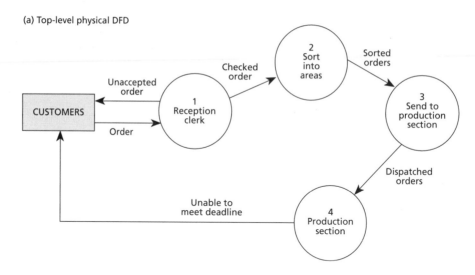

(b) Expanded processes

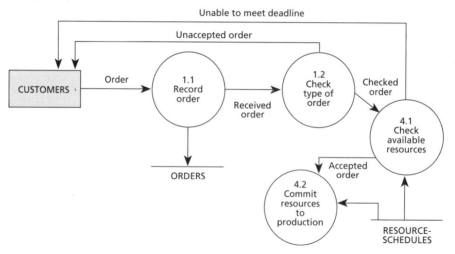

(c) Recombined logical processes

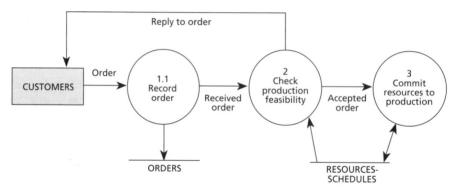

Figure 8.9 *Converting physical DFDs to logical DFDs*

each physical process, find out what it does, and replace it by a leveled DFD of logical functions that represent the physical object's logical activities, or what the object does. All physical processes can be expanded in this way, and their expansion can be combined into a lower-level logical DFD. Thus, in Figure 8.9, the reception clerk is replaced by the two reception clerk functions 'record order' and 'check type of order'. Similarly, 'production section' is replaced by its two functions 'check available resources' and 'commit resources to production'.

This lower-level DFD is then examined. Any common or similar functions are combined, and these combined, higher-level, processes become the higher-level DFD. Examination of the expanded DFD in Figure 8.9(b) shows that 'check type of order' and 'check available resources' have a similar purpose—they determine whether the job can be done. Hence they are combined into a higher-level logical function, 'check production feasibility', in Figure 8.9(c). This process becomes the only process to interface with the users. It replies to the users, either accepting or rejecting their orders, and carries out any other dialog with them. The processes 'record order' and 'commit resources to production' are then also added to the DFD in Figure 8.9(c). The result is that, in Figure 8.9(c), we have the logical DFD equivalent of the physical DFD shown in Figure 8.9(a).

More examples of typical physical DFDs and their conversion to logical DFDs are shown below. Figure 8.10(a) shows a DFD that contains all data in one physical database. This often happens when we are modeling systems with an existing centralized database. The centralized physical database can be replaced by its logical components—that is, the logical data structures stored in the database, as shown in Figure 8.10(b).

Models that include generalized retrieval packages are further examples. In Figure 8.11(a), many processes are combined into the general process inquiry package. Again, the DFD does not show explicit logical processing. Thus, the general inquiry package can be replaced by logical functions, as shown in Figure 8.11(b).

In general, physical DFDs are often characterized by the kinds of structures shown in Figure 8.12. Physical processes usually have many functions and consequently have many flows in and out of them. Whenever you see a structure like the one shown in Figure 8.12, you can be fairly sure that it is a physical process, and you should, if necessary, convert it to a **logical process**.

Let us now return to Text Case D and expand its context diagram.

Logical process Describes any changes of values made by the processes on logical data.

TEXT CASE D:
Construction Company—Expanding the Context Diagram

Figure 8.13 illustrates a first attempt at a top-level DFD. The system is modeled by four obvious physical components: the GRS, the POS, the manual checking system, and the payment section. This top-level DFD is physical in nature. Because it is difficult to go directly from this physical top-level DFD to a top-level logical DFD, we use the idea shown in Figure 8.9 and first level the physical system into its logical components.

The leveled processes are shown in Figure 8.14. This figure shows the leveled diagrams for Processes 1, 2, and 4 of Figure 8.13. Thus, for example, Process 2, which receives parts from suppliers, has been leveled into three parts. Process 2.1

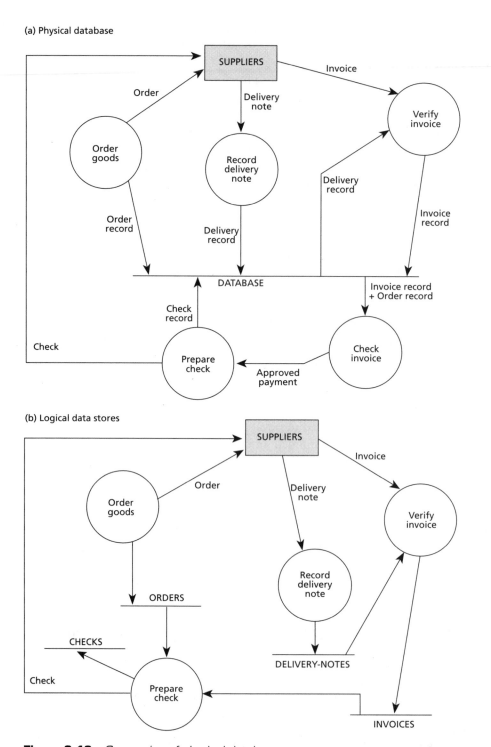

Figure 8.10 *Conversion of physical database*

(a) A physical enquiry package

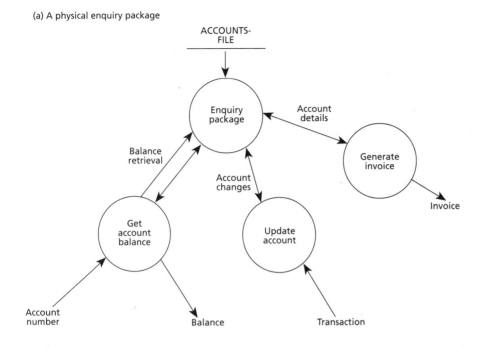

(b) Removing enquiry package

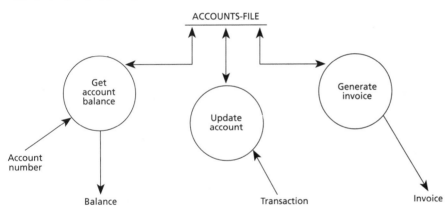

Figure 8.11 *Modeling file access*

simply records a shipment received from a supplier. Process 2.3 generates the shipment list for manual checking (Process 3). Process 2.2 generates delivery advices, as indicated by the matching advices obtained from Process 3.

Note that leveled Processes 1.1, 1.2, 1.3, and 2.1 still represent physical computer system components. Processes 1.4 and 2.3 also represent physical processes, because they describe a particular sort order of data. Process 3 describes a physical processing component (manual processing), and Process 1.2 describes a physical method of data entry. Process 4.1 also refers to a physical component—checks.

Note that Figure 8.14 contains the leveled DFD for a number of top-level processes. Normally, each of these DFDs would appear on a separate page. In Figure 8.14 they appear on the same diagram for illustrative purposes only.

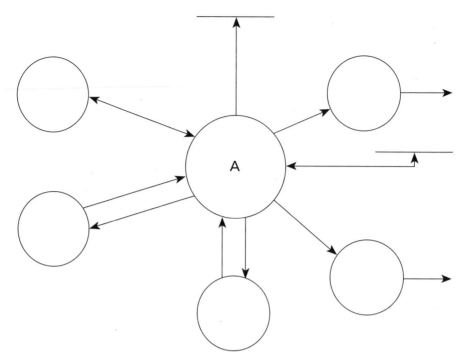

Figure 8.12 *Typical physical process interface*

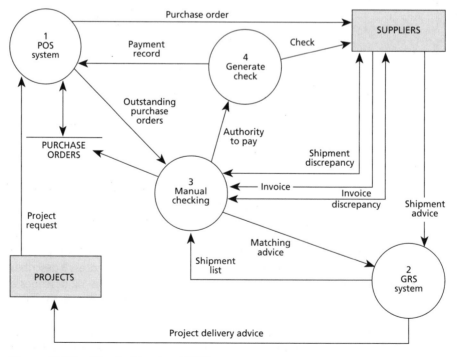

Figure 8.13 *Physical top-level DFD*

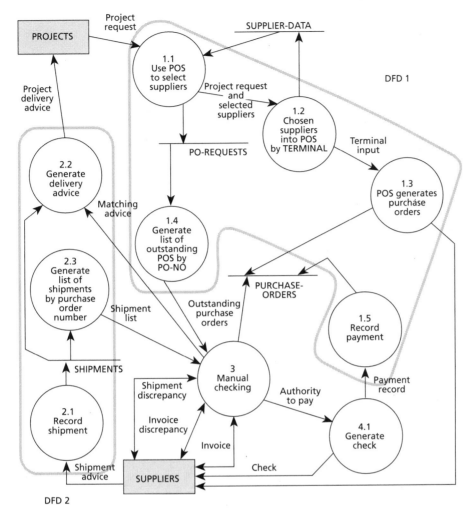

Figure 8.14 *Leveling of physical processes*

To create the top-level logical DFD it is necessary to eliminate references to physical systems and devices and to remove any processes that describe purely physical transformations. It is also necessary to level Process 3 before any recombination into higher-level logical processes. Figure 8.15 shows DFD 3, which is the leveled DFD of Process 3. It describes what is currently done manually. The first step, Process 3.1, is to take the shipment list and match it with the outstanding purchase orders. Any matches are followed by identifying the project involved (Process 3.3) and by an 'advice to store' on where to send the items (Process 3.4). Any discrepancies found in Process 3.1 are brought to the attention of suppliers. All received parts are recorded on the PURCHASE-ORDER file, which is used to authorize payments (Process 3.5). Process 3.5 also checks for any discrepancies between the received part and the invoice and brings them to the attention of suppliers.

Let us examine how to construct a logical top-level DFD from the leveled diagrams. First, purely physical processes like Process 1.2 are removed, and references to physical devices are removed from process descriptions. Thus the

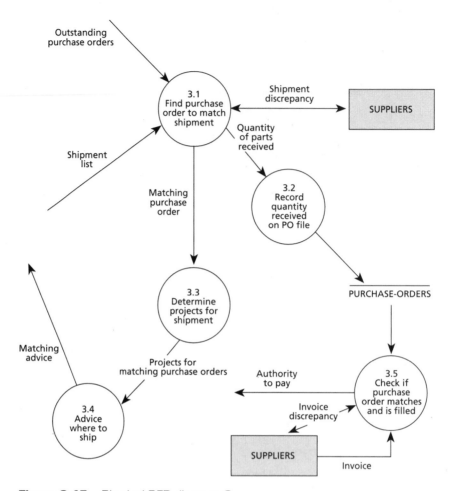

Figure 8.15 *Physical DFD diagram 3*

words 'use POS to' are removed from Process 1.1, and this process becomes Process 1 in the top-level DFD, shown in Figure 8.16. Similarly, the reference to POS is removed from Process 1.3, and this becomes Process 2 in Figure 8.16.

Processes 1.4 and 2.3 are removed because they are purely physical. The leveled Processes 3.1, 3.2, 3.3, and 3.4 in Figure 8.15 are recombined into Process 4 in the logical top-level DFD in Figure 8.15. Note that data store PO-REQUESTS is used by logical Process 4 to find a destination for shipments. The 'shipment destination' is sent to Process 5, which generates the 'project delivery advice'. Process 3.5 in Figure 8.15 becomes Process 6 in the logical top-level DFD. Processes 2.1 and 2.2 become logical Processes 3 and 5 respectively in the logical top-level DFD. Finally, Process 4 is combined with Process 1.5 into logical Process 7.

One question that designers often ask is: 'How can I tell if this is a good DFD?' Unfortunately, this is not an easy question to answer. One part of the answer is whether all the information about the system has been captured. This can be checked only by validating the DFD with the users. The other part of the answer is whether the structure or syntax of the DFD is correct. Again, there is no easy

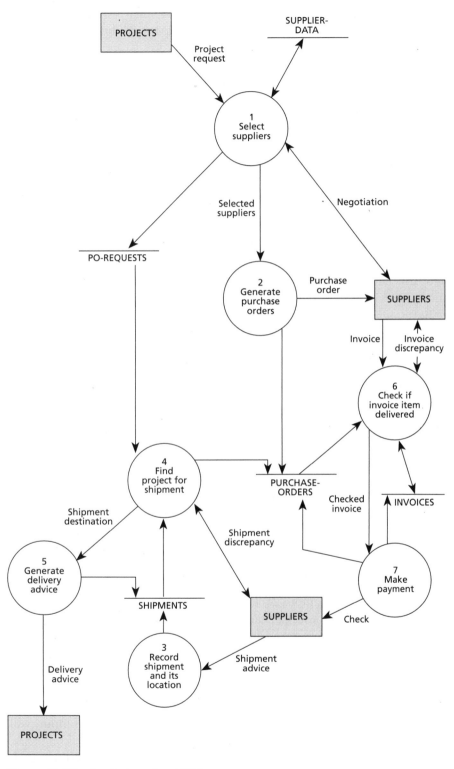

Figure 8.16   *Logical top-level DFD*

answer. However, it is possible to suggest some features of good DFDs, which we will do in the remainder of this chapter. Designers can then determine whether their DFDs violate any of these features. You may postpone reading this part until you have done some detailed diagramming.

WHAT IS A GOOD DATA FLOW DIAGRAM?

DFDs have a number of features that can be used to ensure that they are self-explanatory, complete, and unambiguous. The features are:

- the absence of flowchart structures
- conservation of data
- good naming conventions.

DIFFERENCES BETWEEN FLOWCHARTS AND DATA FLOW DIAGRAMS

You are probably familiar with program flowcharts. They include boxes that describe computations, decisions, iterations, and loops. It is important to keep in mind that DFDs are not program flowcharts and should not include control elements. A good DFD should:

- have no data flows that split up into a number of other data flows
- have no crossing lines
- not include flowchart loops of control elements
- not include data flows that act as signals to activate processes.

Figure 8.17 shows some DFDs that do not conform to these criteria. Figure 8.17(a), for example, shows a splitting data flow that illustrates two outcomes of a computation. In Figure 8.17(b), Process 'compare' has outgoing data flows that are labeled by the conditions under which a data flow occurs, rather than by the contents of the data flow. Figure 8.17(c) illustrates a DFD modeling a program loop. Neither of these constructs is permitted in DFDs. Figure 8.17(d) is another violation that illustrates a signal ('End of month') that activates a process. Structures such as those shown in Figure 8.17 should not appear in your DFDs.

Decisions and repetition, however, do go on in any system and must be modeled. How is this done? Decisions and iterative control are part of the process description rather than data flows. For example, look at Figures 8.18 and 8.19.

Figure 8.18 illustrates a process, 'check item availability'. The input to this process is the data flow 'Item-request'. This data flow requests QTY-NEEDED of an ITEM-NO to be supplied. A check is made by the 'check item availability' process to determine whether a sufficient quantity of the item is held in store. The flows out of the process depend on the outcome of this test. Thus, if enough items are found, a 'delivery note' is output from the process; otherwise an 'unavailable note' is output. Hence, the data flows out of the process, given a certain input, will depend on the outcome of the decision.

Figure 8.19 illustrates repetition. The 'compute daily sales' process looks at each sales docket in a batch of sales dockets. Repetition is defined in the process

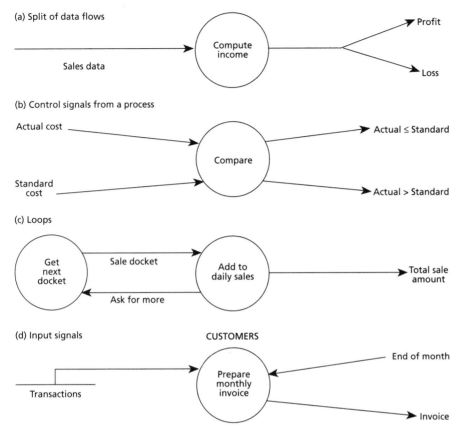

Figure 8.17 *Illegal data flows*

specification by the REPEAT BEGIN . . . END phrase. Again, the repetition control is part of the description of the process and does not appear on the DFD.

Another important criterion used to judge DFDs is conservation of data.

CONSERVATION OF DATA

This useful modeling guideline was originally coined for structured analysis by Gane and Sarson (1979). It applies to processes and data stores. For data stores, conservation of data says that what comes out of the data store must first go in. It is not possible for the data store to create new data elements. An analogous rule exists for processes. A process cannot create new data. It can only take its input data and either output it again or transform it into a new form of data.

Figure 8.20 illustrates some violations of the conservation of data principle. In Figure 8.20(a), the data item QTY is lost by the 'retrieve item price' process. Figure 8.20(b), on the other hand, produces a 'channel use per day' output, but does not have any input information about the number of disk accesses made each day.

NAMING

To make a DFD readable, analysts should avoid using meaningless names. Little information is conveyed by the labels shown on Figure 8.21. Names such as 'needed

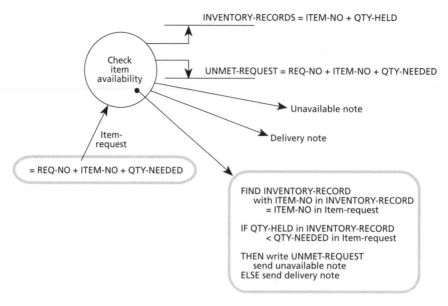

Figure 8.18 *Decisions in DFDs*

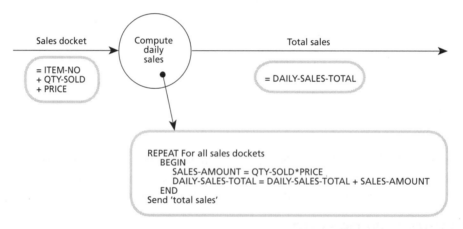

Figure 8.19 *Repetition in DFDs*

data', 'standard operations' and so on should be avoided. Specific names should be used at all parts of the DFD. The choice of these names depends on the type of component named.

NAMING PROCESSES

A process name should be one single phrase, and it should be possible to describe a process in one sentence. The process name should define a specific action rather than a general process. For example, the following names are acceptable:

- 'edit a withdrawal transaction'
- 'compute weekly salary'
- 'compute discount on order'.

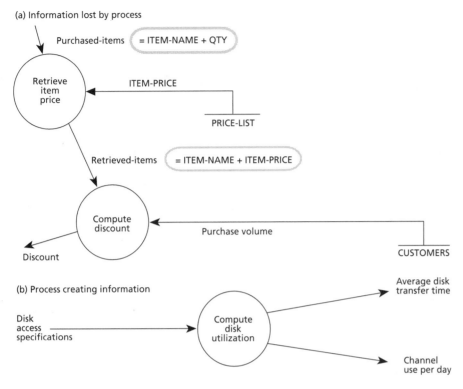

(a) Information lost by process

Purchased-items (= ITEM-NAME + QTY)

Retrieve item price

ITEM-PRICE

PRICE-LIST

Retrieved-items (= ITEM-NAME + ITEM-PRICE)

Compute discount

Purchase volume

Discount

CUSTOMERS

(b) Process creating information

Average disk transfer time

Disk access specifications

Compute disk utilization

Channel use per day

Figure 8.20   *Some data conservation errors*

However, process names should not be general (e.g. 'examine transaction'). Such general terms are necessary only when a process is not a single function. Such a process may, for example, be: 'Read a transaction and process its detail lines if the heading is correct. Also check if the transaction is useful to the sales department.' This process describes more than one function and should be broken up into more than one process. A good guide for evaluating a DFD is to check whether all the process names are single phrases and whether each process can be described in one sentence.

NAMING DATA STORES

The same comment applies to naming data stores. Again, use specific names and avoid general terms such as USER-DATA, CENTRAL-REPOSITORY, or PRODUCTION-DEPT-FILE. Use helpful specific terms like CUSTOMER-ORDER, LAST-YEARS-RECEIPTS, or MACHINE-SCHEDULE.

Needed data

Departmental operation

Information

Figure 8.21   *Meaningless names*

Furthermore, each data store should contain only one specific set of structures and not combinations, such as 'MACHINE-SCHEDULES and PARTS-USED by PRODUCTION-DEPT'. MACHINE-SCHEDULES and PARTS-USED are different kinds of structures and should be modeled by separate data stores.

NAMING DATA FLOWS BETWEEN PROCESSES

Data flows should normally be named as one word (e.g. 'invoice', 'check'). However, there are many instances in DFDs where using only a single word for a data flow can lead to ambiguity and loss of self-description. One frequently occurring instance is where a particular document goes through a number of processes and can be modified by each process. In Figure 8.22, for example, an invoice goes through a number of processes. The 'edit invoice' process checks whether the invoice contains all the needed data. The 'verify invoice' process checks whether the invoiced items have been actually delivered. Finally, the invoice is checked against budgetary provisions and approved for payment.

Figure 8.22 has 'invoice' on all data flows even though it could have been amended or had additional data attached to it at each process. Figure 8.22(b) shows a better naming technique. Now the term 'invoice' is prefixed by a qualifier that specifies the process through which the invoice has just passed.

Figure 8.22(a) has another instance where names should be qualified. Note that in the top part, 'check' appears twice, although obviously the type of check in each case is different. The labeling used in Figure 8.22(b) is preferred. Now each check is qualified as either a 'supplier check' or a 'customer check'.

Some final notes about naming flows between processes and data stores are needed. Often, data flows store or read the whole record in the data store. For example, in Figure 8.22, the 'approve payment' process stores the whole of the 'approved invoice' record into data store APPROVED-INVOICES. By convention, data flows that carry the whole data store record between processes and data stores are not labeled. However, if a process uses only part of a data store record, the data flow must be labeled to indicate the referenced part. Thus, in Figure 8.22(b), the 'make payment' process uses only the data items SUPPLIER and AMOUNT from the data store APPROVED-INVOICES. Note that data flows between data store and processes that contain part of the data store can be labeled by the names in capital letters of the accessed data store items.

Another DFD convention concerns bidirectional flows between processes and data stores. For example, the 'make payment' process reads a SUPPLIER-NAME and AMOUNT from data store ACCOUNTS, updates the AMOUNT, and stores the updated AMOUNT back into data store ACCOUNTS. In this case, the flow is labeled with two arrows to show that it is bidirectional.

SOME OTHER NOTATIONAL SUGGESTIONS

There are many other suggestions for notations on DFDs. You can, for example, allow an external entity to appear more than once in a DFD. Each new appearance has a line added to one of its corners. Another suggestion is to differentiate between accesses, which update a data store, and those that retrieve data from the data store. Updates are indicated by a double arrow.

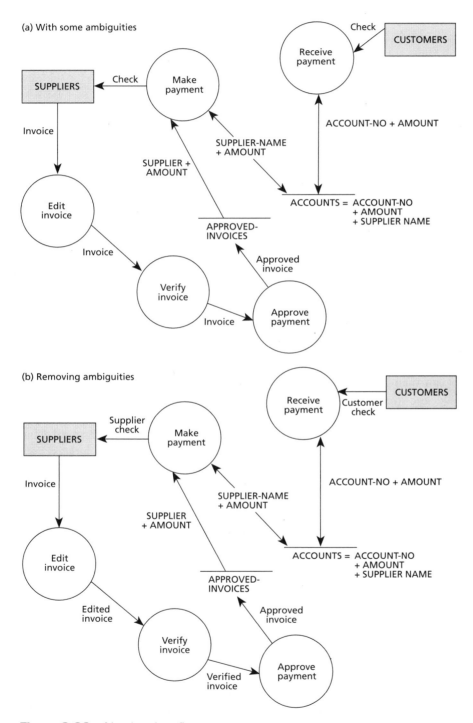

Figure 8.22 *Naming data flows*

MODELING MATERIAL FLOWS

DFDs are intended to model information and not material flows. Sometimes, however, it is desirable also to include material flows. Gane and Sarson illustrate one way of doing this. They include a bottom portion in each process and show the physical location associated with the process—for example, 'store' and 'machine room' in Figure 8.23. Material flow is modeled by a thick line between physical locations. The name of materials is placed inside the thick arrow (see 'parts' in Figure 8.23). Note that Figure 8.23 uses Gane and Sarson's convention to represent processes by rectangular boxes.

SOME MORE ON LEVELING TECHNIQUES

So far this chapter has concentrated on the structure of the DFD and not on the procedures used to develop system models. Structured systems analysis also includes a number of modeling methods and conventions that can help model development. One, called **leveling**, concerns the top-down elaboration of a DFD. Leveling is examined next, and then differences between physical and logical DFDs are described.

Leveling Expanding a process into more detailed processes.

LEVELING DATA FLOW DIAGRAMS

Leveling allows an analyst to start with a top-level function and to elaborate it in terms of its more detailed components. These detailed components are modeled as lower-level DFDs. In Figure 8.2 we started with a context diagram, which was then elaborated in terms of the top-level system functions in Figure 8.3. Each of these functions can then be elaborated further by lower-level DFDs, as shown in Figure 8.4. Finally, the functions are described by process logic.

Apart from supporting a natural, top-down, problem-solving process, leveling improves the readability of the DFD. One ought to be able to understand what the system is doing just by looking at a DFD. If one had a relatively large system and wished to include all the detailed system functions on the DFD, the DFD could become quite large. In fact, it could become so large that it would be difficult to understand and would no longer serve as a communication tool. However, each level of a DFD is small enough to be clearly understood. If someone wishes to know about a particular function at one level, they can refer to it at the next level. This improves the clarity and communication value of the DFD.

SOME LEVELING CONVENTIONS

Although leveling is a very useful tool, a number of conventions must be observed to ensure that no information is lost as a DFD is leveled. These conventions use documentation practices that help analysts maintain consistency between DFDs at different levels.

One important convention is numbering. The context diagram is usually given the number 0. Processes in a top-level DFD are numbered consecutively, starting with 1 and continuing until all processes have been labeled. In Figure 8.24, the top-level DFD has five processes numbered 1, 2, 3, 4, and 5.

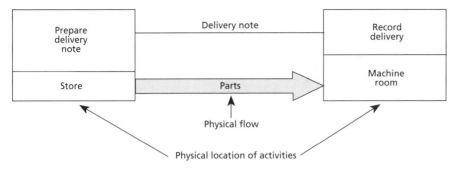

Figure 8.23 *Information and material flows*

As each process is leveled, its DFD is given the same number as the process. Thus, in Figure 8.24, the leveled DFD of Process 1 is named Diagram 1. Each process in the leveled DFD receives a number made up of its diagram number, followed by a stop (i.e. '.'), followed by a number within the leveled DFD. Thus, the processes in Diagram 1 have process numbers 1.1, 1.2, and 1.3.

Another important leveling convention is data flow balancing. Fundamentally, balancing requires that all the data flows entering a process are the same as those entering its leveled DFD. Similarly, all the data flows leaving the process are the same as those leaving its leveled DFD. If you look carefully at Figure 8.24 you will see that this requirement is satisfied for all leveled diagrams. For example, look at Diagram 2, which is the leveled DFD of Process 2. You will see that the only input data flow to Process 2 in the top-level DFD is Y, and its outputs are V and W. Note that the inputs and outputs of Diagram 2 are the same as the inputs and outputs of Process 2 in the top-level DFD.

Leveling can also introduce data stores, which are local to the leveled diagram. In Figure 8.24, data store DS2 contains data that is local to Process 3. Hence, this data store does not appear in the top-level DFD but only in its leveled diagrams. The consequence of this rule is that a data store is used in a DFD only if it is referenced by more than one process in the DFD.

External entities are never part of a process, and analysts should avoid introducing new external entities at lower DFDs. All external entities should appear on the context diagram. They should also appear with a DFD if any process in the DFD has a flow to or from the external entity. Thus, in Figure 8.24, external entity EXT1 has flows to and from Process 1. This external entity appears on the leveled DFD for Process 1.

Note, however, that, if we strictly adhere to data flow balancing, all the detailed flows that occur at the lower-level DFDs must eventually appear at the top-level DFD. If this were to happen, then the top-level DFD could become unmanageable, as it would contain every conceivable flow between the top-level processes. For this reason, a number of modeling techniques are used to apply leveling ideas to data flows. These techniques expand data flows as we level the DFD.

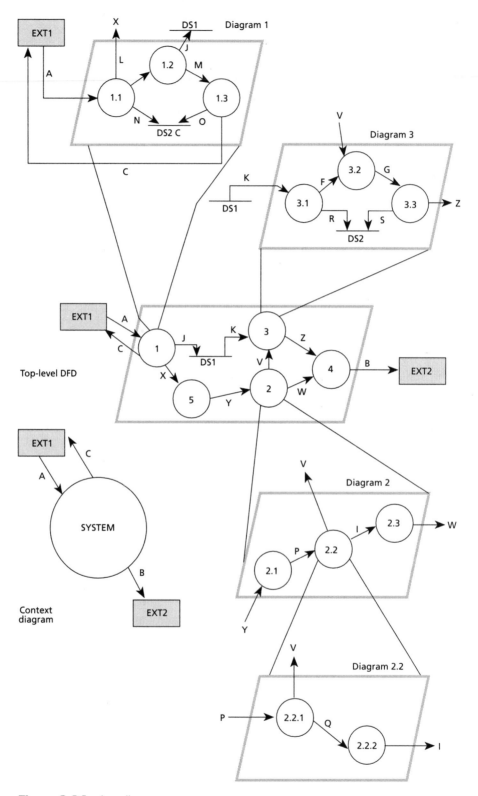

Figure 8.24 *Leveling*

EXPANDING DATA FLOWS

Detailed flows, such as dialog between a process and an external entity, error conditions, and detailed variations of a flow, do not appear at the higher-level DFDs but appear in the more detailed DFDs. Figure 8.25 shows how such detailed flows can be introduced into the lower-level DFDs while maintaining data balancing. In Figure 8.25(a), the top-level DFD models the establishment and maintenance of a loan. Process 1 is used to receive new applications and approve or reject them, Process 2 records payments against established loans, and Process 3 handles any queries from an applicant about a loan. Note that the top-level DFD has a data flow named 'clarification dialog' between Process 1 and APPLICANTS, but details of this flow are not included. The double-arrow on this flow shows that the flow occurs in two directions. Similarly, the data flow between Process 3 and APPLICANTS is labeled 'query and reply', but details of possible queries are not given.

Figure 8.25(b) shows an expansion of Process 1, which includes some data flow expansions. Data flow 'loan application' remains the same, but data flow 'reply to application' is expanded into two flows, 'approved application' and 'rejected application'. Note that the numbering on both the top-level and leveled diagrams is used to check for data conservation. Thus, the data flow 'reply to application' is labeled R in the top-level DFD, and its expansion in Diagram 1 is labeled Rl and R2.

The data flow 'clarification dialog', which is labeled C in the top-level DFD, is expanded in Figure 8.25(b) to show the kind of clarification sought. Thus, Figure 8.25(b) shows that there are two kinds of clarification, to check zipcode, which is labeled Cl, and to check salary, which is labeled C2.

Figure 8.25(b) shows alternative ways of modeling expansions. Zipcode clarification is shown by one flow with two arrows showing the bidirectional nature of this flow. This flow is labeled 'request zipcode' at one end and 'zipcode' at the other. Salary clarification is illustrated by two flows, each in a different direction. The two flows are labeled 'request salary' and 'salary'. The analyst is free to choose either of these alternatives for data flow expansion.

Of course, it is possible to show expansion in another way. This is done by using a process that separates types of input. Thus, Figure 8.25(c) shows Diagram 2, which is an expansion of Process 2 of the top-level DFD shown in Figure 8.25(a). Figure 8.25(c) also illustrates an expansion of data flow 'repayment'. It shows that repayments can be of two types: 'regular payment' or 'advance payment'. Instead of expanding the data flow, however, a process is included in the leveled DFD to distinguish between the two types of flow.

ERROR HANDLING

Error handling, especially in an interactive environment, is very similar to dialog. Details of error handling should be avoided at the higher modeling levels. Where error handling creates an error-correction dialog, it can be modeled in the same way as dialog. Now the dialog is specific to correcting some error and it may be labeled 'error-correcting dialog'.

Sometimes processes do not engage in error dialog but simply reject a given

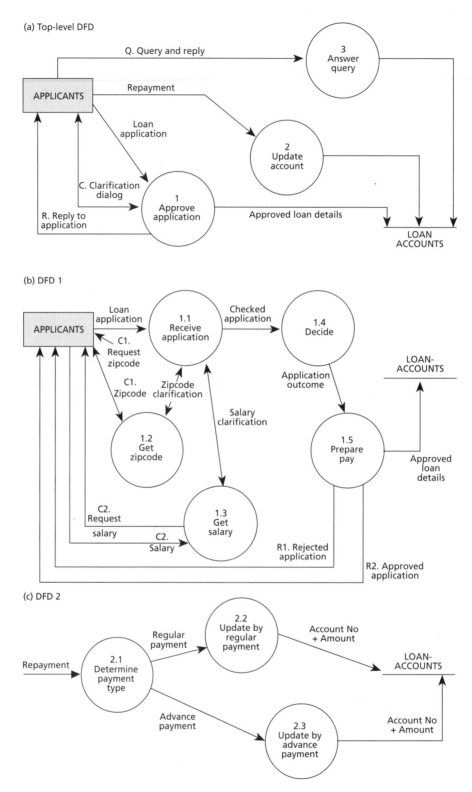

(a) Top-level DFD

(b) DFD 1

(c) DFD 2

Figure 8.25 *Expanding data flows*

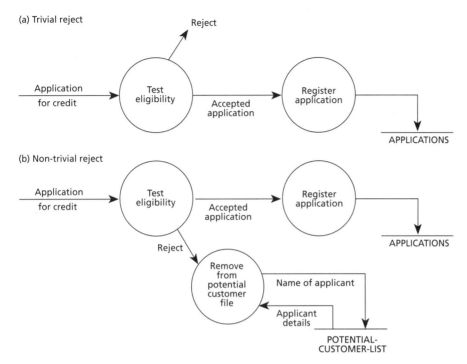

Figure 8.26 *Modeling error conditions*

input. In this case, DeMarco (1978) distinguished two kinds of rejection:

- trivial rejects that do not undo anything that has previously taken place
- non-trivial rejects that need to undo previous work.

The different treatments of these alternatives on the DFD are shown in Figure 8.26. Alternative (a) shows trivial rejects, where we are not interested in what happens to the rejected data. Thus, in Figure 8.26(a), the system does not take any further interest in rejected applications. Alternative (b) shows a rejection that is followed by some action. The rejection is used to remove the rejected applicant from a POTENTIAL-CUSTOMER-LIST.

HOW MANY LEVELS?

It is sometimes difficult to know how far to level down, how many processes to include on a data flow diagram, or even where to start. It is impossible to give precise answers, but some guidelines can be suggested.

Most practitioners say that about seven, plus or minus two, is the ideal number of processes on a DFD. This number can be clearly understood by a visual examination but is not too small to be trivial. A larger number is sometimes too hard to understand; a smaller number often includes too little information to be useful.

However, one should not take a trivial approach and suggest that leveling simply involves taking a higher-level process and partitioning it into seven lower-level processes. Consider, for example, Figure 8.27. It contains two DFDs, each with seven processes. One of these diagrams is clearly less complex and hence easier to understand than the other. Figure 8.27 thus illustrates another requirement of

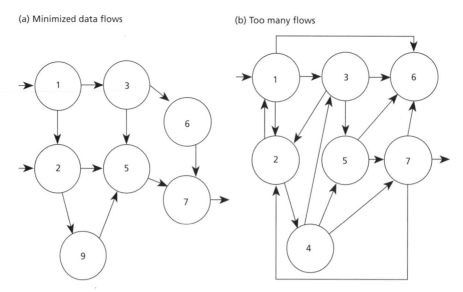

(a) Minimized data flows   (b) Too many flows

Figure 8.27 *Reducing complexity by minimizing flow between processes*

leveling—the interfaces between processes must be minimized. This can be done only by the correct choice of lower-level functions, and this is where the true nature of analysis comes in. The analyst must carefully analyze the top-level process, determine what its self-contained functions are, and represent them on the DFD. If the interconnections are too complex, the leveling should be re-examined and, if necessary, restarted. This may continue over a number of iterations until a satisfactory solution is found.

Minimizing the complexity of interfacing is one leveling guideline. Other informal leveling guidelines are based on the idea of creating functions that are *linguistically modular*. This means that we must actually look at what the higher-level function does or means and describe it in one, or at least a very few, precise sentences. The phrases in these sentences will identify the lower-level functions. For example, the process 'order parts' may be described as: 'For the requisition from a project, select a supplier and then make an order to the supplier.' This immediately suggests that the process 'place an order' should be decomposed into three processes: 'get requisition', 'select supplier', and 'make order'.

For how long does leveling continue? Generally, until you reach a set of processes that can be described by about one page of detailed process specifications.

PROCESSES FOR DEVELOPING DFDs

Beginners often ask how to build DFDs from system descriptions. There are two main approaches. One is to start with data flows and then follow them through the system. The other is to start with processes and connect them.

The technique starting with data flows is illustrated in Figure 8.28. It starts with a context diagram and data flows in and out of the diagram. The design then proceeds in the following steps:

1. Add data stores—Identify the data needed inside the domain of change and construct data stores for this data.

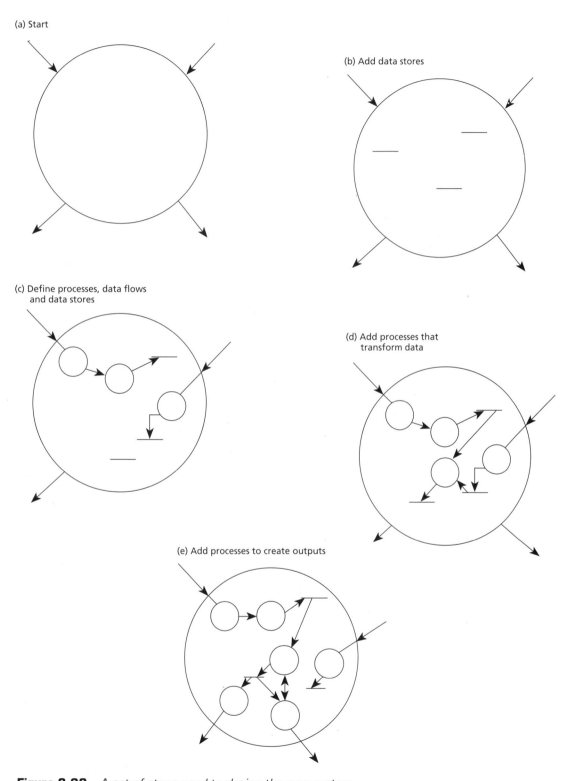

Figure 8.28 *A set of steps used to design the new system*

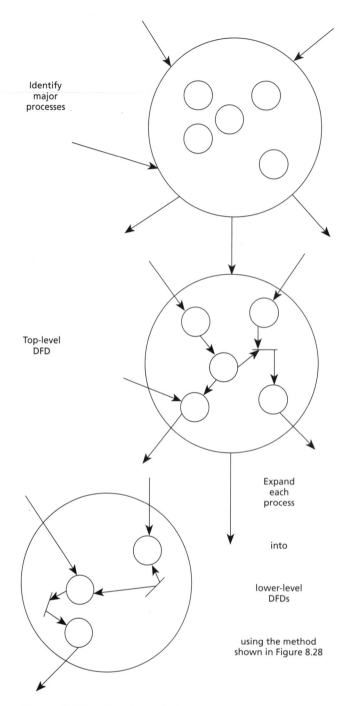

Identify
major
processes

Top-level
DFD

Expand
each
process

into

lower-level
DFDs

using the method
shown in Figure 8.28

Figure 8.29 *Top-down design*

2. Define processes, data flows, and data accepts—Take each input into the domain of change and define processes that the input data flows transform them and place them in a data store.
3. Add processes that transform data—Define processes that use data in the data store and define how the processes use the data.
4. Add processes to create outputs—Look at each output produced and identify the data stores used to produce the output. Define the processes that the data goes through before being output.

Of course, it may not be possible to proceed in this way for a large system. A top-down approach may be preferred instead. This approach starts by defining the top-level processes. Figure 8.29 illustrates how top-level processes are defined. A list of major system functions (such as inventory maintenance and invoicing) is first made and their inputs and outputs are defined. These functions are then linked by data flows to construct a top-level DFD for the domain of change. The method shown in Figure 8.29 is now applied to each top-level function.

SUMMARY

This chapter describes how systems are modeled using data flow diagrams. It describes what a DFD looks like and how DFDs can be developed in a top-down manner by using leveling. The chapter makes an important distinction between modeling what a system does and how a system works. A logical DFD defines what a system does, and a physical DFD defines how a system works. The chapter describes how to develop logical DFDs using conversions between logical and physical DFDs. Some characteristics of good DFDs are also described.

DISCUSSION QUESTIONS

8.1   Where would structured system models be used in the system specification?

8.2   Do you think DFDs are more appropriate to the subject, usage, or system world?

EXERCISES

8.1   Detect any errors in the DFDs in Figure 8.30.

8.2   Consider the DFDs in Figure 8.31. Do they satisfy the conservation of data rules?

8.3   Figure 8.32 is a DFD that describes a system to satisfy user requests for parts. Whenever a parts request is received, a search is made to determine whether the part is available in store. If so, the part is dispatched together with a dispatch notice. Otherwise, a non-availability notice is sent.

    Suppose the system is now changed, as follows:

- Check with suppliers to see whether any parts not in store can be obtained immediately. If so, a purchase order is made out to the supplier and a copy is sent to the user. The supplier then dispatches the parts to the user. The user sends the dispatch back to be matched against the purchase order.

(a) Survey processing

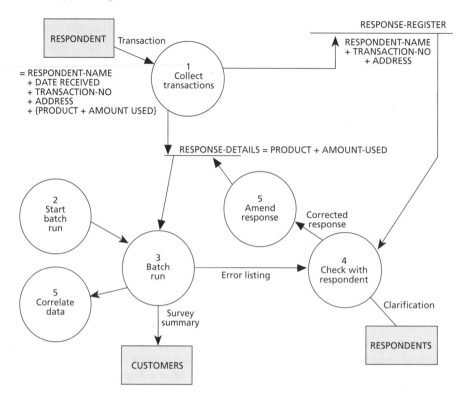

Figure 8.30 *Processing DFDs*

- Regularly check the inventory and place purchase orders whenever the number of parts is below reorder level.

Amend the DFD in Figure 8.32 to include the above changes.

8.4 Can you find any data balancing problems in Figure 8.33?

8.5 Draw a top-level logical DFD for the system made up of the following:

1. Selling customers request the organization to sell various items on their behalf. A record of these requests is kept.

2. Buying customers make requests to buy items.

3. A sale is arranged with a buyer if the item requested by the buyer has been previously put forward for sale by a selling customer.

4. During a sale arrangement:
   - an invoice is prepared for the buyer. A record of the invoice is kept.
   - a notification is sent to the seller whose item was sold. The seller will now hold the item.
   - a commission is computed and debited to the selling customer. This commission is subtracted from the amount sent to the selling customer, and advice of the commission is given in the sale advice.

5. The sale is completed when payment is received from the buyer. A cheque is then sent to the seller together with a dispatch request for items to be sent to the buyer.

(b) Application processing

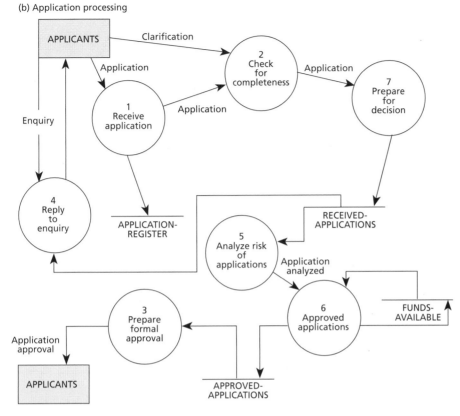

Figure 8.30 (continued)

8.6   Figure 8.34 is a DFD made up of three processes: 'invoice maintenance', 'invoice monitoring', and 'make payment'. These processes have been leveled down one level to illustrate the information flows between them. Rearrange the bottom-level processes into higher-level processes that have fewer data flows between them.

8.7   Figure 8.35 illustrates two top-level DFDs containing a mixture of physical and logical processes. Examine both these DFDs and state which processes are logical and which are physical.

8.8   Develop a top-level DFD for Case Study 1 (Sales/Order System) at the end of the book. Level the top-level processes to obtain detailed system operations.

8.9   Develop a top-level DFD for Case Study 2 (Travels/Arrangements) at the end of the book. Level it as necessary.

(a) Daily sales of items are recorded and totals of each term sold computed:

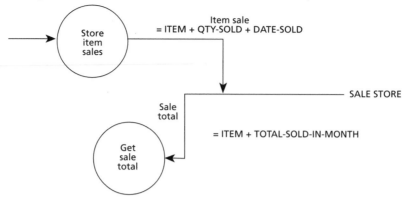

(b) The expenses on a trip are computed:

Litres of petrol consumed

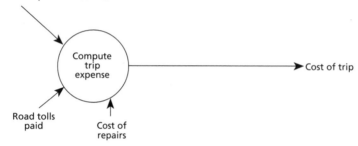

(c) Compute price variation by:

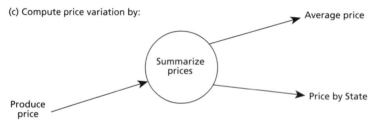

(d) Compute passengers carried per day on a bus-route:

Figure 8.31 *Conservation of data problems*

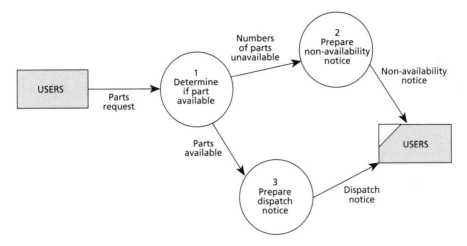

Figure 8.32 *Requests for parts*

(a) Top-level DFD

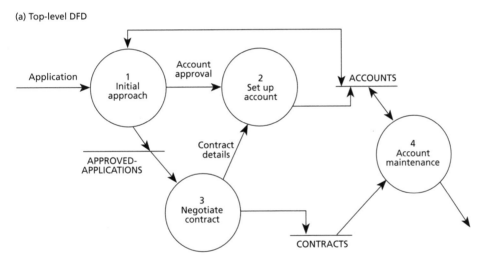

Figure 8.33 *Leveling a DFD*

(b) Leveled DFD

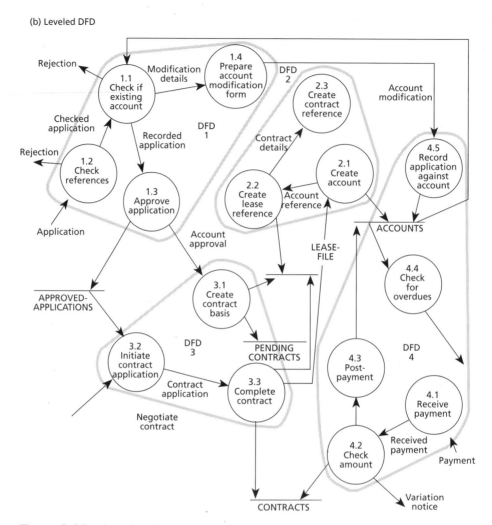

Figure 8.33 *(continued)*

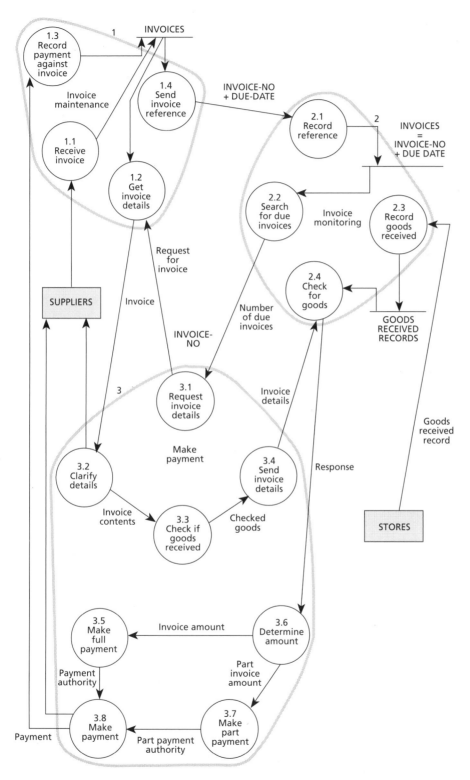

Figure 8.34 *Reducing flows between processes*

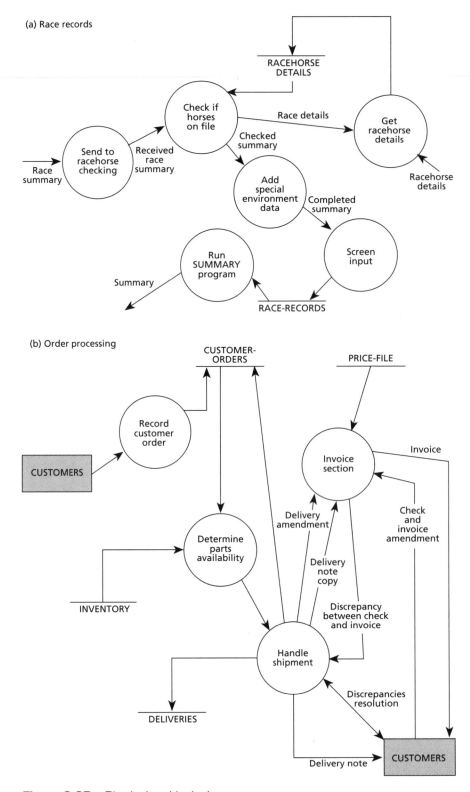

Figure 8.35 *Physical and logical processes*

BIBLIOGRAPHY

DeMarco, T. (1978), *Structured Analysis and System Specification*, Yourdon Press, New York.

Dickinson, B. (1980), *Developing Structured Systems*, Yourdon Press, New York.

Gane, C. and Sarson, T. (1979), *Structured Systems Analysis*, Prentice Hall, Sydney.

Mendes, K.S. (Summer 1980), 'Structured systems analysis: a technique to define business requirements', *Sloan Management Review*, pp. 51–63.

Orr, K.T. (1977), *Structured Systems Development*, Yourdon Press, New York.

Orr, K.T. (1981), *Structured Requirements Definition*, Ken Orr and Associates Inc., Topeka, Kansas.

Page-Jones, M. (1980), *The Practical Link to Structured Systems Design*, Yourdon Press, New York.

Simpson, J.A. (July 1982), 'Impact of structured techniques in the Bank of New South Wales', *The Australian Computer Bulletin*, Vol. 6, No. 6, pp. 18–20.

Teague, L.C. and Pidgeon, C.W. (1985), *Structured Analysis Methods for Computer Information Systems*, SRA, Chicago.

Yourdon, E. (1989), *Modern Structured Analysis*, Yourdon Press, Englewood Cliffs, New Jersey.

DESCRIBING DATA

CONTENTS

KEY LEARNING OBJECTIVES

Why data modeling is important
Semantic concepts of entity, relationship, and attribute
How to represent data using semantic concepts
The relationship between data modeling and data flow modeling
Dependent entity sets and subsets

CHAPTER 9

INTRODUCTION

An important aspect of systems modeling is the analysis of system data. As with a data flow diagram (DFD), an abstract or implementation-independent model of the data is developed as part of the system specification and is later converted to a physical implementation.

Data modeling is a more difficult subject than data flow modeling. You will find that a data abstraction is not as obvious as a DFD abstraction. A DFD looks almost like a system. It has boxes that you can actually envisage as physical operations, and its flows can be imagined just as easily. A data model, however, is often more abstract and difficult to relate to actual system components—for example, data associations, which are not visible as physical things in the system.

Data modeling is part of the development process. In the linear development cycle, it is used during the system requirements phase to construct the data component of the analysis model. This model represents the major data objects and the relationships between them. It should not be confused with data analysis, which takes place in the system design phase. System design organizes data into a good shape. Usually this means removing redundancies, a process often called normalization, which uses ideas from relational theory. This normalized model is then converted to a physical database.

Most designers develop only the high-level conceptual model in the system specification phase. The more detailed analysis using normalization is carried out during design. This chapter describes one commonly used set of techniques for developing the conceptual data model: entity–relationship modeling. Normalization is described in later chapters when we turn to design.

As in a DFD, a model of data consists of a number of symbols joined up according to certain conventions. We will describe conceptual modeling using symbols from a modeling method known as entity–relationship *analysis. This method was first introduced by Chen in 1976 and is now widely used.*

ENTITY–RELATIONSHIP ANALYSIS

Entity–relationship (E–R) analysis uses three major abstractions to describe data. These are:

Entity–relationship (E–R) model A model that represents system data by entity and relationship sets.

- *entities*, which are distinct things in the enterprise
- *relationships*, which are meaningful interactions between the objects
- *attributes*, which are the properties of the entities and relationships.

The idea of E–R analysis is illustrated by Figure 9.1. Similar objects or things are grouped into entity sets. In Figure 9.1, all the persons have been put into a PERSONS entity set and all the projects into a PROJECTS entity set. You should note the difference between the terms *entity* and *entity set*. Each individual object is called an **entity**. A collection of such entities is an **entity set**. Thus, each person or each project is an entity. A collection of persons is the entity set PERSONS, and a collection of projects is the PROJECTS entity set.

Entity A distinct object in a system.

Entity set A component in an E–R diagram that represents a set of entities with the same properties.

We then model all interactions between the things in entity sets by relationships and relationship sets. Just as with entities, there is a difference between the terms

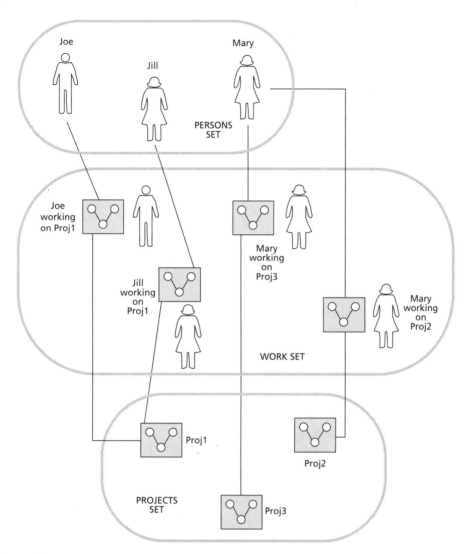

Figure 9.1 *E–R modeling*

relationship and *relationship set*. If a person works on a project, then there is a **relationship** between that person and the project. A **relationship set** is a collection of such relationships. In Figure 9.1, a collection of relationships of persons working on projects is modeled by the WORK relationship set. This set is a relationship set because it represents a collection of relationships between things in two different entity sets. Each relationship in the WORK relationship set is one person working on one project.

Relationship sets are harder to perceive than entity sets. We can actually see entities, but we cannot see relationships. This makes data analysis difficult, because it is necessary to create models of things that do not physically exist as single objects. However, if you stretch your imagination you should perceive a relationship occurrence as a person working on a project—for example 'Jill' working on 'Proj1' in Figure 9.1. A collection of such occurrences of persons working on projects

Relationship One interaction between one or more entities.

Relationship set A component in an E–R diagram that represents a set of relationships with the same properties.

becomes the relationship set WORK. Once you understand the difference between occurrences and sets, you will find data analysis much easier.

Of course, diagrams such as that shown in Figure 9.1 are not particularly useful for modeling. Imagine having to represent each person in a large organization by a personal symbol! A more concise representation is to use entity–relationship (E–R) diagrams. The E–R diagram does not represent individual entities and relationships, only the entity and relationship sets. It uses rectangular and diamond-shaped boxes to do this. As shown in Figure 9.2, the rectangular boxes represent entity sets, with the name of the entity set placed inside the box. In Figure 9.2 there are two entity sets, named PERSONS and PROJECTS. The diamond-shaped boxes represent the relationships between the entities, and the name of the relationship set is placed inside the box. The only relationship set in Figure 9.2 is WORK.

TERMINOLOGY

It is particularly important to use consistent terminology to understand data analysis terms—for instance, those that distinguish the kinds of situations depicted in Figures 9.1 and 9.2. Figure 9.1 illustrates actual entities and relationships. Thus, it shows actual people, in this case 'Joe', 'Jill', and 'Mary', and actual projects, in this case 'Proj1', 'Proj2', and 'Proj3'. Diagrams such as that shown in Figure 9.1 are often called **occurrence diagrams** and represent individual entities and relationships. Figure 9.2 does not show individual persons or projects or work relationships, only the set of people, set of projects, and set of work relationships. This diagram is called an *E–R diagram* and it represents entity and relationship sets. The entity sets in the E–R diagram in Figure 9.2 are called the PERSONS set and the PROJECTS set. The relationship set is called WORK.

Note that we use the word *set* rather than *type* for a reason. The word *type* is often associated with computer terms such as record type or variable type. However, data analysis is not concerned with computers but with the structure of system data, and uses terms more natural to data properties than to the definition of data in computer terms. Data analysis avoids computer terms as much as possible to prevent any preconceptions about computer structures slipping into the analysis. Of course, sets are converted to record types later during computer system design. This conversion is described in Chapter 15.

> **Occurrence diagram** A diagram that represents entities and relationships.

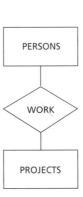

Figure 9.2   *An E–R diagram*

ENTITY–RELATIONSHIP STRUCTURES

Figure 9.2 illustrates the simplest E–R construct. It shows two entity sets with one relationship set between them. It is also possible to have more complex structures. There can be a large number of entity and relationship sets in an E–R diagram. Each entity set in the E–R diagram can be linked to more than one relationship set. Figure 9.3, for example, contains the entity set PERSONS, which is associated with two relationship sets. One relationship set is WORK-ON with projects. This relationship set shows that persons work on projects. The other relationship set, ARE-IN, shows that persons are in departments. A number of other entity and relationship sets are also shown in Figure 9.3. Entity set PARTS, for example, is associated with three relationship sets: USE with PROJECTS to show that projects use parts, SUPPLY with SUPPLIERS to show that suppliers supply parts, and HOLD with WAREHOUSES to show that warehouses hold parts.

Figure 9.3 shows each relationship set linked to two entity sets and, at most, one relationship set between any two entity sets. This is not, however, a general requirement of E–R diagrams. There are, in fact, very few restrictions placed on drawing E–R diagrams. For example, it is possible to have two or more relationship sets between the same two entity sets. One example of this is illustrated in Figure 9.4. Here there are two relationship sets, OWN and LEASE, between COMPANIES and VEHICLES. It is necessary to have these two different relationship sets because there are two different ways in which companies can interact with vehicles. Each of these ways has different properties. Relationship OWN occurs between a company and a vehicle if that company purchases the vehicle. This relationship may have properties such as DATE-PURCHASED or PRICE-PAID. Relationship LEASE exists when a company rents a vehicle. In that case, the properties may be LEASE-TERMINATION-DATE and MONTHLY-RENTAL.

It is also possible for more than two entity sets to be associated with the same relationship set. Figure 9.5(a), for example, shows the relationship set BUY, which models an interaction between CUSTOMERS, STORES, and PARTS. Each relationship in this set includes a person, a part bought by the person, and the store where the purchase was made. Relationship sets that include more than two entity

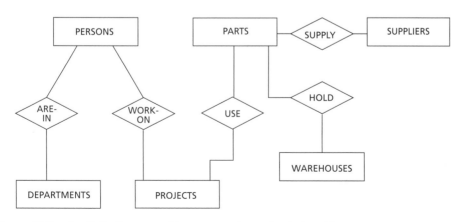

Figure 9.3 *An E–R diagram with many entity and relationship sets*

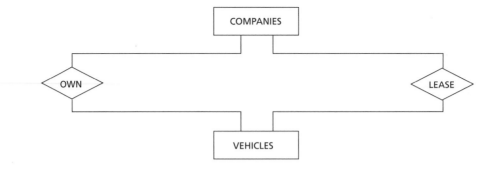

Figure 9.4 *Two relationship sets between the same two entity sets*

N-ary relationship A
relationship that
includes entities from
more than two entity
sets.

**Binary
relationship** A
relationship that
contains entities from
at most two entity
sets.

sets are known as **N-ary**. Relationship sets that include only two entity sets are known as **binary**. However, N-ary relationship sets should be avoided in E–R diagrams, because they include more than one concept. For example, the relationship in Figure 9.5(a) describes both the store where a customer bought the goods and the details of the purchase. It is always possible to remove an N-ary relationship by replacing it with an entity set. Relationship sets are then created between this new entity set and those entity sets that participated with the N-ary relationship set. In this way, the E–R diagram in Figure 9.5(a) can be converted to the E–R diagram in Figure 9.5(b). The E–R diagram Figure 9.5(b) contains only binary relationship sets.

There are some things you should avoid. For instance, the system itself should not appear as an entity set in the E–R diagram. Thus, for example, we should not have the entity set BUSINESS in the E–R diagram in Figure 9.6(a) if we are modeling the business. The entity set BUSINESS would include only one entity. The correct E–R diagram here is shown in Figure 9.6(b). This E–R diagram does not include the entity set BUSINESS but includes only those entity sets that are of interest to the business.

You should also avoid derived relationship sets in an E–R diagram. Figure 9.7 includes one such derived relationship set. The E–R diagram there contains three relationship sets. Relationship set HAVE describes the sections in each department, and relationship set EMPLOY describes the persons employed in each section.

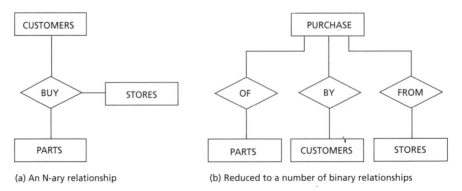

(a) An N-ary relationship   (b) Reduced to a number of binary relationships

Figure 9.5 *Reducing N-ary relationship sets*

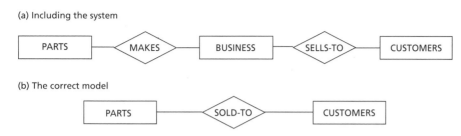

(a) Including the system

(b) The correct model

Figure 9.6 *Excluding the system from the E–R diagram*

Relationship set ARE-IN then models the persons in each department. We can find all the persons in a department through relationship sets HAVE and EMPLOY. Once we know all the sections in a department through relationship set HAVE and then find all the persons in each section through relationship set EMPLOY, we will know all the persons in the department. The relationship set ARE-IN is thus not needed, because it contains information that can be derived from the other relationship sets. Relationship sets that include derived information should not appear in an E–R diagram.

MORE ABOUT ENTITY–RELATIONSHIP MODELING

After the entity and relationship sets are identified, the next step is to determine the **attributes** (or properties) of objects in the sets. For example, what is of interest to us about persons? This may be a person's NAME, their ADDRESS, or their unique identifier, PERSON-ID, in the organization. These properties are written alongside the boxes of the E–R diagram.

Attribute in an E–R diagram A property of a set in an E–R model.

There are alternative ways of showing attributes on an E–R diagram. You will find such alternatives in practice. One alternative places the attributes in circles and attaches these circles to boxes in the E–R diagram. Another lists the attributes in a table next to the E–R diagram. In this book, attributes are placed next to the boxes on the E–R diagram to give the complete picture of the data on one diagram. However, if you come across some of the other alternatives, remember that they mean the same thing.

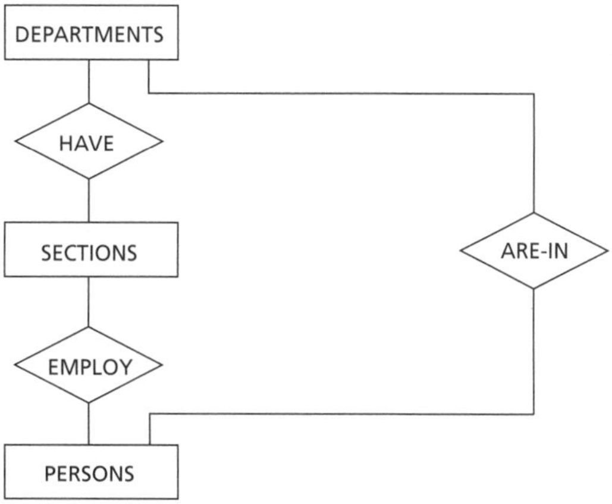

Figure 9.7 *The derived relationship set, ARE-IN*

Identifier A set of properties whose values identify a unique object in an object set.

In addition, we usually choose one of the attributes of an entity or relationship set to be the **identifier**. The identifier has one important property: its values identify unique entities in the entity set. The identifier attributes are underlined in the E–R diagram.

Thus, in Figure 9.8, PERSON-ID is the identifier of the PERSONS entity set. Each person in the PERSONS entity set has a unique value of PERSON-ID. Similarly, PROJECT-ID is the identifier of the PROJECTS entity set, as each project has a unique value of PROJECT-ID. Choosing identifiers for relationship sets is somewhat more complex.

The convention used in this book is to use the identifiers of the entities that participate in the relationship as the relationship identifiers. In most cases it is necessary to know the value of both of these identifiers in order to identify a unique relationship. Thus, in Figure 9.2, we need to know both the value of the person identifier, PERSON-ID, and the value of the project identifier, PROJECT-ID, to identify a particular relationship in relationship set WORK. A value of the person identifier on its own would not be enough, because that person can work on more than one project and hence appear in more than one relationship. Similarly, a value of the project identifier is not enough, because there may be more than one person working on a project. Thus, we must know both the person and the project identifiers to identify one relationship. In some cases, however, one identifier is sufficient to identify a relationship—for example, when entities in one entity set are restricted to one relationship. Thus, if a person could work on only one project at the most, the value of the person identifier would be enough to identify a unique WORK relationship.

The convention used in this book, however, is to use all the entity identifiers to make up the relationship identifier. Some people may point out that such identifiers do not correspond to file keys, but you should remember that data analysis does not imply any computer implementation, and identifiers have nothing to do with file keys. It is true that if sets are converted to files, the identifiers can become the file keys. However, at this stage, relationship identifiers are not file keys but are the identifiers of entities that participate in the relationship.

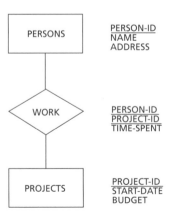

Figure 9.8 *Adding attributes*

On the E–R diagram, we show the number of relationships in which an entity can appear. This value is sometimes known as the relationship cardinality.

MODELING RELATIONSHIP CARDINALITY

One property often shown on E–R diagrams is relationship **cardinality**, which specifies the number of relationships in which an entity can appear. An entity can appear in:

- one relationship
- any variable number (N) of relationships
- a maximum number of relationships.

Cardinality The number of relationships in which one entity can appear.

If you refer to Figure 9.1 you will see that a person can appear in more than one WORK relationship, and so can a project. This is shown by the letters N and M on the E–R diagram links (see Figure 9.9). These letters stand for variables and can take any value. If there were a limit to the number of times an entity can take part in the relationship, then N or M would be replaced by the actual maximum number. N and M are used (rather than N to N) to show that entities in the different sets may participate in a different number of relationships.

Figure 9.10 illustrates a 1:N relationship. Here a project has one manager, whereas a manager can manage any number (N) of projects. Thus, the occurrence diagram shows that manager 'm1' manages two projects: 'p1' and 'p2'. Thus manager 'm1' appears in two MANAGE relationships. Every project in Figure 9.10, however, appears in one MANAGE relationship only. Note that in the E–R diagram, '1' appears opposite the MANAGERS set, and the diagram indicates that a project is managed by one manager.

MODELING RELATIONSHIP PARTICIPATION

E–R diagrams often specify the manner of participation of entities in a relationship set. The participation of entities in a relationship set can be mandatory, optional, or conditional. If all entities in a given set must appear in at least one relationship in a relationship set, then their participation in the relationship set is mandatory. If each entity need not appear in the relationship set, then its participation in the

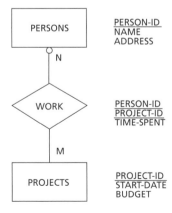

Figure 9.9 *Adding cardinality*

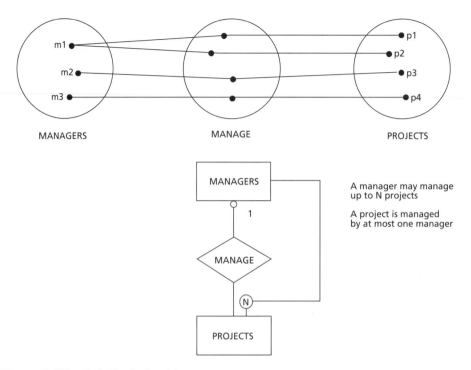

Figure 9.10 *A 1:N relationship*

relationship set is optional. Thus, suppose that a project must have at least one person working on it. The project must appear in at least one WORK relationship, and the participation of each project in the WORK relationship set is mandatory. A person, however, need not work on a project. The participation of each person in a WORK relationship set is optional. The notation used to indicate optional participation is shown in Figure 9.9. An optional participation has a small circle, O, on the link next to the entity set. There is no mark on the link for mandatory participation.

BUILDING ENTITY-RELATIONSHIP MODELS

Although it is relatively easy to describe what an E–R diagram looks like, it is much harder to describe how one goes about developing an E–R diagram for a particular system. There are some important points to consider here: how to choose entities, relationships, and attributes; how to choose names; and what steps should be followed in analysis.

EACH SET TO MODEL ONE CONCEPT

One of the most important things to remember is that in analysis we are trying to capture the precise semantics of a system. To do so it is necessary to identify the fundamental components of a system and relationships between them. For this reason, each set in the E–R diagram should model only one concept. Thus, each entity set models only the one concept—that is, entities with some common properties. Each relationship set models the one concept of two entities, each from the same set interacting in the same way. If you adhere to this idea you can place only things with the same properties into the same entity set. Thus, all persons would fall into one entity set named PERSONS, all parts into one entity set named PARTS, and so on. In any case, you would not place persons and parts into the one entity set.

What kinds of things are to be modeled as entity sets? The most common system entities are distinct physical things in the organization, things such as persons, parts and invoices. However, other things that are not so clearly visible are also modeled as entities. The most common of these are organizational entities such as projects, departments, and budgets. Finally, things that happen, such as deliveries, faults, or examinations, may also be modeled as entities.

CHOOSING ATTRIBUTES

Attributes, just like entity and relationship sets, should express simple concepts. For this reason, attributes should also be simple. Any attribute in an E–R model should therefore take a simple value. E–R diagrams should not contain multivalued or structured attributes, as shown in Figure 9.11(a).

In Figure 9.11(a), entity set PERSONS has four attributes. Attribute PERSON-ID is the person's identifier, and attribute DATE-OF-BIRTH is the person's birthdate. Both of these attributes take one value for each person entity. Attribute QUALIFICATION is another, but this may take many values for one person because a person can have many qualifications. The asterisk (*) indicates that QUALIFICATION is a multivalued attribute. Attribute ADDRESS, on the other hand, is itself made up of other attributes—SUBURB, STREET, and NUMBER. Attribute ADDRESS is known as a structured attribute.

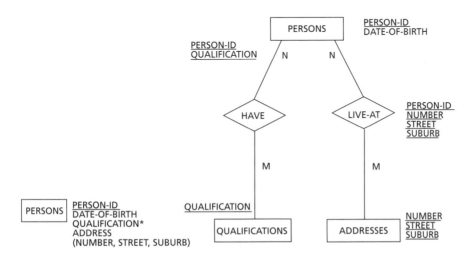

(a) Non-simple attributes

(b) Removing multivalued and structured attributes

Figure 9.11 *Reducing to simple attributes*

Any multivalued or structured attributes should not appear in the final E–R diagram. They can be removed by converting them to relationship sets, as shown in Figure 9.11(b). Here there are two new entity sets: QUALIFICATIONS and ADDRESSES. The multivalued attribute QUALIFICATION now becomes an attribute of the new entity set QUALIFICATIONS. The structured attribute ADDRESS has now become the entity set ADDRESSES, and its components have become the attributes of this new entity set. Entity set PERSONS is associated with these two new entity sets through relationship sets HAVE and LIVE-AT. These relationship sets are used to find a person's qualifications and address.

CHOOSING OBJECT SET NAMES

Proper choice of names often helps in E–R modeling. Remember that one goal of E–R modeling is to produce a model that is easily understood by both users and computer personnel. To do this we must choose set names that make the diagram readable. In an E–R diagram, entity sets are usually labeled by nouns, whereas relationship sets are labeled by verbs. This is quite natural in E–R modeling, because entities are usually distinct objects that are best described by nouns. Relationships are actions and are better described by verbs. Often, however, relationship sets are named by prepositions. This is particularly so when modeling structural relationships, like that shown in Figure 9.5(b). This shows that each entity in entity set PURCHASES is made up of a part, a customer, and a store. Such structural relationships often appear in E–R diagrams. For example, a building can consist of rooms, or parts can be made up of other parts.

AN ANALYSIS SEQUENCE

Most analysts suggest that E–R modeling should start by identifying the system entity sets. The question is how we actually choose the entity sets. Previous information is often useful for identifying initial entities. Data stores and external entities in a DFD often suggest entity sets. Sometimes people also go through the descriptions of a system and underline all the nouns, which suggests an initial set of entity sets. Once there is an initial set of sets, some refinement usually takes place.

After defining the initial entity sets, we look at how entities in the entity sets can interact with each other and model this interaction by relationship sets. Thus, once we find out that persons work on projects, we would add a relationship set between the PERSONS and PROJECTS entity sets and give this relationship an appropriate name—in this case, WORK. Often this is derived from a statement like 'persons work on projects' somewhere in the system description. We then go on identifying further interactions between entities and model them by relationships with appropriate names. Then we can add cardinality to the system. As we go into more detail, attributes are added to the E–R diagram and identifiers are chosen. However, the guideline does not mean that we must choose all the entity sets before we start to look for relationships, or have all the relationships before starting with attributes. It simply means that we should have an entity to use in a relationship and not propose a relationship set first or define attributes and have no sets to attach them to. Thus, we can define some entity sets, then some relationship sets between them. We can then add more entity sets, then perhaps add some attributes to existing sets, then add some more entity sets, and so on. The model thus grows in an evolutionary manner.

This evolution calls for a continuous refinement of the model. We may initially draw a top-level diagram that may include N-ary relationships or even multivalued and structured attributes. We may then refine the model by replacing such components by using the techniques shown in Figures 9.5 and 9.11. We may at the same time remove any derived relationships. Thus, E–R modeling is a continuous and evolutionary process, and a model gradually evolves that correctly represents the semantics of a system.

We now develop an E–R diagram for our Text Case D.

TEXT CASE D:
Construction Company—Drawing the E–R Diagram

The E–R diagram for Text Case D is developed in two steps, the first of which is to develop a top-level E–R model. This model is shown in Figure 9.12. It will then be refined further. To develop the top-level model, we first define the entity sets, then the relationship sets, and then add attributes to these sets.

The DFD diagram in Figure 8.16 can serve as a guide here. Thus, the two external entities, SUPPLIERS and PROJECTS, become entity sets. Then each data store in Figure 9.12 also becomes an entity set. As a result, we have the entity sets PURCHASE-ORDERS, SHIPMENTS, and INVOICES. A dependent entity set PROJECT-REQUESTS is defined to model project requests. It is defined as a dependent entity set because a project request cannot arise without there being a project. Its identifier is made up of the identifier of projects, PROJECT-NO and REQ-NO, which identifies a request within a project. The entity set PROJECT-DELIVERIES is also included to model deliveries made to projects.

Next, relationship sets are defined. A number of relationship sets are first defined between entity set SUPPLIERS and the documents that suppliers use. Thus, there is a relationship set TO between entity sets SUPPLIERS and PURCHASE-ORDERS, because purchase orders are sent to suppliers. Similarly, there are relationship sets SEND and FROM between entity sets SUPPLIERS and SHIPMENTS and entity sets SUPPLIERS and INVOICES, because suppliers send these documents to the organization. A number of relationship sets between documents are then added. The relationship set OF between entity sets SHIPMENTS and PURCHASE-ORDERS models the fact that shipments are made to satisfy purchase orders. The relationship set INCLUDE between entity sets INVOICES and SHIPMENTS models the fact that invoices are sent for items delivered in shipments. The relationship set FOR shows that invoices are issued for payments against purchase orders. The relationship set MET-BY between entity set PROJECT-DELIVERIES and dependent entity set PROJECT-REQUESTS models the fact that the project deliveries are made in response to project requests.

Now, attributes are added to the entity and relationship sets. The attributes added to the sets can be found by examining Figure 9.12. For simplicity, some fields are modeled as multivalued fields in the top-level E–R diagram. Thus, for example, look at PURCHASE-ORDERS in Figure 9.12. This is modeled by the entity set PURCHASE-ORDERS. Document PURCHASE-ORDERS has three heading fields, PO-NO, SUPPLIER-NAME and DATE-ORDERED. PO-NO and DATE-ORDERED

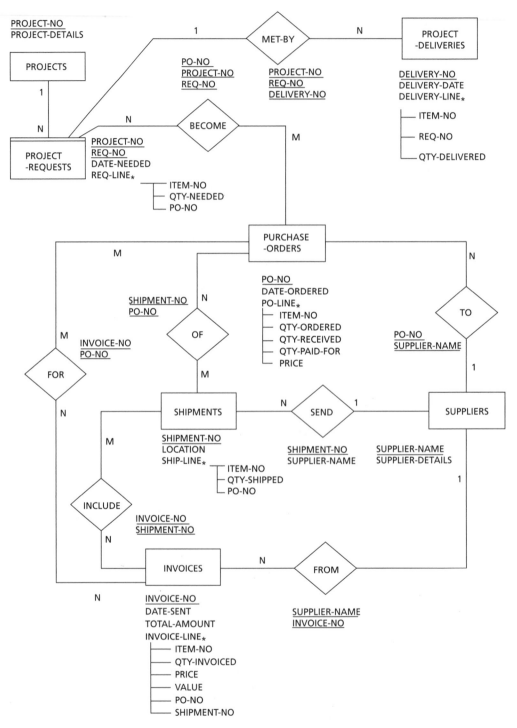

Figure 9.12 *A first attempt at an E–R diagram*

become attributes in entity set PURCHASE-ORDERS. SUPPLIER-NAME does not become an attribute in set PURCHASE-ORDERS but is associated with entity set PURCHASE-ORDERS through relationship set TO. Document PURCHASE-ORDERS has a number of lines, each made up of the five fields ITEM-NO, QTY-ORDERED, QTY-RECEIVED, QTY-PAID-FOR, and PRICE. These lines are modeled by the multivalued attribute PO-LINE, which is made up of these five fields. The asterisk next to PO-LINE indicates that there may be more than one PO-LINE for each PURCHASE-ORDER. You should note that a new attribute, INVOICE-NO, has been added to entity set INVOICES. This attribute is an internal identifier added to distinguish invoices received from different suppliers. You may wish to examine all the documents in Figure 9.12 and study how they have been converted to the E–R diagram.

SOME ADDITIONAL CONCEPTS

The most common extensions to the E–R model are dependent entity sets and subsets. This book introduces these extensions here. For more elaborate descriptions, you should go to more advanced database texts (e.g. Hawryszkiewycz, 1991).

DEPENDENT ENTITIES

The first E–R model extension is the concept of the **dependent entity set**. Dependent entity sets are also sometimes known as **weak entity sets**. You should recall what is meant by the word *set*. An entity or relationship set is made up of objects that have the same properties. All objects in the same dependent entity set have not only the same properties, but also another property in common: their existence depends on the existence of a parent entity in another set. If that parent is not of interest to the system, its dependent entities also cease to be of interest to the system.

The dependent entity set is shown in the E–R diagram by a rectangular box with a second line drawn across the top. Figure 9.13 shows two E–R models that include dependent entities. Figure 9.13(a) models INVOICES and INVOICE-LINES. An invoice line obviously cannot exist without there being a corresponding invoice, and appears in the system only if a corresponding invoice exists. This dependence is modeled by an arrow directed from the parent entity set (in this case, INVOICES) to its dependent entity set (in this case, INVOICE-LINES). Figure 9.13(b) is another example of dependence. Here, project tasks are set up for projects

Dependent entity set A set of entities whose existence depends on other entities.

Weak entity set Another term for a dependent entity set.

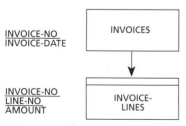

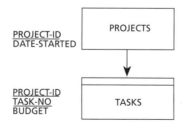

(a) There cannot be an invoice line without an invoice

(b) Tasks can only be created for projects and cannot stand on their own

Figure 9.13 *Dependent entities*

in the system. Thus, a project to build a computerized payroll system may include tasks such as analysis, design, and implementation.

A task cannot be set up without there being a project. In an E–R model, tasks are modeled by dependent entity set TASKS, projects are modeled by entity set PROJECTS, and the dependence is shown by the arrow from PROJECTS to TASKS.

DEPENDENT ENTITY SET

Note that dependent entities have composite identifiers. Their identifier consists of the identifier of their parent entity, together with another attribute that uniquely identifies the dependent entity within the parent. Thus, an invoice line is made up of the attribute INVOICE-NO (which identifies the invoice in which the invoice line appears) and the attribute LINE-NO (which identifies the particular line in the invoice). Similarly, a task identifier is made up of the attribute PROJECT-ID (which identifies the task's project) and the attribute TASK-NO (which identifies a task within the project).

USING DEPENDENT ENTITY SETS

Dependent entities are very useful for modeling historic information. Historic information must be kept whenever two entities interact more than once. Each such interaction may happen at a different time, and each interaction may have different property values. An example of historic information is the amount of machine time used by projects each day. The amount of time used by the same machine on the same project may be different each day. The obvious way to model machine use by projects without using dependent entity sets is shown in Figure 9.14(a). The machines are modeled by entity set MACHINES and projects by entity set PROJECTS. The use of machines by projects is modeled by the relationship set USE.

The occurrence diagram in Figure 9.14(a) shows that a machine in entity set MACHINES interacts more than once with the same PROJECTS entity in relationship set USE. Machine 'Mach1', for example, appears in two USE relationships, 'u1' and 'u2', with project 'Proj1'. There is one such relationship between the same machine and the same project for every day that the machine uses that project. Relationships between the same two objects in the same relationship set are sometimes known as **multiple relationships**.

Multiple relationship set A relationship set where the same two entities can appear in more than one relationship.

Without historic data, the identifiers of relationships in relationship set USE would be the identifiers of MACHINES and PROJECTS—namely, MACHINE-NO and PROJECT-ID. To model historic information, relationship USE contains an additional identifier, DATE-USED, to distinguish the different relationships between the same machine and project. The attribute TIME-SPENT-ON-PROJECT in relationship set USE is the amount of machine time used by a project on one day. Relationship set USE also has the attribute MACHINE-CONDITION-FOR-DAY to store the machine condition for the day. The machine condition is the same for all projects on a given day.

Relationship sets with multiple relationships often include redundancy, because they model more than one concept. Thus relationship set USE, in Figure 9.14(a),

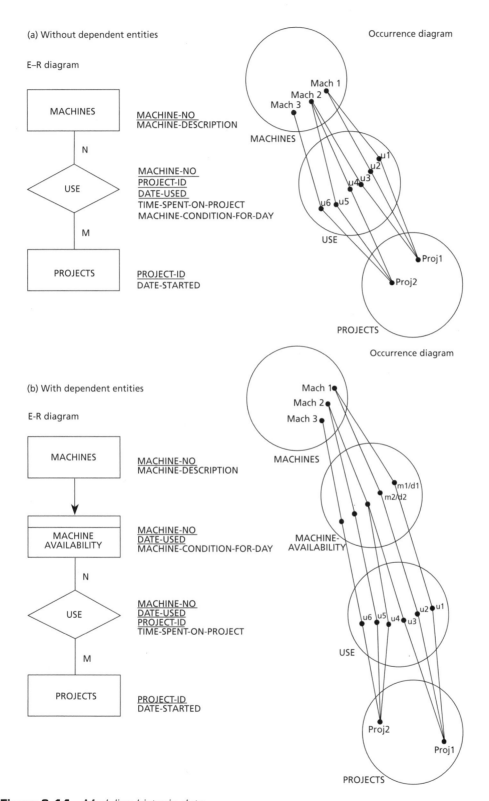

Figure 9.14 *Modeling historic data*

describes both the condition of a machine and the projects that used the machine. If you look carefully at the E–R model in Figure 9.14(a), you will see that a value of MACHINE-CONDITION-FOR-DAY will appear as many times as the number of projects that use the machine for the day. This is because the machine condition for the day is independent of the projects that use the machine.

A better way to model this system is to use a separate dependent entity set to model the machine condition of the day. In Figure 9.14(b), the dependent entity set MACHINE-AVAILABILITY models the availability of machines and their condition each day. The occurrence diagram illustrates the difference from Figure 9.14(a). The occurrence diagram in Figure 9.14(b) no longer has any multiple relationships between the same machine and project. Instead, the machine has a dependent entity in dependent entity set MACHINE-AVAILABILITY for each day the machine is available. For example, 'Mach1' has two dependent entities, 'm1/d1' and 'm1/d2' in dependent entity set MACHINE-AVAILABILITY, to show that it is available on two days.

There are no multiple relationships between the same project and machine in Figure 9.14(b). They have been replaced by relationships between a project and a number of machine availabilities. Thus, the multiple relationship 'u1' and 'u2' between the same machine and project in Figure 9.14(a) now become relationships between the same project but different machine availabilities, 'm1/d1' and 'm1/d2', in Figure 9.14(b).

The attribute MACHINE-CONDITION-FOR-DAY now becomes an attribute of dependent entity set MACHINE-AVAILABILITY. The redundancy of Figure 9.14(a) no longer exists, because the value of MACHINE-CONDITION-FOR-DAY will appear once only for each machine and each day. You should note that the identifier of MACHINE-AVAILABILITY has a composite identifier made up of the parent identifier, MACHINE-NO and DATE-USED, which is the date on which the machine is available. Machines available on a given day can be used by projects on that day. Relationship set USE models the use of machines by projects and contains attributes such as TIME-SPENT-ON-PROJECTS, which are specific to the use of the machine by the project.

SUBSETS

There are many modeling situations where we wish to treat entities from one entity set in the same way in some cases and differently in other cases. For example, all loan applications may be treated in the same way at the beginning. They are all filed, given a number, and so on. Later their treatment may depend on the kind of application (e.g. whether they are for a personal loan or a home loan).

Subsets Some of the objects from one entity set.

To model different methods of treatment of entities in an entity set, it is necessary to show the division of entity sets into **subsets**. Figure 9.15 shows how such a division can be modeled using the E–R model. The occurrence diagram in Figure 9.15 shows the idea of subsetting. It shows five persons in entity set PERSONS. Some of these persons, such as 'Jill', are teachers and appear in subset TEACHERS; and some, like 'Mary', are students and appear in subset STUDENTS. It is also possible for some persons to appear in both subsets—for example, 'John', who is both a teacher and a student.

E–R diagram

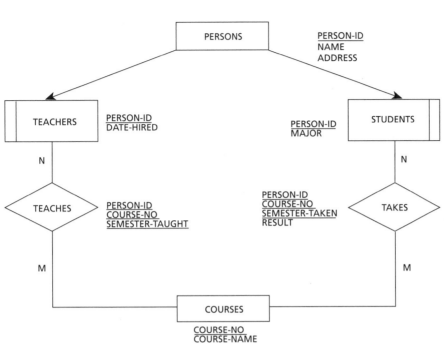

Occurrence diagram for entity set PERSONS and its subsets

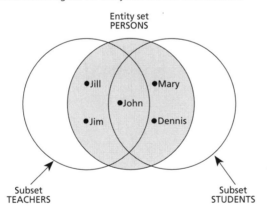

Figure 9.15 *Subsets*

To extend our terminology to subsets, we say that entities of a set can be entities in subsets of that set. Thus, in Figure 9.15, 'Mary' is an entity in set PERSONS. 'Mary' is also an entity in subset STUDENTS of entity set PERSONS.

The E–R model in Figure 9.15 shows how subsets are modeled on an E–R diagram. The E–R diagram includes the entity set PERSONS and its attributes PERSON-ID, NAME, and ADDRESS. PERSON-ID is the identifier of persons in entity set PERSONS. The subsets are modeled by rectangular boxes with a second line on the left-hand side of the box. Two subsets of entity set PERSONS, namely TEACHERS and STUDENTS, are modeled by the E–R model in Figure 9.15.

The directed arrows from PERSONS to TEACHERS and STUDENTS show that STUDENTS and TEACHERS are subsets of PERSONS.

It is also common for entities in different subsets to have some different attributes and some common attributes. The common attributes are shown as attributes of the parent entity set, whereas attributes particular to a subset are shown only as attributes of that subset. Thus, a person who is a teacher may have the attribute DATE-HIRED, and a person who is a student may have the attribute MAJOR. Attributes, which depend on the type of entity, are stored in the subset rather than the entity set. Thus, in Figure 9.15, DATE-HIRED is an attribute of TEACHERS and not PERSONS. Similarly, a MAJOR is an attribute of STUDENTS. However, ADDRESS is an attribute of PERSONS, because both TEACHERS and STUDENTS have the attribute ADDRESS.

SUBSET IDENTIFIERS

The next question is how to identify members of a subset. The simplest way is to use the same identifier in a subset as in the parent of the subset. Thus, if PERSON-ID identifies a unique person in the PERSONS set, then it will also identify that person in a subset of entity set PERSONS. In Figure 9.15, PERSON-ID is an identifier of entity set PERSONS and the identifier of two subsets, TEACHERS and STUDENTS. In some cases, however, it is useful to create a new identifier to identify members of a subset. This is necessary when subsets from two different entity sets are combined.

COMBINING SUBSETS FROM A NUMBER OF ENTITY SETS

Figure 9.16 illustrates yet another kind of subset. Subset CLUB-MEMBERS contains some members of entity set STUDENTS and some members from entity set STAFF. These members use courts at the club. Subset CLUB-MEMBERS is

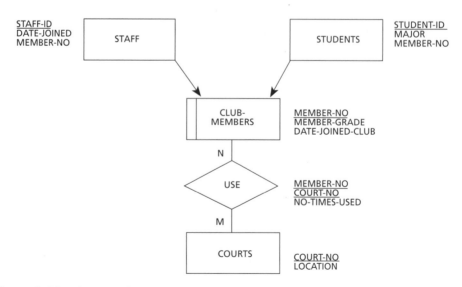

Figure 9.16 *A non-uniform subset*

called non-uniform, or merged, because it contains entities from more than one entity set.

Subset CLUB-MEMBERS must also have its own identifier, MEMBER-NO. We cannot use either STAFF-ID or STUDENT-ID in subset CLUB-MEMBERS, because the same value of these identifiers can identify different persons. STAFF and STUDENTS include the attribute MEMBER-NO to show the membership numbers allocated to staff and students. You might argue why not make STAFF and STUDENTS subsets of CLUB-MEMBERS. Because those staff and students who are not club members would not appear in the system.

TEXT CASE D:
Construction Company—Refining the E–R Diagram

An initial E–R diagram for Text Case D was illustrated in Figure 9.12. The next E–R modeling step is to refine this diagram. The refined E–R diagram is shown in Figure 9.17. Each multivalued attribute in the top-level E–R diagram is replaced by a dependent entity set in the refined E–R diagram. For example, PO-LINE of PURCHASE-ORDERS is replaced by the dependent entity set ORDER-LINES. All the fields of the multivalued attribute become attributes of the newly created dependent entity set. In addition, a new identifier is added to the dependent entity set to identify the dependent entities within their parent. The attribute PO-LINE-NO is thus used to identify lines within a PURCHASE-ORDER.

Another refinement is to add entity sets ITEMS to the E–R diagram. This entity set is added because a reference to items appears in most documents. Relationship sets are added between the entity set ITEMS and all the dependent entity sets for document lines that refer to items.

Relationship sets that previously existed between documents are now replaced by relationship sets between the document lines. This calls for a detailed analysis to precisely determine the relationships between document lines. The refined E–R diagram shows that:

- one purchase order line is related to many shipping lines, because items in one purchase order line may be delivered in more than one shipment
- one shipment line is related to many invoice lines, because items in one shipment line may be spread across a number of invoices
- one purchase order line is related to many invoice lines, because items in a purchase order line may be invoiced over a number of invoices
- one order line is related to many request lines, because a purchase order line may combine requests from more than one project request
- one request line is related to many delivery lines, because items in one project request line can be delivered in a number of deliveries to the project site.

E–R DIAGRAMS AND DFDs

So far, E–R analysis and data flow analysis have been discussed as two separate modeling methods. However, in practice, both these techniques are used together in system modeling. Data flow analysis is used to model system flows, whereas E–R analysis is used to model system data. There are two implications in this. First, as both models describe the same system, they must be consistent in their use of system names. Second, each model can be used to help develop the other, and to check

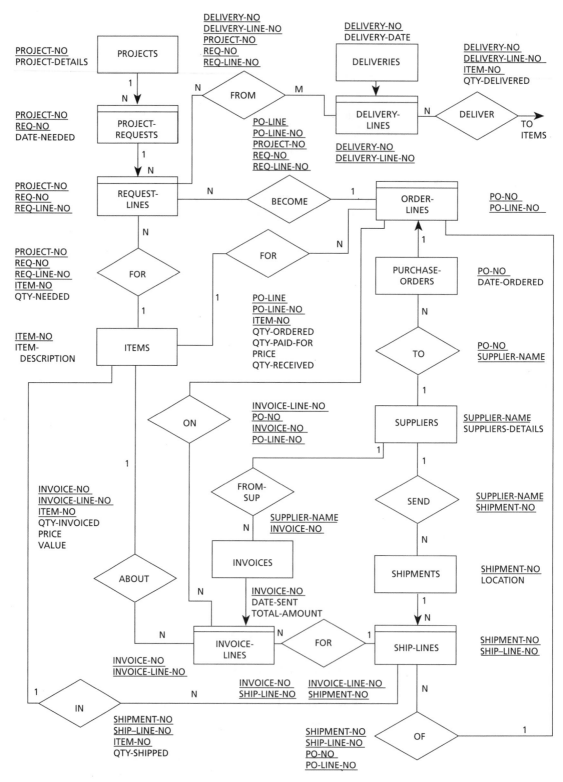

Figure 9.17 *A refined E–R diagram*

that the other model is complete, so that if something is missed in one model it can be found in the other, and vice versa.

However, one cannot derive one model from the other. Hence, it is possible to start modeling with either the DFD or the E–R diagram, or to use both together. Most methodologies start with the DFD and then use the data stores or flows as indicators of the major system entities. However, it is possible to use the information gathered in developing one model to help develop the other. The DFD can also suggest what to include in an E–R diagram. Typical guidelines here include the following:

- Each external entity is modeled usually by an entity set. We often need information about such external entities—for example, their addresses—and this would be modeled as part of this entity set.
- Data stores often indicate possible entity sets. A detailed examination of a data store often suggests entity sets and their attributes.
- Processes often suggest relationships. A process that uses two components often suggests a relationship between these components. Thus, a process that uses two stores often suggests that a relationship exists between the entities in those stores. A process that uses flows with an external entity and a store often suggests a relationship between the external entity and entity sets in the store.

In contrast, data flows often do not suggest entity sets. They often move information from one source to another. Such information may be collected from more than one entity set and need not be restricted to one set. However, some data flows may sometimes carry information related to one entity, such as an order or a delivery.

As an example, consider the DFD in Figure 8.3 and how it can be used to suggest an E–R diagram. The first step may be to create an entity set for each of the three external entity sets. These three entity sets are shown in Figure 9.18 as entity sets DEPARTMENTS, MANAGERS, and SUPPLIERS. Then we note that there are data stores for budgets and accounts and set up the ACCOUNTS and BUDGETS entity sets. A relationship is then set between BUDGETS and DEPARTMENTS to show the departments to which a budget applies.

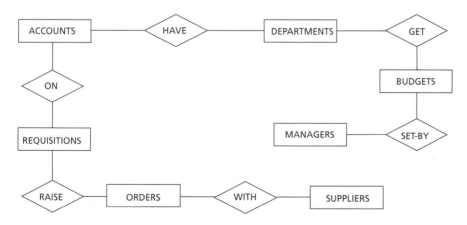

Figure 9.18 *An initial E–R diagram*

REQUISITIONS is the next entity set that we might set up. You might ask why have REQUISITIONS as an entity set. There are no data stores for requisitions in Figure 8.3. However, the flow from the external entity DEPARTMENTS does suggest a requisition, and there might be a data store that contains requisitions, probably in one of the processes, such as 'Check funding'. We may also add entity set ORDERS, because this would be part of process 'Organize supplies' and is sent to suppliers. Again, there is no data store of orders, but the flow to SUPPLIERS suggests that ORDERS might be stored in process 'Organize supplies'. We then assume that process 'Organize supplies' uses requisitions to create orders and thus set up a relationship between ORDERS and REQUISITIONS and another between ORDERS and SUPPLIERS, because orders are sent to suppliers. Finally, we might set up a relationship between REQUISITIONS and ACCOUNTS because of the checking in process 'Check funding'. The DFD in Figure 9.18 may then be a good starting point for E–R modeling.

The idea of using a DFD to suggest an initial E–R model does not in any way imply that E–R models should be derived from DFD diagrams. They are, in fact, developed in parallel, and one usually suggests improvements to and completeness of the other. Thus, it is possible to check model consistency and ensure that no information is missed during analysis. Names in the E–R and DFD models are often checked to ensure that both contain all the information. Documentation is frequently designed to check for such consistency. Many CASE tools also provide consistency checks.

The question, of course, is how tight to make the link between the two models. One way is to keep a fairly loose link. Here we ensure that there is a consistency between the names used in each model and that each model can provide leads for detailed searches or corrections in the other. Alternatively, there may be a more comprehensive set of link requirements. For example, we may require each data store to correspond to an entity set, a relationship set, or a combination of the two.

SUMMARY

This chapter describes ways to describe data as part of a system and methods used to develop data models during systems analysis. The goal at this early stage is to capture the important characteristics of data. This is done by using a high-level semantic model. The chapter describes how entity–relationship (E–R) modeling techniques are used to develop such a model. Later in design, some further refinements are made to the E–R model. This model is then refined by converting it to a relational model. The relational model provides a number of criteria to construct normal form relations. The chapter also discusses the relationship between the E–R model and DFDs and how this can be used to ensure that correct and complete models are developed during analysis.

EXERCISES

9.1 Develop E–R diagrams for the following:

1. Customers make orders.

2. People work in departments.

3. Customers buy items.

4. Vehicles are owned by persons.

5. Athletes take part in events.

6. Deliveries of parts are made to customers.

9.2 Draw E–R diagrams showing the cardinality for the following:

1. An invoice is sent to one customer, and many invoices can be sent to the same customer.

2. A part is used in many projects, and many projects use the part.

3. A person works in one department, and there are many persons in a department.

4. A vehicle is owned by one person, and a person can own many vehicles.

5. Students take subjects. Each subject can be taken by many students, and each student can take many subjects.

6. Persons apply for loans. Each loan must be made to one person, but each person can make many applications.

7. An operator can work on many machines, and each machine has many operators. Each machine belongs to one department, but a department can have many machines.

9.3 Draw E–R diagrams with attributes, cardinality, and identifiers for the following:

1. Customers identified by a CUSTOMER-NAME and with an ADDRESS buy items. Items are identified by an ITEM-NO and have a COLOR. The QTY-BOUGHT of an item by each customer is recorded. An item can be bought by many customers.

2. Departments are identified by a DEPT-NO and have a budget. A department can manage many projects, but each project is managed by one department. Projects are identified by a PROJECT-NO and have a START-DATE.

3. An order with a unique ORDER-NO and ORDER-DATE can be made for any number of parts (identified by ITEM-NO and with a COLOR). QTY-ORDERED is the amount of each part ordered. Each order is made to one supplier (who has a unique SUPPLIER-NAME and one ADDRESS).

4. A fault occurs on one item of equipment. A logbook contains FAULT-NO, FAULT-DATE, and FAULT-DESCRIPTION. Each item of equipment has a unique EQUIP-NO and an EQUIP-DESCRIPTION and TYPE. Each such item is located in one building, which has a unique BUILDING-NAME and one ADDRESS.

5. A student (with STUDENT-ID, ADDRESS, and SURNAME) takes any number of subjects, which have a unique SUBJECT-NAME and a SUBJECT-DESCRIPTION. The student is enrolled in a major that has a unique MAJOR-NAME and LENGTH. The date on which a student started a major is recorded. A subject is taught by one teacher, who is identified by a TEACHER-ID and has a TEACHER-ADDRESS.

9.4 Convert the N-ary relationship set shown in Figure 9.19 to a set of binary relationship sets.

9.5 Draw an E–R diagram for the following:

• A customer reserves a date for maintenance or repairs to a vehicle. The reservation is given a RESERVATION-NO. The customer and vehicle details are recorded with the reservation. The TIME-OF-RESERVATION is also recorded.

• Information stored about customers includes CUSTOMER-NAME, ADDRESS, and TELEPHONE-NO.

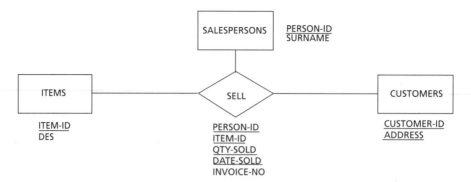

Figure 9.19   *A N-ary relationship set*

- Information kept about the vehicle includes MAKE, REG-NO, and DATE-OF-MANUFACTURE.

- After examination, a number of jobs are recorded for the vehicle. Each job has a JOB-NO within the booking, and reasons for carrying out the job are recorded as WHY-NEEDED.

- The parts used for each job and the TIME-SPENT on each job are recorded.

- The information about parts incudes PART-NO and PRICE.

9.6   Draw E–R diagrams for the following, using dependent entities if necessary. Where required, make sensible assumptions and carefully specify those assumptions.

1. A department has a number of sections. Each department has a unique DEPT-NO and BUDGET, and each section has a unique SECTION-NO within the department.

2. An office (identified by OFFICE-NO) in a building (with a unique BUILDING-NAME) is occupied by a person (identified by PERSON-ID and with a SURNAME). The building has a unique ADDRESS and the office has an OFFICE-SIZE.

3. A vehicle is identified by a REGISTRATION-NO and is of a given MAKE. Vehicles are used every day to make deliveries. A driver is assigned to a vehicle for the whole day, and that driver can make any number of deliveries during the day. Each delivery is identified by a DEL-NO and is made to a given ADDRESS.

4. A student identified by a STUDENT-ID and with a given ADDRESS can enrol for any number of semesters. The enrolment in a semester can be of a given TYPE ('part-time' or 'full-time'). The student can take any number of subjects in the semester and get a GRADE for each subject taken. The subject is identified by SUBJECT-NAME and has a DESCRIPTION.

9.7   EMPLOYEES in an organization can be 'part-time' or 'full-time' employees. The organization keeps the PERSON-ID and SURNAME for each employee. In addition, for each part-time employee, the organization keeps that employee's START-TIME and END-DATE.

The DATE-JOINED is kept for each full-time employee. In addition, each full-time employee occupies one POSITION. Each position is identified by a POSITION-NO and is in one DEPARTMENT. The date on which the full-time employee was appointed to the position is also stored.

9.8   An organization gets requests from clients to carry out in-depth market studies. It needs to find the most qualified employees to carry out these studies. It often also draws on the experience of some of its external consultants.

1. Each client request in the organization is given a REQUEST-NO, and the REQUEST-DATE, CLIENT-NAME, and CLIENT-ADDRESS are also recorded. The STUDY-FIELD for each request is also recorded.

2. The organization keeps a record of its internal employees, including their skills. This includes EMPLOYEE-NO, EMPLOYEE-NAME and number of expert skills. The skills are used to match the employees to the request study field.

3. It also keeps information about its external consultants, including their CONSULTANT-NAME, SKILLS and ADDRESS.

4. A record is kept of people assigned to each client request. This includes the TIME-SPENT by the person on the request.

5. The documents produced by each study are also numbered and classified by their NAME and any number of STUDY-FIELDs. These documents may then be retrieved for reference when a request with similar study fields is received.

9.9   Draw an E–R diagram for the following:

An exhibiting organization keeps information about paintings and sculptures. Each painting has a PAINTING-NAME, PAINTER-NAME, and PAINTING-DESCRIPTION. Each sculpture has a SCULPTOR-NAME, SCULPTURE-NAME, and SCULPTURE-DES.

Paintings and sculptures may appear in the same gallery. For the purpose of keeping track of the location of items, each painting and sculpture is given a unique identifier, ART-NO.

Each gallery has an identifier, GALLERY-NO, and a size. Each gallery can store any number of art objects. Each art object appears in one gallery only. The DATE-PLACED-IN-GALLERY is kept for both paintings and sculptures.

Note that PAINTING-NAME is unique within PAINTER-NAME, and SCULPTURE-NAME is unique within SCULPTOR-NAME.

Some more advanced exercises follow. These often requiring starting with N-ary relationships, then reducing them and adding new entity sets and identifiers.

9.10   You have gathered the following data: you are required to draw an E–R diagram from the data.

* Persons (identified by PERSON-ID) work on machines (identified by MACHINE-NO) to produce garments.

* Various GARMENT-KINDS can be produced. Each GARMENT-KIND has a description (GARMENT-DESCRIPTION) and is made up of a variety of materials (identified by MAT-KIND). A record of the QTY-NEEDED of each MAT-KIND for each GARMENT-KIND is stored.

* The production of each garment is recorded as a job identified by JOB-NO. Each JOB-NO has a START-TIME and an END-TIME and is performed by one person on one machine. A number of garments of the same kind can be produced on one job.

* Other information of interest is:
  * the NAME and DATE-OF-BIRTH of each person
  * the DATE-PURCHASED of each machine
  * the DESCRIPTION of each MAT-KIND
  * the TIME-SPENT by a person on a job
  * the NUMBER-OF-GARMENTS produced on one job.

9.11   Below are some statements about order processing in an organization. You are required to construct an E–R diagram from these statements.

* Persons in the organization are identified by a PERSON-ID and have a SURNAME, FIRST-NAME, and DATE-OF-BIRTH.

- The persons are responsible for orders, which are identified by an ORDER-NO and have an ORDER-DATE, DESCRIPTION, and QUOTED-PRICE. Each order is from one customer. Only one person is responsible for a given order, but a person may be responsible for many orders.

- The organization manufactures the order in a series of jobs. A person responsible for an order makes formal requests to sections to carry out these jobs. The requests are identified by a REQUEST-NO. They nominate a START-DATE and an END-DATE for each request.

- A number of jobs can be created by a section in response to a request. Each job is identified by a JOB-NO and has a COST. All jobs for one request go to the same section, which is identified by SECTION-ID and has one MANAGER.

- Each job uses a QTY-USED of one or more materials. Materials are identified by MAT-ID and have a MAT-DESCRIPTION.

9.12 Draw an E–R diagram that describes the following message transmission system:

Messages are sent in an organization. Messages have a DATE-SENT and LOCATION-MAILED. Messages may be of bulletin or letter type. A bulletin is identified by a unique BULLETIN-NO, whereas a letter is identified by a SENDER-NAME and TIME-SENT.

Persons (identified by a unique PERSON-ID and with NAME and DATE-OF-BIRTH) in the organization belong to groups (identified by GROUP-NO and with a given GROUP-LOCATION and FUNCTION). Each person is assigned to a group on a given DATE-ASSIGNED, and a person can belong to more than one group. Similarly, a group is made up of more than one person.

Letters may be addressed to either groups or individual persons. Only individual persons can send bulletins or letters. Bulletins are addressed to groups only.

BIBLIOGRAPHY

Chen, P.P. (March 1976), 'The entity–relationship model—toward a unified view of data', *ACM Transactions on Database Systems*, Vol. 1, No. 1, pp. 9–36.

Davenport, R.A. (1979), 'Data analysis for database design', *Australian Computer Journal*, Vol. 10, No. 4, pp. 122–37.

Flavin, M. (1981), *Fundamental Concepts of Information Modeling*, Yourdon Press, New York.

Hawryszkiewycz, I.T. (1991), *Database Analysis and Design* (2nd edn), Macmillan, New York.

Howe, D.R. (1983), *Data Analysis for Data Base Design*, Edward Arnold, London.

Nijssen, G.M. and Halpin, T.A. (1987), *Conceptual Schema and Relational Database Design: A Fact Oriented Approach*, Prentice Hall, Sydney.

Robinson, H. (1981), *Database Analysis and Design*, Chartwell-Bratt, Bromley, Kent.

PROCESS
DESCRIPTIONS

CONTENTS

KEY LEARNING OBJECTIVES

How to specify processes in detail
How to describe processes in data flow diagrams
Decision trees and tables

CHAPTER 10

INTRODUCTION

The modeling methods described in the previous two chapters all used a graphical representation. This has been done for a reason. Graphs are easier to understand, and they often give a total picture of the system. However, graphical models become unwieldy when used to specify processes in detail. Instead, such detailed process descriptions tend to use languages or scripts rather than diagrams. This chapter describes a number of ways of describing processes in detail. These range from natural language descriptions to scripts that must satisfy structure rules. The chapter then concentrates on the way in which detailed processes are defined in structured systems analysis.

Process descriptions have to satisfy a number of desirable properties. These depend on the level of the specification. At the usage level, they must be sufficiently rich to capture detailed user operations. At the system level, they must be more precise, as they become the system specification. Processes in the system specification should be specified in such a way that they can be converted to a computer program. Such specifications are often called executable specifications. *An alternative to executable system specifications are specifications that can be converted to a computer program using a specific set of rules. A third desirable property of process descriptions is that they should be easy to read and understand.*

NATURAL LANGUAGE SPECIFICATIONS

Most early design methods used natural language to specify system requirements. Such natural language specifications were generally found unsuitable because they often led to ambiguities. For example, consider the description: 'Add the expense to the travel budget if the trip exceeded two days or was longer than 250 km and a company vehicle was used.' What do we do if the trip was 300 km long, took three days, and used a private vehicle? It depends on whether we group 'was longer than 250 km' with 'exceeded two days' (in which case, the answer is 'do not add') or we group it with 'company vehicle was used' (in which case, the answer is 'add'). The process description is therefore ambiguous.

For this reason, natural language is generally not used for detailed process specification but is sometimes used at higher specification levels to define broad process goals.

SCRIPTING

Script A description of a process.

Basically, **scripting** means a structured written description of something. This book describes a number of ways of scripting. One method uses scenarios together with rich pictures to describe processes in the usage world. Ethnographic scripts were used for a similar reason. Both of these are relatively informal methods, describing a set of steps followed by people in their work. They both describe specific instances of work practice in organizations. Use cases tend to introduce more formality, as they are generalizations of a number of scenarios. Examples of use case specifications are given in Chapter 11.

Computer specifications, however, often require even more structured scripts. Such scripting can be anything from a program to a structured specification. The important thing is that such scripts have a structure that encourages precision. They must leave no doubts in the mind of the reader as to what they mean. Often this means that the scripts follow a logical structure. They may also use key words to shorten the amount of script. The scripts may also have to follow rules, be limited to predefined constructs, use specific instructions, or closely approximate computer programs themselves. There are some advantages in using scripts that follow precise rules, as they can be directly converted to program code.

One example of scripts that follow a predefined structure comes from the work of Schank and Abelson (1977), who defined scripting as a method of defining knowledge. Here, scripts are defined as scenes that describe the interaction between roles in an environment. Such scripting methods can be adapted in defining processes in systems analysis.

TEXT CASE A:
Interactive Marketing—A Script for Arranging a Sale

We assume that the following are involved in marketing:

      C—Consumer

      V—Vendor

      T—Trader.

We now define a script made up of four scenes—initiating a purchase, agreeing on a price, arranging a purchase, or aborting it.

Scene: initiating purchase

      C—ASK (for PRODUCT) from T

      T—PRESENTS (list of product)

      C—ASK (for price) from T

      do bargaining.

Scene: bargaining

      C—ASK (for price) from V

      V—PRESENTS (price) to T

      T—PRESENTS (price) to C

      C—REPLIES (accept)—do arrange purchase

      C—REPLIES (reject)—do abort purchase

      C—REPLIES (ask again)—repeat bargain.

Additional scripts would be written for arranging a purchase, which would create a sales record, and abandoning a purchase, which would delete any record of negotiation.

In the above example, scripting described the interaction between three roles. One important requirement of scripting is that it must use a precise syntax to be unambiguous. In the example above, each line of the script has a VERB, such as ASK, as well as some arguments and the source and target of each interaction. The important thing to remember about scripts is that they must be precise and describe unambiguously the steps followed in a system. Precise methods for scripting have also been developed for DFDs and are described below. Later the book describes scripting methods used in object modeling known as use cases.

SCRIPTING PROCESSES IN DATA FLOW DIAGRAMS

All processes in a DFD must have a process description. Different methods can be used to describe processes in a DFD. The scripts used to describe top-level processes, for example, differ from those used to describe detailed processes. Top-level processes are usually brief and can use natural language. The process is usually described by one sentence that states what the process does. For example, Process 1 in Figure 8.3 may be described as 'Check whether there are sufficient funds in the department's budget to meet the spending request'.

What is needed at detailed levels are specifications that are unambiguous and can be easily understood by both users and programmers. The main techniques proposed for this purpose are:

- structured English
- tight English
- decision tables
- decision trees.

Structured English and tight English put verbal descriptions into a logical structure, which removes logical ambiguities. This provides the best of two areas—the logical structure removes logical ambiguities—but English narrative can still be used to describe activities. The logical structures used to remove ambiguity are the logic constructs of structured programming. Process descriptions use ideas from **block structured languages** to express process logic in a clear and unambiguous way.

Block structured language A way of programming that clearly expresses process logic.

The next two methods—*decision tables* and *decision trees*—are preferred where one of a large number of actions is to be selected. The action selected depends on a large number of conditions. Structured or tight English is not usually used for this purpose because the logic structure would become repetitive.

The differences between the four methods are illustrated in Figure 10.1, where there is a description of a process that determines whether a customer is to be given credit. The process is described using each of the four methods. The process chooses one of the three possible actions ('allow credit', 'refuse credit', or 'refer to manager'). The chosen action depends on whether customers have exceeded their current credit limit, on the size of the purchase, and on the customer's payment history. Note that credit is allowed if the credit level has not been exceeded. If the credit limit has been exceeded, then a customer's credit history and amount of credit are examined. Customers with a bad credit history are refused credit. Customers with a good credit history can be allowed a further credit of up to $200 at the discretion of a manager.

You will note that in structured English these conditions are expressed by logic on the IF . . . THEN . . . ELSE construct. The first statement checks the condition 'credit limit exceeded'. Statements that follow THEN specify what is to be done if the condition is true. Statements that follow ELSE (which falls directly under the THEN) specify what is to be done if the condition is false. Further conditions are tested if the condition 'credit limit exceeded' is true. These tests are indented to the right to be within the first THEN clause. The first test is condition 'customer has bad payment history'. The THEN and ELSE clauses indented under this test

(a)   Using structured English
　　　IF credit limit exceeded
　　　　　THEN
　　　　　　　IF customer has bad payment history
　　　　　　　　　THEN refuse credit
　　　　　　　　　ELSE
　　　　　　　　　　　IF purchase is above $200
　　　　　　　　　　　　　THEN refuse credit
　　　　　　　　　　　　　ELSE refer to manager
　　　　　ELSE allow credit

(b)   Using tight English
　　　5.1 IF credit level exceeded
　　　5.1.1 THEN (credit limit exceeded)
　　　　　IF customer has bad payment history
　　　　　　　5.1.1.1 THEN refuse credit
　　　　　　　5.1.1.2 ELSE (customer has good payment history)
　　　　　　　　　IF purchase is above $200
　　　　　　　　　5.1.1.2.1 THEN refuse credit
　　　　　　　　　5.1.2.2.2 ELSE (purchase is below $200)
　　　　　　　　　　　refer to manager
　　　5.1.2 ELSE (credit level not exceeded) allow credit.

(c)   Using a decision table

Conditions	Credit limit exceeded	Y	Y	Y	Y	N	N	N	N
	Customer with good payment history	Y	Y	N	N	Y	Y	N	N
	Purchase above $200	Y	N	Y	N	Y	N	Y	N
Action	Allow credit					X	X	X	X
	Refuse credit	X		X	X				
	Refer to manager		X						

Key:
Y = Yes, condition true
N = No, condition not true

(d)   Using a decision tree

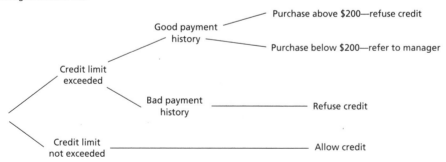

Figure 10.1 *Specifying processes*

specify what is to be done for the 'true' and 'false' outcomes of this test. A 'false' outcome leads to a test of condition 'purchase above $200', and the outcomes of this test appear as further indentations. Too many indentations often lead to confusion in a structured English specification and should be avoided. The term *nesting* refers to indentation levels.

Tight English uses a numbering scheme to simplify indentation. Statement 5.1 checks whether the credit limit has been exceeded. Statement 5.1.1 specifies what has to be done if the credit limit has been exceeded. Statement 5.1.2 specifies what is to be done if the credit level has not been exceeded. Tests within other tests use a further level of numbering. Thus 5.1.1.1 and 5.1.1.2 specify the outcomes of condition 'customer has bad payment history' within level 5.1.1.

A decision table shows each possible set of conditions in one column and the corresponding actions in the same column. The decision tree defines the conditions as a sequence of left to right tests, commencing with credit limit, then, depending on the outcome, looking at the payment history and, finally, at the size of the purchase.

You should also note that tight English is similar to structured English and that decision trees use the same ideas as decision tables. Structured English and decision tables are described in detail in the remainder of this chapter.

STRUCTURED ENGLISH

Structured English syntax is very similar to block structured languages. It provides the keywords to structure the process specification logic as a block structured language while leaving some freedom when describing the activities and data used in the process. As with block structured languages, process specification logic consists of a combination of sequences of one or more imperative sentences with decision and repetition constructs.

IMPERATIVE SENTENCES

An imperative sentence usually consists of an imperative verb followed by the contents of one or more data stores on which the verb operates. For example:

add PERSONS-SALARY to TOTAL-SALARY

It is important that imperative sentences use verbs that are clear and unambiguous. Verbs such as 'process', 'handle', or 'operate' should not be used. Instead, verbs should define precise activities, such as 'add' or 'compute average'. Sometimes computer general words such as 'edit' or 'convert' have a specific meaning within a given context and are used in that context. Adjectives that have no precise meaning, such as 'some' or 'few', should also not be used in imperative sentences because they cannot be used later to develop programs. These adjectives are redundant in the process description and should not be used, as they create unnecessary confusion.

Additional standards are often used in imperative statements. For example, data flow names often appear in lower case between quotes, and specific data items in the data flows are capitalized. Data store names, as well as specific data store items, appear in capitals in imperative statements. If necessary, specific item names can be qualified by their data flow names or data store names to avoid ambiguity.

Boolean and arithmetic operations can be used in imperative statements. The exact operators used will depend on the standards adopted by a particular organization and usually include the following:

Arithmetic	multiply (*)
	divide (/)
	add (+)
	subtract (−)
	exponentiate (**)

Boolean	add
	or
	not
	greater than (>)
	less than (<)
	less than or equal to (<=)
	greater than or equal to (>=)
	equals (=)
	not equal to (•)

STRUCTURED ENGLISH LOGIC

Structured English uses certain keywords to group imperative sentences and define decision branches and iterations. These keywords are:

BEGIN	REPEAT	IF
END	UNTIL	THEN
CASE	WHILE	ELSE
OF	DO	FOR

GROUPING IMPERATIVE SENTENCES

A sequence of imperative statements can be grouped by enclosing them with the BEGIN and END keywords. For example, Process 2.3 in Figure 10.2(a) can be defined as:

```
BEGIN
     Receive 'sale report'.
     Get SALES record for PART-NO in 'sale report'.
     TOTAL-QTY = TOTAL-QTY + QTY-SOLD.
     SALE-VALUE = QTY-SOLD * UNIT PRICE.
     TOTAL-VALUE = TOTAL-VALUE + SALE-VALUE.
     Write SALES record.
     Send 'summary advice'.
END
```

This sequence defines what happens when a 'sale report' is received. Each 'sale report' reports a sale of one part-kind to a customer. The part-kind is identified by PART-NO; QTY-SOLD is the number of parts sold; UNIT-PRICE is the sale price of one such part. The item CUSTOMER in 'sale report' identifies the customer.

Process 'record sale' maintains data store SALES, which keeps an accumulated total of the number of parts sold and total moneys received for each part-kind. To do this, the process uses PART-NO from 'sale report' to select a record about the

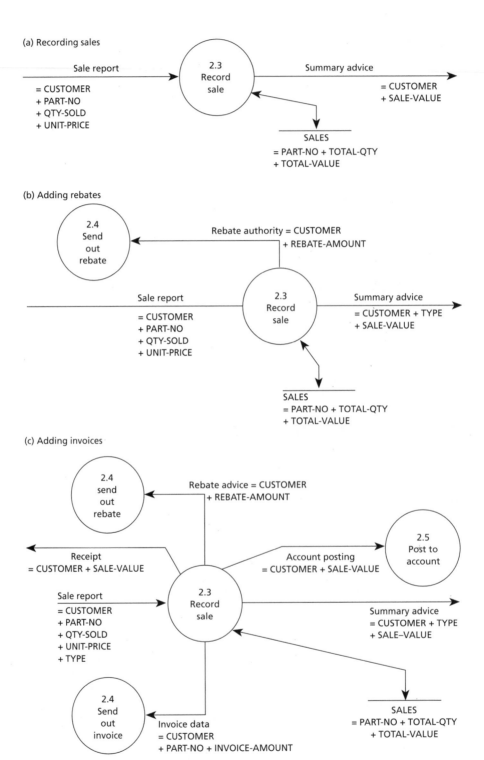

Figure 10.2 *Editing records*

part-kind from data store SALES. It updates the value of TOTAL-QTY by QTY-SOLD and the value of TOTAL-VALUE by the SALE-VALUE of the sale, and writes the record back to data store SALES. The process also sends out a 'summary advice' for each sale. The 'summary advice' contains the total value of a sale, SALE-VALUE, and the customer identifier, CUSTOMER.

Remember that structured English is not a programming language and that imperative statements are as brief as possible. There are no unnecessary move statements such as:

MOVE CUSTOMER in 'sale report' to CUSTOMER in 'summary advice'

which would normally be found in a programming language. It is assumed that the value of CUSTOMER remains the same as it passes through the process. Data items are not qualified unless there is ambiguity. Thus, TOTAL-QTY is not qualified as TOTAL-QTY in SALES because the qualification does not add any useful information but makes the specification longer.

DECISIONS

Two types of decision structures usually appear in structured English. These are as shown in Figure 10.3. Figure 10.3(a) shows a structure that allows a choice between two groups of imperative sentences. The keywords IF, THEN, and ELSE are used in this structure. If a condition is 'true', then GROUP A sentences are executed. If it is false, then Group B sentences are executed.

The structure shown in Figure 10.3(b) allows a choice between any number of groups of imperative sentences. The keywords CASE and OF are used in this structure. The value of a variable is first computed. The group of sentences executed depends on that value. In Figure 10.3(b), the value of TEST is first computed. If that value is 'A', Group A sentences are executed. If it is 'B', Group B sentences are executed, and so on.

Suppose Process 2.3 in Figure 10.2(a) is amended to compute rebates for sales above a certain amount. The amended process is shown in Figure 10.2(b), and its process description now becomes:

```
BEGIN
    Receive 'sale report'.
    Get SALES record for PART-NO in 'sale report'.
    TOTAL-QTY = TOTAL-QTY + QTY-SOLD.
    SALE-VALUE = QTY-SOLD * UNIT-PRICE.
    TOTAL-VALUE = TOTAL-VALUE +- SALE-VALUE.
    Write SALES record.
    Send 'summary advice'.
    IF SALE-VALUE > 500.00
        THEN
                BEGIN
                    REBATE-AMOUNT = SALE-VALUE * .02.
                    Send 'rebate authority'.
                END
END
```

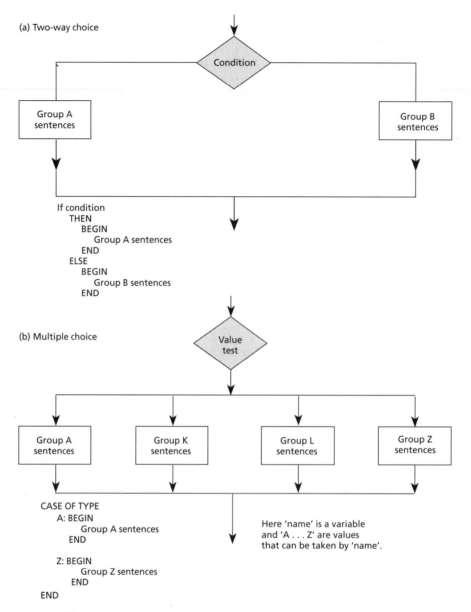

(a) Two-way choice

```
If condition
    THEN
        BEGIN
            Group A sentences
        END
    ELSE
        BEGIN
            Group B sentences
        END
```

(b) Multiple choice

```
CASE OF TYPE
    A: BEGIN
            Group A sentences
        END

    Z: BEGIN
            Group Z sentences
        END
END
```

Here 'name' is a variable and 'A . . . Z' are values that can be taken by 'name'.

Figure 10.3 *Decision structure*

An IF clause has now been added. It checks whether the total amount of the sale exceeds $500 and sends out an authority for a 2% rebate to the customer.

A further change is made to process 'record sale' to illustrate the CASE structure. Figure 10.2(c) shows the changed process. Now each 'sale report' contains the field TYPE. The value of TYPE shows how payment is to be made. A payment can be 'cash', 'check', 'credit', or 'account'. Process 'record sale' is now changed to cater for these types of payment.

An invoice must be sent for each 'credit' sale. An 'account' sale must be posted to an account, and only 'account' sales attract a rebate. A receipt is returned for all 'cash' and 'check' sales. A 'summary advice' is still sent for all sales.

The changed process is shown in Figure 10.2(c), and its process description is as follows:

```
BEGIN
    Receive 'sale report'.
    Get SALES record for PART-NO in 'sale report'.
    TOTAL-QTY = TOTAL-QTY + QTY-SOLD.
SALE-VALUE = QTY-SOLD − UNIT-PRICE.
TOTAL-VALUE = TOTAL-VALUE + SALE-VALUE.
Write SALES record.
Send 'summary advice'.
CASE TYPE OF
    'account':
            BEGIN
                Send 'account posting'.
                IF SALE-VALUE > 500.00.
                    THEN
                        BEGIN
                            REBATE-AMOUNT = SALE-VALUE* .02.
                            send 'rebate advice'.
                        END
            END

    'cash', 'check':
        BEGIN
            Send 'receipt'.
        END
    'credit':
        BEGIN
            INVOICE-AMOUNT = SALE-VALUE + SALE-VALUE * .01.
            send 'invoice data'.
        END
END
```

Note that each sentence sequence for a CASE outcome is enclosed by a BEGIN and END statement.

REPETITION

Figure 10.4 shows two ways of specifying iterations in structured English. One way is to use the WHILE . . . DO structure. Here a condition is tested before a set of sentences is processed. The other way uses the REPEAT . . . UNTIL structure. Here the group of sentences is executed first, and then the condition is tested.

As an example, consider process 'finalize order' in Figure 10.5. This process computes the value of an order for parts. The 'preliminary order' is identified by

(a) Using WHILE

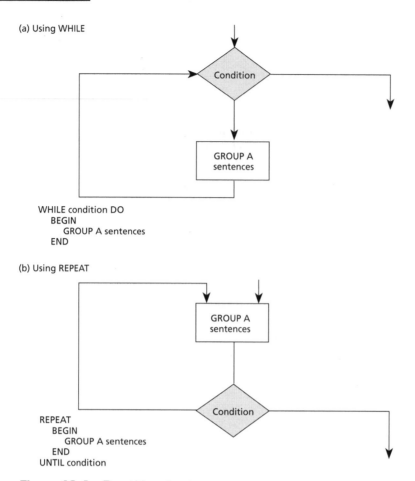

```
WHILE condition DO
    BEGIN
        GROUP A sentences
    END
```

(b) Using REPEAT

```
REPEAT
    BEGIN
        GROUP A sentences
    END
UNTIL condition
```

Figure 10.4 *Repetition structures*

ORDER-NO and is made up of a number of lines. Each line contains PART-NO to specify a needed part and QTY-NEEDED to specify the quantity of that part ordered. The process looks up the price of each part in file PRICES and computes the total value of the order. The process is specified as:

```
Get 'preliminary order'.
ORDER-VALUE = 0.
WHILE there are more order lines DO
    BEGIN
        Get next 'order line'.
        Get PRICES record for PART-NO in 'order line'.
        PART-VALUE = QTY-NEEDED * PRICE.
        ORDER-VALUE = ORDER-VALUE + PART-VALUE.
        Create order line in 'finalized order'.
    END
Send 'finalized order'.
```

Figure 10.5 *Finalizing orders*

An alternative to the WHILE ... DO structure is to use the FOR structure. The same process is now specified as:

Get 'preliminary order'.
ORDER-VALUE = 0.
FOR each order line in an order DO
    BEGIN
        Get next 'order line'.
        Get PRICES record for PART-NO in 'order line'.
        PART-VALUE = QTY-NEEDED * PRICE.
        ORDER-VALUE = ORDER-VALUE + PART-VALUE.
        Create order line in 'finalized order'.
    END
Send 'finalized order'.

A small change is made to process 'finalize order' illustrating the REPEAT ... UNTIL construct. The amended process is shown in Figure 10.6. Each order now has an ORDER-LIMIT. Order values must fall below the limit. Process 'finalize order' is now specified as:

Get 'preliminary order'.
ORDER-VALUE = 0
REPEAT
    BEGIN
        Get next 'order line'.
        Get PRICES record for PART-NO in 'order line'.
        PART-VALUE = QTY-NEEDED * PRICE.
        ORDER-VALUE = ORDER-VALUE + PART-VALUE.
        IF ORDER-VALUE < ORDER-LIMIT
            THEN create order line in 'finalized order'.
    END
UNTIL (ORDER-VALUE > ORDER-LIMIT) or
        (there are no more order lines).
Send 'finalized order'.

Figure 10.6 *Finalizing values with order limit*

Order lines are now priced only until the accumulated total for the value of the order is below the value of ORDER-LIMIT. Once this limit is exceeded, no more lines are added to 'finalized order'.

SOME COMMENTS ON USING STRUCTURED ENGLISH

Those familiar with programming (especially with using block structured languages) may ask whether we can use structured English to nest decisions within iterations and iterations within decisions, and have a number of levels of nesting. This can be done but should be avoided as much as possible. Remember that what we are trying to achieve are process specifications that can be readily understood by those reading them. Complex levels of nesting should not be necessary, because we try (through leveling) to reduce processes to their most fundamental function. The guideline is that the structured description of a procedure should not exceed one page. If it does, you should try to level the process down further.

DECISION TABLES

Although structured English can describe most processes, such process descriptions can become clumsy if we try to specify a process that selects one of a possible set of actions using a set of complex rules. This particularly applies if it is necessary to repeat a decision or to do the same process more than once to retain the logic structure. For example, consider Figure 10.7. The structured English specification in Figure 10.7 specifies two tests (one for type of account and one for type of transaction) and different combinations of the same actions (additions to or subtraction from accounts and counters), depending on the tests. You will find that either the tests or the actions must be repeated in the specification.

A better way of describing such logic is to use decision tables. A decision table for this process is illustrated in Table 10.1.

A decision table is first divided into two parts: the conditions and the actions. The conditions part states all the conditions that are applied to the data. The actions are the various actions that can be taken depending on the conditions. The table is constructed with columns, so that each column corresponds to one combination of conditions.

The entry in the column indicates the existing condition. In Table 10.1, condition 'account type' can take two values: P for 'private' and T for 'trade'.

```
IF a trade account
THEN
        BEGIN
          IF withdrawal
              THEN
                  BEGIN
                      make an account deduction
                      add 1 to trade withdrawal
                  END
              ELSE (a deposit)
                  BEGIN
                      make an account addition
                      add 1 to trade deposits
                  END
        END
  ELSE (a personal account)
        BEGIN
          IF withdrawal
              THEN
                  BEGIN
                      make an account deduction
                      add 1 to personal withdrawal
                  END
              ELSE (a deposit)
                  BEGIN
                      make an account addition
                      add 1 to personal deposits
                  END
        END
```

Figure 10.7 *Multiple conditions described by structured English*

Condition 'activity type' can also take two values: W for withdrawal and D for deposit. In many decision tables there are only two values for a condition (i.e. T for true or F for false). It is possible for conditions to take more than two values. Condition 'activity type' could, for example, take the additional value B to get the balance of an account. Decision tables where conditions take more than two values are sometimes called *extended decision tables*.

Table 10.1 *Multiple conditions with decision tables*

		P	P	T	T
Conditions	Account type				
	Activity type	W	D	W	D
Actions	Add to account		X		X
	Subtract from account	X		X	
	Add 1 to trade withdrawal			X	
	Add 1 to personal withdrawal	X			
	Add 1 to trade deposits				X
	Add 1 to personal deposits		X		

Key:
P = Personal
T = Trade
W = Withdrawal
D = Deposit

The actions taken for the combination of conditions in the column are given by the crosses in the column. If the action line is crossed, then that action is taken, given the set of column conditions. For example, if 'account type' is P and 'activity type' is W, then two actions follow: 'subtract from account' and 'add 1 to personal withdrawal'.

SOME ISSUES IN PROCESS DESCRIPTION

The two most important issues in process description in structured systems analysis are when to abandon graphical modeling and begin scripting, and what scripting method to use. Only guidelines can be provided here. One is to use graphical techniques as long as possible, as these tend to give a total picture of a system, but not to use them to extremes to describe detailed computations. Thus, as a guideline, any graphical object should be represented by a fairly substantial script, otherwise it is a graphical script.

DESCRIBING DATA STRUCTURES

Process descriptions use data and require a formal description of data structure. These include descriptions of both the data flows and the data stores, as well as the detailed data elements and structures that make up these flows or stores. These descriptions can then be referenced by the DFD or the E–R diagram. Figure 10.8 illustrates typical data structure description.

DESCRIBING DATA ELEMENTS

There is one entry for each data element. One such data element entry is illustrated in Figure 10.8. This entry describes the data element PRODUCT-CODE. The entry describes any aliases or alternative names for the data element. For example, PRODUCT-NO can be used to mean the same thing as PRODUCT-CODE. The entry also includes the data element description, which includes the kind of values the element can take and the range of these values.

DESCRIBING STRUCTURES

Data structures are combinations of data elements that appear in various parts of the DFD. Most data descriptions describe structures by hierarchies. For example, the data structure described in Figure 10.8 is the INVOICE. The structure INVOICE is made up of a number of lower-level structures, namely INVOICE-HEADING, INVOICE-LINE, and SUPPLIER-DATA. INVOICE-HEADING is made up of two data elements; SUPPLIER and ORDER-NO.

Similarly, INVOICE-LINE is made up of three data elements, PRODUCT-CODE, QTY, and PRICE. The asterisk next to INVOICE-LINE indicates that this structure may be repeated any number of times in INVOICE.

Data element: PRODUCT-CODE
Alias: PRODUCT-NO
Description: A five-character code. The first two characters are alphabetic to
        indicate class. The last two characters are a number within the class.

Where used: INVOICE

Data structure: INVOICE
INVOICE-HEADING
   SUPPLIER
   ORDER-NO
INVOICE-LINE*
   PRODUCT-CODE
   QTY
   PRICE
SUPPLIER-DATA

Description: Standard format constructed from supplier invoice

Where used:
Data stores:
Data flows:
   SUPPLIER-INVOICES

Data flow: SUPPLIER-INVOICE

Source:
   External entity suppliers or process:
Destination:
   External entity or process: 1.3.1

Data structure: Depends on supplier

Volume: 10/day

Physical description: Paper invoices, format depends on supplier

---

Data store: INVOICES

Contents: INVOICE + DATE-RECEIVED
               + DATE-PAID

Processes used by:
3.7   STORE INVOICE
7.2.1   RECORD PAYMENT

Physical description: Computer file

Size: average of 20,000 records

Figure 10.8 *Data entries for structured systems analysis*

The illustration in Figure 10.8 uses a hierarchical description of structures. A hierarchy is described by listing the data elements that make up a structure as a hierarchy. For example:

ORDER
   SUPPLIER-NO
   DATE-ORDERED
   DATE-REQUIRED
   ORDER-NO

Here ORDER is the structure name. The order is made up of four data elements: SUPPLIER-NO, DATE-ORDERED, DATE-REQUIRED, and ORDER-NO. The indentation of the item names under ORDER implies that they are components of ORDER. Of course, most data structures are more complex than the ORDER structure shown above. Item values may be repeated, and there may be optional or alternative values, and structures within structures.

An example showing structures within structures and repetition follows:

```
ORDER
    SUPPLIER-NO
    DATE-ORDERED
    DATE-REQUIRED
    ORDER-NO
    ITEMS-ORDERED*
        ITEM-NO
        QTY-ORDERED
```

Now ORDER contains the structure ITEMS-ORDERED. This structure is made up of two data elements. ITEM-NO specifies the item ordered, and QTY-ORDERED specifies the quantity of the ordered item. Because an order can contain many items, the structure ITEMS-ORDERED can be repeated many times. The asterisk after the structure name indicates that a structure or a data element can be repeated.

The notation also allows specification of alternative or optional items in a structure. This is done as follows:

```
ORDER
   { SUPPLIER-NO   }
   { SUPPLIER-NAME }
    DATE-ORDERED
    DATE-REQUIRED
    ORDER-NO
    [ORDER-STATUS]
    ITEMS-ORDERED*
        ITEM-NO
        QTY-ORDERED
```

The braces { } indicate alternative structures. Thus, an order may contain SUPPLIER-NO or SUPPLIER-NAME, but not both. The brackets [ ] indicate an optional component. Thus, an ORDER may or may not contain the data element ORDER-STATUS.

There are, of course, alternative ways of describing structures. One such alternative has been described by DeMarco (1978). In this notation, structure components are described by using the plus sign (+) instead of indentations. Thus, ORDER is now described as:

ORDER = SUPPLIER-NO + DATE-ORDERED + DATE-REQUIRED
    + ORDER-NO

Repetition is described by placing the repeating structure within brackets; alternative structures are defined by square braces and optional structures by parentheses. Thus, the previous example, using DeMarco's notation, becomes:

$$\text{ORDER} = \left[\begin{array}{l} \text{SUPPLIER-NO} \\ \text{SUPPLIER-NAME} \end{array} \right] + \text{DATE-ORDERED}$$
$$+ \text{DATE-REQUIRED} + \text{ORDER-NO}$$
$$+ (\text{ORDER-STATUS}) + \{\text{ITEM-NO}$$
$$+ \text{QTY-ORDERED}\}$$

Note that structures within structures do not have to be explicitly named using this notation. The structure {ITEM-NO + QTY-ORDERED} was given the name ITEMS-ORDERED using the hierarchical notation, but no such name is needed using DeMarco's notation.

USING CASE TOOLS FOR DOCUMENTATION

Documentation, especially in large systems, can be a time-consuming process. Not only is it necessary to create the initial documents, it is also necessary to change them as requirements change. Such changes can be almost impossible to maintain, especially in the presence of tight delivery requirements. Most methodologies are now supported by CASE tools to help documentation. There are a large number of CASE tools available on the market and information on many of them can be found on the WWW. As an example we briefly describe a suite of tools provided by the Oracle corporation.

Oracle, a well-known vendor of database products, supports them with a suite of tools. The major tools are Designer/2000 and Developer/2000. The architecture of the Oracle system centers around a repository that stores models and is broadly shown in Figure 10.9. Many more details can be found from the Oracle Worldwide Web site at http://www.oracle.com.

The tools provided by Oracle are broadly divided into two major suites: one known as Design/2000 that contains modeling tools, and another known as Developer/2000 that contains a set of development tools. For more detail, you should refer to the Oracle Worldwide Web site. In summary, Designer/2000 begins with tools for business process re-engineering for redesigning business processes. For analysis, it supports entity relationship modeling, dataflows, functional hierarchy and matrix modeling techniques. System design supports the generation of design models from the analysis model that includes data structure diagrams and relational tables. It also supports the generation of module structures. CASE support is continually changing and readers are advised to visit the Oracle site at http://www.oracle.com for the latest information.

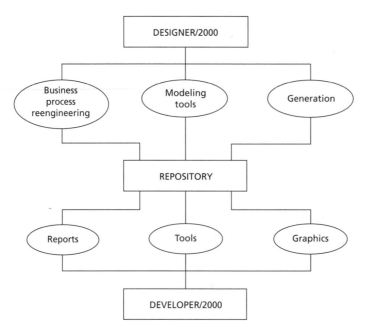

Figure 10.9 *Oracle CASE tools*

SUMMARY

This chapter describes methods used for process descriptions. It begins by describing why precise process specifications are needed and continues by presenting some of the main description methods used in structured systems analysis.

The first method is structured English. This is a mixture of natural language prose and key words used to define logic. Structured English thus tries to get the best of two worlds: program logic to define precision and natural language to get the convenience of the spoken word.

The chapter describes another method that is often used to describe processes: decision tables. Decision tables are used where a large number of conditions must be tested to select one of many possible actions.

DISCUSSION QUESTIONS

10.1   Why is natural language not useful as a system-level specification?

10.2   What is the difference between scripts at the usage and system levels?

10.3   Would structured English be useful in describing informal interactions between people?

EXERCISES

10.1   Use structured English to describe the following system. The system receives a batch of transactions, each of which has a key value.

The system examines each transaction in the batch and, depending on the type, does the following:

- For Type X transactions, the system stores the transaction as a Type X record, but only if there is a Type Y record with the same key value. Otherwise, the system writes out an error line. If there is a Type W record with the same key value as the Type X transaction, the Type W record is converted to a Type Z record.

- For Type Y transactions, the system stores the transaction as a record of Type Y in the main file.

- For Type Z transactions, the system stores the transaction Type Z record, but only if there is a pair of Type X and Type Y records with the same key value as the Type Z transaction. If there is no such pair, then the Type Z transaction is stored as a Type W record, but only if its date is before 1982. If none of these conditions hold, then an error line is output.

The system ensures that no two records of the same type have the same key value.

10.2 Develop a decision tree and a decision table for the following:

The gatekeeper at an amusement park is given the following instructions for admitting persons to the park:

- If the person is under three years of age, there is no admission fee.

- If a person is under 16, half the full admission is charged. If the person is accompanied by an adult, the admission is reduced to a quarter of full admission (the reduction applies only if the person is under 12).

- Between 16 and 18, half the full admission fee is charged if the person is a student; otherwise the full admission is charged.

- Over 18, the full admission fee is charged.

- A discount of 10% is allowed for a person over 16 if they are in a group of 10 or more.

- There are no student concessions during weekends. On weekdays, under-12s get one free ride.

10.3 Convert the DFD in Figure 10.8 into structured English. You should note that, dependent on transaction type, one or more of the four processes are applied to the transaction. Thus, for one transaction we need only determine the material; for another, the personnel and computing cost; and for another still, some other combination of processes. Do you feel that structured English is a better way of describing this process than the DFD in Figure 10.8? Alternatively, do you think there is a better way to represent the system in Figure 10.8?

10.4 An invoice clerk receives invoices from suppliers. Each invoice contains information on:

- an order and supplier number

- items delivered

- quantity of each item delivered

- price of each item

- total invoice amount.

The invoice clerk examines the invoice and compares it with both the order and a stock report. The stock report contains data on the goods received in the organization's store from various suppliers. This data includes the order number and the supplier who delivered the items.

If the items on the order, the invoice, and the stock report match, then the invoice clerk checks the total invoice amount. If the amount is correct, the invoice clerk sends an authority to the accounts department to issue a check for the invoice. If the amount is incorrect, the invoice clerk adjusts the invoice and authorizes the accounts department to issue a check for the adjusted amount. At the same time, the invoice clerk prepares and dispatches a vendor memo advising of the adjustment.

If the items on the invoice do not match the stock report but do match the order, the invoice clerk first checks the correctness of the report. If it is correct, the invoice clerk first makes an adjustment to the invoice amount, authorizes accounting to prepare a check for the adjusted amount, and prepares a vendor memo advising of the adjustment. At the same time, a stock memo is sent advising of further items to be received against the order and issuing a supplementary order number to both the store and the supplier.

1. Prepare a decision table to illustrate the activities of the invoice clerk.

2. Is any further information required to completely describe all possible situations that the invoice clerk may meet? If not, what other information would you need to completely describe the activities of the invoice clerk?

BIBLIOGRAPHY

DeMarco, T. (1978), *Structual Analysis and System Specification*, Yourdon Press, New York.

McDaniel, H. (1978), *An Introduction to Decision Logic Tables*, Perocelli Books, Princeton, New Jersey.

Pollack, S.L., Hick, H.T. Jr. and Harrison, W.F. (1971), *Decision Tables: Theory and Practice*, Wiley-Interscience, New York.

Schank, R.C. and Abelson, R.P. (1977), *Scripts, Plans, Goals and Understanding*, Lawrence Erlbaum Associates, Hillsdale, New Jersey.

OBJECT MODELING

CONTENTS

KEY LEARNING OBJECTIVES

The object-oriented paradigm

Object classes and class diagrams

Relationships between objects

Use cases

Methods to model behavior

Sequence diagrams

Transition diagrams

CHAPTER 11

INTRODUCTION

This and the next chapter describe object-oriented (OO) methods and their use in analysis. Object-oriented methods differ from structured systems analysis and design because they result in models that integrate data and processes. Thus, rather than modeling process flows by data flows, data by E–R diagrams, and processes by process descriptions, all of these three components are integrated together into the one object analysis model.

This chapter first introduces the object-oriented methods and their notations and how object models are used to represent systems at the system level. It outlines some of the characteristics of object structures—encapsulation of data and methods—and explains how these characteristics can be used to advantage in system development. The chapter then continues by describing how behavior can be modeled.

A growing number of software products directly support object models at the implementation level. Such systems are either object-oriented languages or object-oriented database management systems. They provide another advantage for using the object-oriented approach by making it possible to directly, or seamlessly, convert a system model to an implementation model.

THE OBJECT ENVIRONMENT

Object modeling combines processes, data, and flows into the one modeling paradigm, thus allowing objects to be modeled as independent entities and flexibly combined into cooperating systems.

This idea is illustrated by the simple client–server example shown in Figure 11.1. Here we have two objects, a client and a server. These objects communicate with each other through messages. The client object can send a message that requests a service from the server object. A message activates a process, or method, in the receiving object. Often the message has the same name as the method. The message causes the execution of the method program, which carries out any processing required by the message, and returns the response. A server object can provide a service such as computing a statistical average or looking up a file. It is also possible to model other kinds of communication between objects, such as one object issuing a command (e.g. update a database) to another object.

This flexibility is made possible by the **encapsulation** of data and processes into the one object. Processes are often called **methods**, which are implemented as

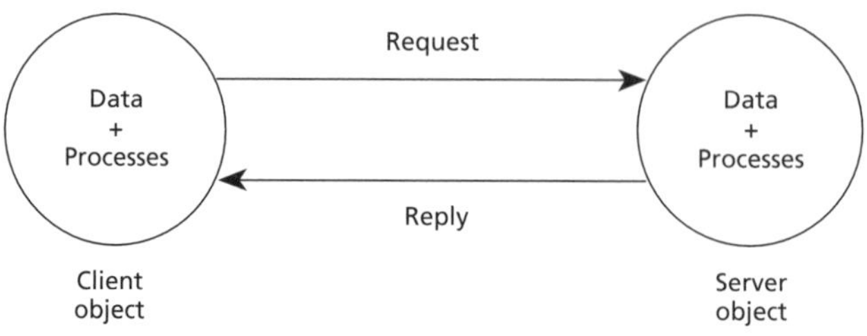

Encapsulation Inclusion of many features in the one object.

Methods A feature that describes programs within an object.

Figure 11.1 *Objects*

programs within the object. Such encapsulation means that it is not necessary to think in terms of building one large system. Instead, we identify objects as independent entities with their local goals. Such objects can then exchange messages between themselves to achieve a global goal of the large system.

The encapsulation of data and processes means that objects can be designed independently of each other, with (usually) only their interfaces specified. Interface specifications are the messages that can be accepted by each object. Using the same paradigm to represent these two features eliminates some of the disadvantages of storing data, processes, and flows separately. Thus, the analyst need no longer maintain separate models for data and processes but can include everything in one system model. It is not necessary, therefore, to develop and validate links between models based on these different techniques. Greater autonomy between objects also increases the flexibility in building and changing systems. Thus, we can model the system as a collection of objects that communicate with each other. Changes can often be localized to a particular object or to the links between them.

OBJECT STRUCTURE

We often talk about objects as having features. Objects can have other features in addition to properties and methods. Object-oriented modeling does not have a standard set of features, and objects can have as many different features as we like. Thus, we can have features such as:

- object states that can initiate methods
- constraints maintained between objects and object properties
- checks on pre- and post-conditions to be satisfied for messages and replies
- triggers that activate messages for given data conditions.

The question is to select those features that allow autonomous objects to be modeled in the most natural yet simple way. There is no agreement on what these features should be, although there is recognition that properties and methods must be included in any model. Each programming language or implementation, however, will support only some of the features.

Figure 11.2 shows a simple object that includes only two kinds of features: properties and methods. It also shows that properties themselves can be of different kinds. One kind takes simple values—for example, PROJECT-NO, START-DATE, BUDGET—while another kind can be a reference to another object. Thus, MANAGER in Figure 11.2 does not include a person name but is a reference to the object that represents the project manager. Another example of a reference is PEOPLE-ASSIGNED, where the N indicates that this property can contain a number of references rather than one reference. TASKS are another kind of property that defines objects included in projects. These objects, like dependent entities, cannot exist on their own. The term *composite* is sometimes used to indicate that one object, in this case PROJECTS, contains other objects. So far this is similar to an entity set. Thus, PROJECT-NO, START-DATE, and BUDGET can be seen as entity attributes. MANAGER and PEOPLE-ASSIGNED can be seen as relationships, and TASKS as dependent entities. There is, however, one important

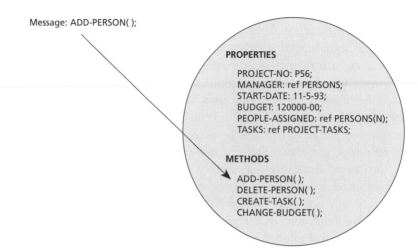

Message: ADD-PERSON( );

PROPERTIES

    PROJECT-NO: P56;
    MANAGER: ref PERSONS;
    START-DATE: 11-5-93;
    BUDGET: 120000-00;
    PEOPLE-ASSIGNED: ref PERSONS(N);
    TASKS: ref PROJECT-TASKS;

METHODS

    ADD-PERSON( );
    DELETE-PERSON( );
    CREATE-TASK( );
    CHANGE-BUDGET( );

Figure 11.2 *An object with encapsulated data and methods*

difference from E–R modeling: the object also contains methods. The methods shown in Figure 11.2 are ADD-PERSON, DELETE-PERSON, CREATE-TASK, and CHANGE-BUDGET. These methods are activated by messages outside the object and can change object properties. Thus, an ADD-PERSON message would add another reference.

Methods are programs that can change attribute values. The object interfaces to its environment through messages, where each message activates a method within the object. Thus, in Figure 11.2, the message 'add-person' would activate method ADD-PERSON, which would add a new person, X, to the project. Thus, both the properties and programs are encapsulated in the one object.

UML (Unified Modeling Language)

Figure 11.2 describes an object using a notation that shows the object as a circle. In the early days of object modeling, its proponents used a variety of techniques and notations to represent object models. Gradually, however, there has been an evolution towards a standard known as UML (Unified Modeling Language). This standard arose through a combination of techniques from the methodologies defined by Rumbaugh, Jacobson, and Gooch. It combines the strengths of these early methodologies to form a standard methodology that supports all development process phases. UML is a comprehensive language, and many users use only subsets of this language. The remainder of this chapter will use a subset of UML and follow it as closely as possible to give you an idea of the UML and, especially, its principles. However, students should consult relevant manuals or specialized text for a complete treatment of UML.

CLASSES AND OBJECTS

One important aspect of object modeling notation is that we represent an *object class* as well as individual objects or *object instances* of that class. Thus, we can have an object class, PROJECTS, and each project is an instance of that class. It is possible in an implementation to store the class, PROJECTS, as a separate object. Individual projects become instances of the class, in much the same way as entities were instances in entity sets. Each project instance is stored as a separate object. Object methods are often stored once in the class object rather than being duplicated in each instance of that class. Thus, all object instances inherit their methods from their class object.

A common way to represent classes and object instances is to use the UML notation. The object instance is identified by a value of one of its attributes, in this case a87, which is underlined. Using this notation, the object class is represented as a square box with subdivisions for each of the object features. Two subdivisions in Figure 11.3 describe the properties and methods. Figure 11.3 also illustrates another possible feature, constraints. Here, the constraint 'ensure' places a limit on the number of people in a project. Although not part of UML, general object principles do not prohibit such extensions to object classes.

The general syntax of the attributes here is:

<attribute-name>: type = <initial value> for object class, or
<attribute-name> = <value> for the object instance.

When N follows an attribute, it means that the attribute can take a number of values in the one object instance. Note that some attributes, known as class-value attributes, take a class as their value. This means that the value of this attribute is an instance of that class. Thus, PEOPLE-ASSIGNED is a multi-valued class-value attribute. Its value is the task instance, in this case 'task20' and 'task31'. The general syntax used for methods is:

<method-name>(<input parameter list>) : <return parameter list>

where each item in the list has the general syntax of the attribute. The parameter lists are often omitted, especially in early analysis steps.

OBJECT CLASS DIAGRAMS

A system usually has a large number of object classes. A representation that shows all these classes and any relationships between them is known as the **object class diagram**. UML defines a number of relationships between object classes, but only three are illustrated in Figure 11.4. These are association, containment, and

Object class diagram A diagram that shows object classes and relationships between them.

CLASS NAME	PROJECTS	a87 : PROJECT
PROPERTIES	PROJECT-ID: string; MANAGER: string; START-DATE: date; BUDGET: real; PEOPLE-ASSIGNED(N): PERSONS; TASKS(N): PROJECT-TASKS;	PROJECT-NO: a87; MANAGER: Melissa; START-DATE: 10-7-98; BUDGET: 45000.00; PEOPLE-ASSIGNED(N): {Mary, Lin}; TASKS(N): {task20, task31};
METHODS	Add-project(project-no); Delete-project(project-no); Create-task(task-name, task-details); Change-budget(new-budget);	Add-project(project-no); Delete-project(project-no); Create-task(task-name, task-details); Change-budget(new-budget);
	Object Class	Object Instance

Figure 11.3 *A notation for object modeling*

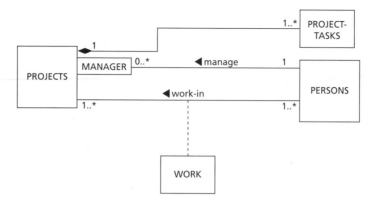

Figure 11.4 *Object class diagram*

inheritance. Just for comparison, these are closely related to relationships, subsets, and dependent entities in E–R diagrams.

ASSOCIATIONS

The most common relationship between object classes is the association relationship, which is represented by a line between the classes. For example, in the object class diagram in Figure 11.4, there is an association between PERSONS and PROJECTS. The line is labeled by the name of the association, in this case 'manage'. The arrow next to the line indicates the direction of reading of the name. Here the direction is right to left; that is, PERSONS manage PROJECTS.

An association is usually based on object class properties. This property is defined by the qualifier, which is one of the object attributes. In Figure 11.4, 'MANAGER' is the qualifier of the 'manage' association. This means that the value of MANAGER in an instance of object class PROJECTS determines a unique instance of an associated PERSON, who of course appears as a person in the PERSONS class.

The notation in Figure 11.4 also shows cardinality. Thus, '1' next to PERSONS on the 'manage' association indicates that the manager can be only one person. The '0..*' next to MANAGER indicates that a person can manage between 0 and many projects. The '1' next to PROJECTS on the PROJECTS–PROJECT-TASKS link says that each project-task is associated with one project. The '1..*' next to PROJECT-TASKS states that a project must have at least one task but can have many.

Figure 11.4 also shows an association class, WORK. Note that the 'work-in' association is such that a person can work in many projects, and a project can have many people assigned to it. Often such associations carry information; for example, number of hours that a person worked on a project. The information in such a relationship is stored in an association class. Thus, WORK is an association class. It has a dotted line to the 'work-in' association to indicate that it is an association class.

COMPOSITION

Figure 11.4 also shows another important relationship, composition, or known earlier as containment and often referred to as aggregation. Here one object can be composed of objects from other objects. Thus, in Figure 11.4, the PROJECTS object is composed of a number of PROJECT-TASKS objects. A composition relationship is also represented by a line from a property to an object class with a diamond at the end of the line. Association relationships are often temporary, whereas composition relationships are permanent. Thus, project managers can change. However, once a task is allocated to a project, it stays with the project—just like a dependent entity in E–R diagrams.

DETAILED OBJECT CLASS DIAGRAM

Figure 11.4 illustrates an object class diagram that shows only the object classes and relationships between them. Another representation is a detailed object class diagram, which also shows all the object features (Figure 11.5).

INHERITANCE

Inheritance is very similar to the idea of subsets in E–R models. An object can inherit features from another object. It can also have some additional features or, if needed, replace some of the object features. Figure 11.6 illustrates a very simple inheritance structure. It shows both the object class diagram and the detailed object class diagram. The object class diagram shows that there are two kinds of TASKS: one carried out locally and another that is outsourced. Both these kinds of tasks inherit the features of TASKS and also have additional properties and methods. For example, both OUTSOURCED-TASKS will have all the properties of TASKS and two additional properties, ORGANIZATION and CONTRACT-NO. It also has

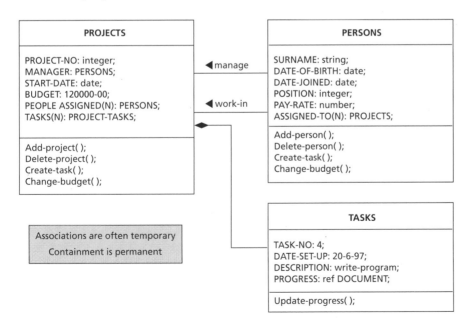

Figure 11.5 *A detailed object class diagram*

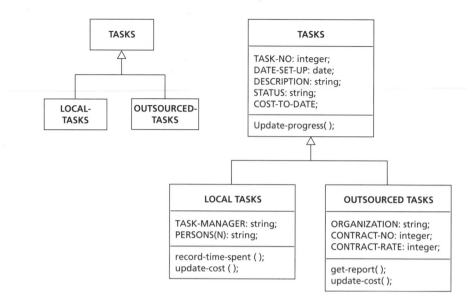

Figure 11.6 *Inheritance*

the additional method 'get-report'. The term *specialization* often comes up in discussions of inheritance. We can say that LOCAL TASKS and OUTSOURCED TASKS are a specialization of TASKS. They inherit all the features of TASKS but are also specialized with additional features. The notation for the inheritance association is the arrow pointing from the specialized classes to the parent class.

Inheritance is important because it gives designers the ability to specialize objects while at the same time not requiring a change to calling messages. We may, for example, decide to set up yet another kind of task, say, a CONTRACTED-TASK, where a special contractor is hired to complete it. This could be added as yet another specialized object of TASKS.

POLYMORPHISM

Polymorphism
Selecting the method appropriate for the class of object called.

Polymorphism is closely related to inheritance and means the ability of the one construct to take many forms. In object-oriented programming it particularly applies to the ability of a message to change its effect depending on the instance of the object called. Thus, a message 'update-cost' addressed to TASKS will select the appropriate method depending on the type of offer being considered. Furthermore, adding a new specialized object with a new specification of 'compute-value' will mean that the message 'compute-value' will select that method if addressed to an instance of the new class.

MULTIPLE INHERITANCE

The subject of inheritance is crucial to object orientation. What we have shown so far is objects inheriting their properties from only one other object. It is also possible to have structures where an object inherits properties from more than one other object. This is generally known as *multiple inheritance*.

Figure 11.7 shows an example of multiple inheritance. Here we make SALES,

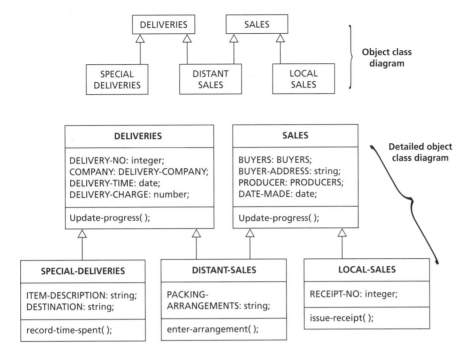

Figure 11.7 *Multiple inheritance*

some of which are over-the-counter sales, others of which require delivery to the buyer. These two kinds of sales are modeled as specializations of SALES. They have all the features of SALES and additional features. Thus, LOCAL-SALES has a receipt, as receipts are issued at the time of sale. DISTANT-SALES inherits the properties of SALES and the properties of another object, called DELIVERIES, because every distant sale results in a delivery. A DISTANT-SALE is a union of all the properties of SALES and DELIVERIES. It inherits properties concerned with selling from SALES and properties of delivering from DELIVERIES. Thus, processes such as making a transportation arrangement come from DELIVERIES, whereas arranging a sale comes from SALES.

DISTANT-SALES thus illustrates the idea of multiple inheritance—that is, inheritance of features from more than one object. It would, of course, be possible to include the properties and methods concerned with transportation in DISTANT-SALES. However, DELIVERIES is modeled as a separate object, because there is more than one kind of delivery in the system. It is also possible to arrange deliveries of special items that are not part of the sales system but which use the same process as deliveries made through the sales system. Hence, all features common to both these kinds of deliveries are abstracted into the object class DELIVERIES.

Multiple inheritance introduces a problem when both parents have a feature with the same name. The question is which of these two features to inherit. This is usually specified to be the sequence in which the inheritance links are defined.

MODELING BEHAVIOR

Behavior is described in different ways in different worlds. For example, in the implementation world, behavior is described as a program. This program precisely defines the steps to be followed in a given computation. However, when we begin modeling, it is often best to start by describing what happens in the usage world. In many object-oriented methodologies this is done through use cases. A use case describes the steps that people follow in carrying out a system activity. In fact, the first thing we often do is to identify the major activities and show them in a *use case model*.

USE CASE MODEL

As an example of a use case model, let's look at automatic teller machines (ATMs). A number of activities take place through an ATM. These include withdrawing cash, transferring funds between accounts, depositing funds, and getting reports. We show all of these activities by a use case model as shown in Figure 11.8. Here the system is the rectangular box, and each oval shape in this box represents an activity. Actors also take part in each activity. Actors are shown outside the box with links to those activities in which they take part. Each activity in the system is described by a use case.

A use case model shows all the use cases and the actors in each use case. Each use case has two major components: a script and actors. The script describes the steps followed in the activity in detail. The actors describe what active elements are included in the use case. The actors can be users themselves, like customers, or they can be other systems, like a bank system.

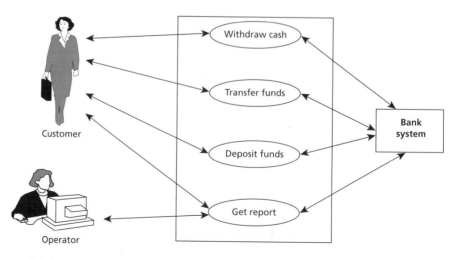

Figure 11.8 *A use case model*

USE CASES

The use case script is usually quite informal and uses a set of statements to describe the steps followed in an activity. It describes the usual set of events that take place in the activity. A use case for cash withdrawal would then look like the following set of statements:

> **Use case—Using the ATM**
> A greeting message is waiting on the ATM.
> The customer inserts their card into the machine.
> The ATM asks the customer for their PIN.
> The customer enters their PIN.
> The ATM displays the customer's accounts and requests a selection.
> The customer selects an account.
> The machine displays the account limits.
> The customer keys in the withdrawal amount.
> The bills are dispensed.
> The card is ejected.
> The receipt is printed out.

This use case describes the most straightforward set of steps in withdrawing cash from an ATM. Now ask yourself what else can happen that can lead to variations of these steps. What alternatives are there to the usual set of events?

Examples here include:

- What if the card is not accepted?
- What if the customer enters an incorrect PIN?
- What if the withdrawal is too large?
- What if the customer stops the process and withdraws the card?

You must now write a use case for each of these.

The idea of doing this—that is, writing only simple use cases—is to separate a complex system into as many simple cases as possible. Each simple case describes a flow of events without any loops or decisions. The complete set of cases and their alternatives describe all the possibilities. This seems to be a lengthy process, but it ensures that all possibilities are precisely defined. Later we will describe how all of these separate simple cases are brought together to build a system that includes all possibilities.

USE CASE RELATIONSHIPS

Sometimes when we are modeling large systems we find that there are many use cases and that many of them include common elements. Some methods provide ways of modeling such common elements. This is done by *use case relationships*. There are two common use case relationships: the uses relationship and the extends relationship.

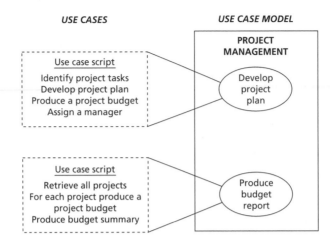

Figure 11.9 *Use case model*

USES RELATIONSHIP

Suppose we have the use case model shown in Figure 11.9. This shows a use case model with two use cases and a very broad script for each use case. When you examine the two scripts, you will find that they both contain a similar activity, 'Produce project budget'. What we can do is create a separate use case for this common activity.

This common activity is now shown as a separate use case on a modified use case model, shown in Figure 11.10. Here, the 'produce budget' activity becomes a separate use case, which is used by the other two use cases.

EXTENDS RELATIONSHIP

Suppose now in some use cases we wish to produce a more detailed project budget, one that classifies expenditure by kind of expense. The idea of extension is then used. It is shown in Figure 11.11.

Let us now look at a more complex case.

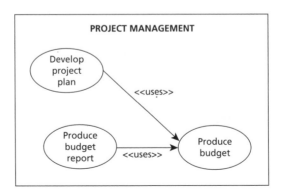

Figure 11.10 *Modified use case model*

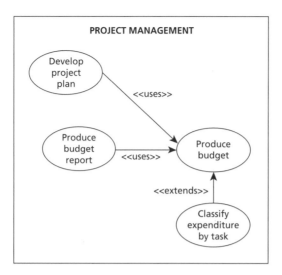

Figure 11.11 *Extending with a specialized action*

TEXT CASE A:
Interactive Marketing—A Use Case Model

The use case model for interactive marketing is shown in Figure 11.12. It includes a number of actors. First of all there are the buyers and sellers. There is a facilitator to make sure the system is working properly. There are also transporters, who pick up sold items from sellers and deliver them to buyers.

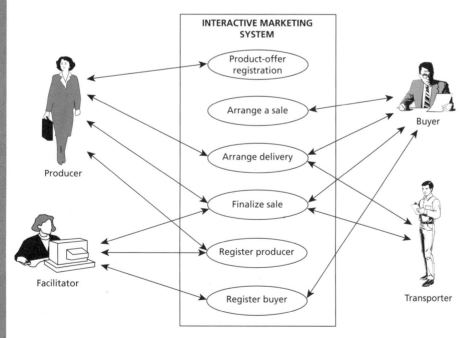

Figure 11.12 *Use case model for interactive marketing*

Each use case is now described by a script. Some of these are described below. The first use cases concern the registration of buyers and sellers. The use case describing this is as follows:

USE CASE—REGISTER BUYER
1. A buyer calls the facilitator by phone.
2. The facilitator asks the buyer for their details, address, and so on.
3. The facilitator registers the buyer in a buyer file.

USE CASE—REGISTER PRODUCER
1. A producer calls the facilitator by phone.
2. The facilitator asks the seller for their details, address, and so on.
3. The facilitator registers the seller in a buyer file.

The next important user case is where a seller wants to offer a product for sale through the system. This is described as follows:

USE CASE—PRODUCT-OFFER REGISTRATION
1. A producer enters a new offer for an existing product into the system. This includes asking-price and quantity-available.
2. The offer is given a unique OFFER-NO and recorded in the system.
3. The producer is notified that the offer has been recorded and given an offer-no.
Exception: the product is not recorded as a standard product.
1. The facilitator is informed that a new product is being offered.
2. The facilitator makes an entry for a new standard product.

The next important use case describes how buyers purchase products. A buyer first examines whether a suitable product can be found. A sale is arranged when the buyer finds a suitable offer.

USE CASE—ARRANGE A SALE
1. A buyer requests a stated amount of a selected product.
2. The system searches product-offers to find those that satisfy the request and displays these offers to the buyer.
3. The buyer selects one or more offers.
4. The selection is recorded against the offers as a commitment.
5. The system creates a sales record and records these quantities as sold on a sales record.

Once the sale is arranged, the facilitator arranges for a transporter to pick up the sold items from the producer and deliver them to the buyer. Then the facilitator finalizes the sales by issuing invoices. You should note that the system is not responsible for ensuring that all the parties are paid. This is left for the parties to arrange themselves. They have all been issued with appropriate invoices and make payments on these invoices.

USE CASE—ARRANGE DELIVERIES
1. The facilitator gets a sales-record.
2. The facilitator gets a quote from a number of transporters.
3. The facilitator selects a quote and informs the successful transporter of the selection.
4. The facilitator contacts the producer and transporter to get their agreement on the quoted price.
5. When the producer agrees on a time and charge, the facilitator records agreement on the delivery docket.

Once all arrangements are made, the facilitator prepares invoices and distributes them to the parties concerned. A product invoice is sent to the buyer to pay the producer, and a copy is sent to the producer. A delivery invoice is sent to the producer to pay the transporter. The payments happen outside the system.

FINALIZING THE SALE
1. Once the sales record is created and delivery is arranged, the facilitator prepares a product invoice for the buyer.
2. The invoice is sent to the buyer, and a copy is sent to the seller.
3. A delivery invoice for transportation costs is prepared and sent to the seller with a copy to the transporter.

ALTERNATIVE COURSES
For special deliveries, create a special delivery docket.

A complete description would also include all the alternatives and exceptions to each use case.

FROM USE CASE TO OBJECT

Remember that the eventual goal is to produce an object class diagram. The question then becomes what to do next. As shown in Figure 11.13, the whole idea is then to reduce use cases to a set of object classes. To simply go directly from use cases to an object class diagram is often too complex, especially in large systems.

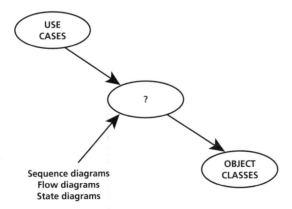

Figure 11.13   *Going to object class diagrams*

For this reason, a number of intermediate modeling techniques are provided to simplify this task. UML contains a large number of such techniques and includes a rich notation. This book does not provide the complete notation, which requires a book of its own, but provides a simplified description to make readers aware of the principles needed to identify and describe object classes. One of these is a sequence diagram.

SEQUENCE DIAGRAMS

Sequence diagram A diagram showing dynamic relationships between objects.

One way to view **sequence diagrams** is as messages exchanged between objects in a use case. Each use case is first carefully examined to determine the objects in the use case. Each object is drawn as a vertical line with the object name next to it. Then messages between objects are shown in their time sequence, going from top to bottom. As an example, the sequence diagram for our ATM is shown in Figure 11.14. The three objects here are the customer, the ATM, and the bank system. The sequence of messages starts when the customer inserts a card. The ATM recognizes the card and sends a message to the customer requesting the PIN. The customer then enters the PIN, which is then checked by the bank system. Accounts are displayed to the customer, who then selects the account and withdrawal amount.

Sequence diagrams often show additional information. In particular, they can show the parameters that are part of each message and that describe the information carried by the message.

Some sequence diagram rules:

- As a general rule, each actor becomes an object in the sequence diagram.
- When drawing a sequence diagram, remember that time moves downward. No decisions are shown. Every alternative in a use case has its own sequence diagram. Thus, a use case can result in many use case diagrams, one for each exception or alternative.

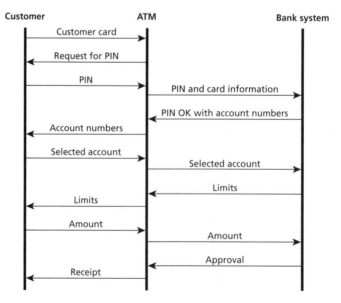

Figure 11.14   *Sequence diagram for the ATM*

- Remember that the lines between the objects are messages and should be given nouns as names. They also include parameters that state the data carried by each message, although this is not the case in Figure 11.15, as the parameters are self-explanatory. For example, the message 'PIN' between Customer and ATM would include a value of the PIN.

USING SEQUENCE DIAGRAMS

Why do we draw sequence diagrams? They are often a good first step to object class diagrams. Figure 11.15 is a preview of how we get object classes and methods. This shows one use case, which describes customers making inquiries about prices of store items. The inquiry is made to a manager, who sends a price-request to the store. When the store-price is received, the manager sends a sales price to the customer. Each message in Figure 11.15 states the message parameters. For example, 'customer' and 'item-name' are the parameters of message 'price-inquiry'. Each of the objects in the sequence diagram eventually becomes an object class. The messages coming in to the object identify the methods in the object class; there must be a method to manage each object class. The parameters carried by the incoming and outgoing messages identify the properties of the object class.

Of course, an object class often has additional methods, sometimes called internal methods. Thus, for example, there may be an internal method to compute any commission when an item is sold, which is added to the item price before passing it on to the customer.

Now, if we go back to Figure 11.15, we can also identify the methods that must be provided by the ATM. The first method is the one that deals with the incoming card. This method checks to see whether the card is a legal card. It simply reads the card and asks for a PIN. Then there must be a method that receives the PIN. This method sends the card and PIN information to the bank system. Another method receives the account numbers from the bank system and displays the

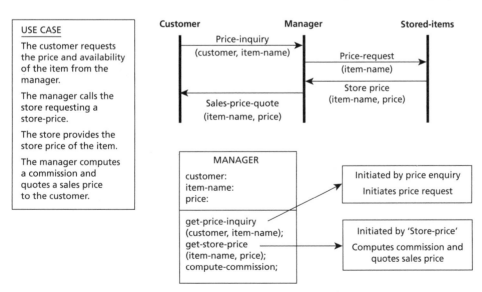

Figure 11.15 *From use cases to object classes*

accounts to the customer. Additional methods then receive the selected account from the customer and pass it to the bank system. Still another receives limits from the bank and passes them to the customer. One more receives the amount from the customer and passes it to the bank system, and another receives approval from the bank system and issues a receipt to the customer.

TRANSITION DIAGRAMS

There is yet another way to model use case dynamics. This is to use transition diagrams. Figure 11.16 shows the transition diagram for the ATM. Here each circle or ellipse represents the state of the object. States can change as the result of messages or events. The change is shown by the arrows, which are labeled by the message that caused the change of state. Thus, normally the ATM is 'idle'. When a customer inserts a card, the ATM responds with a 'request for PIN' and enters the state 'waiting for PIN'. Once this is received, the ATM 'requests a security check' and goes to state 'waiting for authority from bank', and so on.

COMPARING TRANSITION DIAGRAMS AND SEQUENCE DIAGRAMS

A transition diagrams shows all the exceptions on the one diagram, as in Figure 11.17. This is shown by having more than one transition from each state. For example, Figure 11.17 shows two outcomes while the ATM is waiting for a PIN. One is that the customer inputs the number; the other is that the customer removes the card, in which case the ATM goes back to the idle state.

You need as many transition diagrams as there are objects. One question that arises is whether it is better to use sequence diagrams or transition diagrams. One answer is that transition diagrams tend to be better when we are modeling mechanical systems, such as the ATM. Sequence diagrams are often preferred for modeling business processes. Sometimes it may be useful to do both to verify each other and ensure that nothing is missing from the model.

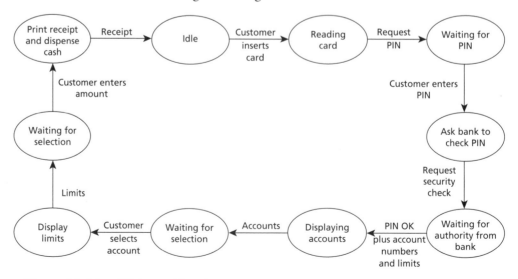

Figure 11.16 *ATM transition diagram*

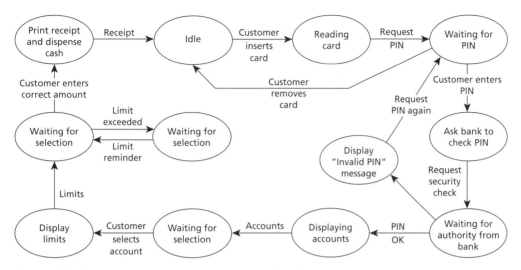

Figure 11.17   *Adding alternatives to a transition diagram*

MANAGING COMPLEXITY

So far we have described ways of drawing sequence and transition diagrams for individual activities. Our goal, however, is to build a system that supports all the activities, their exceptions, and alternatives. The remainder of this chapter describes ways of integrating information gathered about the various use cases and sequence diagrams. One approach is to use event flow diagrams, which originated with many of the early methodologies. Another approach is to use collaboration diagrams, which were developed as part of UML.

EVENT FLOW DIAGRAMS—BRINGING IT ALL TOGETHER

The idea now is to bring all the sequence diagrams together.

The effect of this is that everything to do with one object is now shown in that one object. Before, the same objects could appear in many sequence diagrams. This idea is illustrated by Figure 11.18, which shows two sequence diagrams, each with their objects and messages. They are combined into the one event trace diagram. The collaboration diagrams shown here are simplified and do not show all the features of the collaboration diagrams defined in UML.

The following are important to remember when constructing the event flow diagram:

- All objects in each sequence diagram appear only once in the event flow diagram. Thus, the event flow diagram in Figure 11.19 includes OBJ1, OBJ2, and OBJ3, which appear in each sequence diagram. It also includes OBJ4, which appears only in one sequence diagram.
- All messages are shown in the event flow diagram. The messages are grouped by the direction of flow. Thus, m1 and m5 appear together, as they are all the messages that flow from OBJ1 to OBJ2.

The next step now is to get a first cut at an object class. As shown in Figure 11.20, a routine approach here is for the message arguments to become the object

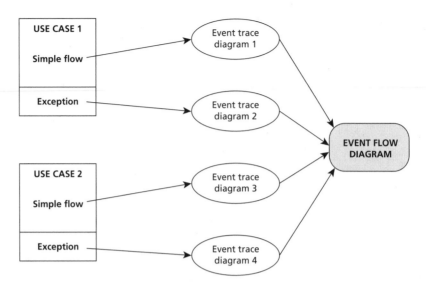

Figure 11.18 *Bringing the use cases together*

class properties and the incoming messages to become the methods. Thus, the object class must have a method to receive an incoming message. This method can then produce an outgoing message. Alternatively, it is possible to add methods to send out messages. These methods would be activated by methods that receive incoming messages.

Figure 11.22 shows an event flow diagram for the ATM. It shows all the messages between the objects. The incoming messages then identify the methods to be provided in the objects, which are also shown in Figure 11.21 for illustrative purposes.

The event flow diagram combines all the sequence diagrams and can be used to provide a first cut at an initial object class diagram. It does not, however, include the sequence of these flows. UML, however, proposes a collaboration diagram that models both the flows between objects and the sequence of these flows.

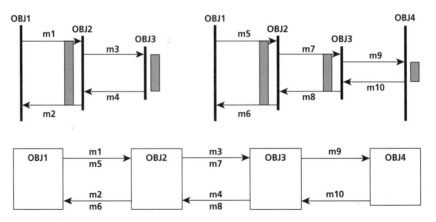

Figure 11.19 *Combining sequence diagrams*

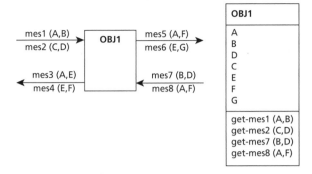

Figure 11.20   *A first cut at designing an object class*

COLLABORATION DIAGRAMS

Another technique used in UML is the collaboration diagram (Figure 11.22). It follows the idea of the event flow diagram but differs in two ways. First, there are usually a number of collaboration diagrams, usually one for each use case. Second, the collaboration diagram also shows the sequence of message flows.

The difference now is that each message has a number. This number defines the sequence of the flows. Each object has a number—say 'n'—and messages originating from the object are numbered 'n.i', where i is the message sequence number. In Figure 11.22, the flow starts with message 1.1 originating from CUSTOMER. The response is message 2.2 from the ATM. The number sequence '1.1/2.1' says that flow 2.1 occurs in response to message 1.1. You can then follow the remainder of the flows by observing this sequence, observing that message 1.2 follows message 2.1 and so on.

OBJECT MODELING IN THE DEVELOPMENT PROCESS

Just like structured systems analysis, object methods also follow a development process. Many of the early object modeling methods concentrated on the representations of data and usually applied to only a subset of the development stages. Thus, for example, Coad and Yordon (1990) concentrated on using objects

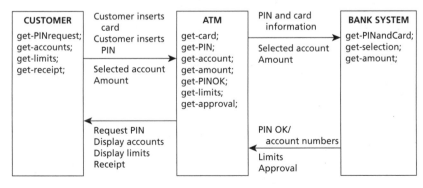

Figure 11.21   *A simplified event flow diagram for the ATM*

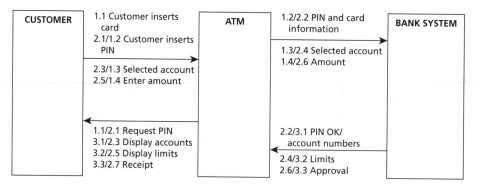

Figure 11.22 *Collaboration diagram for the ATM*

for developing analysis models, whereas Grady Booch (1994) primarily emphasized the design model. A development process requires models at each of the phases. As object-oriented methods matured, development processes like OOSE (Object-Oriented Software Engineering) covered increasingly more phases of the development process.

OBJECT MODELING AND THE LINEAR CYCLE

Figure 11.23 illustrates one way of using object modeling in the development process. It shows the four major stages from Figure 5.4. Concept formation is similar to any other methodology, and object methods usually start to be used from the

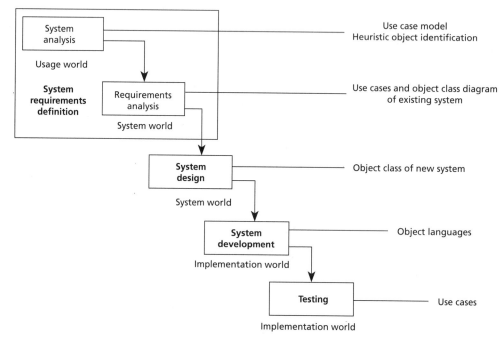

Figure 11.23 *Object orientation in the development cycle*

system requirements definition. During the system specification phase, object methods are used to create a requirements model, which is a collection of a number of use cases, and an analysis model, which describes the system in terms of object classes, very similar to that shown in Figure 11.23. A design model is then developed in terms of object classes, which can then be implemented by using object-oriented languages.

There are, of course, other approaches. For example, rather than using an object-oriented implementation model, conventional systems, including relational databases, can be used in the implementation.

The early object-oriented development phases create an object class diagram. The main activity here is to identify the object classes. Such identification can prove to be quite difficult, as different people can perceive a system differently, leading to their identifying different object classes.

THE OOSE METHOD

The OOSE (Object-Oriented Software Engineering) method was proposed by Jacobson (1992) and later became part of UML. The OOSE method is use case driven and uses the five different models shown in Figure 11.24, each of which has a strong relationship with the use case model.

The five models are as follows:

- The domain object model, which defines the standard terms used in an application.
- The analysis model, which is itself made up of three components, the entity, interface, and control objects.
- The design model, which adopts the object model to the implementation environment.
- The implementation model to implement the system.
- The test model to verify the system.

Each of these models is used at different stages of the development process, and formal conversion rules are provided to go from one stage to the next. Figure 11.25 shows where each of the models is used in the OOSE development cycle described in Chapter 5.

The analysis model that is produced at the time of requirements definition is general in the sense that it can be implemented using either traditional database

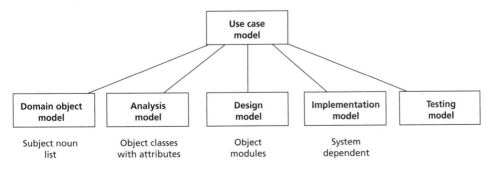

Figure 11.24 *Major models used in OOSE*

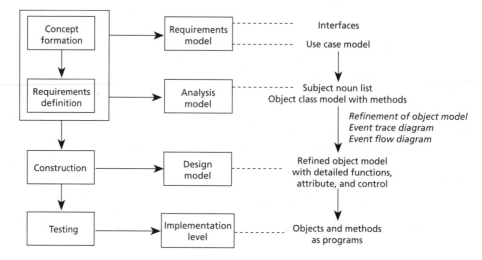

Figure 11.25 *The OOSE development process*

management systems or object-oriented implementations. Methods are provided to develop system models for either of these two alternatives.

EMPHASIS ON USE CASES

The OOSE development process is basically use case driven. Thus, use cases are initially used to create the analysis model as follows:

1. Identity the most obvious objects (or entities) from use cases—for example, the roles themselves—then artifacts that result or are used by the use cases—for example, product list or sale.
2. From the use cases, analyze how the system will be used and the interfaces that are needed by the users.
3. Identify any additional lower-level objects that may be reused in a number of use cases—for example, update account.

Step 3 can also be initiated directly from the use cases, as suggested by Jacobson. The approach here is to begin to decompose some of the use cases using interaction diagrams (an earlier name for sequence diagrams) to create the requirements model.

OBJECT ORIENTATION AND RAPID APPLICATION DEVELOPMENT

Object methods are particularly suitable in prototyping and rapid application development. Here we can start with some objects, connect them into a system, and experiment with the system. We can begin to add new functionality to a design by allowing objects to be easily put together to form systems. Alternatively, we can begin to specialize objects and store them in a library, and then reuse them or reuse their specialized versions. New methodologies are slowly evolving to support cycles that combine both evolution and synthesis.

TOOLS FOR OBJECT-ORIENTED DEVELOPMENT PROCESSES

There is also an evolving set of tools for development processes that use object modeling. Perhaps the best known set of products is Rational Rose, which supports the UML, an evolving standard described later. Detailed descriptions of UML and Rational Rose can be found at http://www.rational.com.

SUMMARY

This chapter describes the main concepts used in object modeling. It begins by describing object class diagrams and uses cases and then describes methods used to describe behavior. The chapter concludes by describing development methodologies that use object modeling. The next chapter describes ways of using these techniques to identify objects during analysis.

EXERCISES

11.1—Terminology
Figure 11.26 is a detailed object class diagram.

PART A

1.   What is 'change-dept()'?

2.   What does DEPT on the link between PERSONS and DEPARTMENTS mean?

3.   Suppose the relationship between PERSONS and DEPARTMENTS is such that a person can be moved between departments. Should the relationship be an association relationship or a composition relationship?

4.   Is the association between DEPARTMENTS and PERSONS actually needed?

5.   In how many departments can one person be?

PART B

Draw an object instance diagram showing some object instances (e.g. Bill works in accounting and is an accountant, Mary works in sales and is a sales representative, Lin manages the accounts department).

11.2

Figure 11.27 is an object class diagram. It describes faults reported to the organization about its products. Here clients can report any number of problems, each related to one purchased item. Each

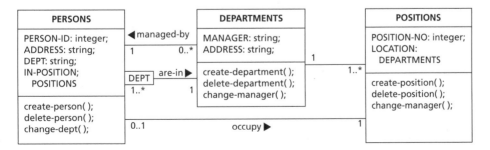

Figure 11.26   *A detailed object class diagram*

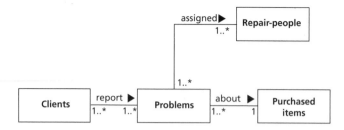

Figure 11.27 *Reporting problems*

problem has a unique identifier and is about one purchased item. There can, of course, be many problems reported about the same item. Any number of repair-people can be assigned to solve the problem.

Further analysis shows that there is additional information in the system. In particular, a problem solution is divided into a number of tasks, and any number of parts are used in each task. Each task is allocated to one person.

Complete the object class diagram by showing:

•   the tasks carried out in repairing the problem

•   the person carrying out the task

•   the parts used in each repair.

11.3

An object class diagram for people carrying out jobs on projects is given in Figure 11.28.
Amend the diagram (if necessary):

•   so that it shows the job each person on a project works on and the time that a person spends on the job. Assume that more than one person can work on a job

•   to show the time a machine is used on each job.

11.4

Figure 11.29 is a simplified object class diagram for a supply chain. It shows that suppliers supply parts that are used in production orders that produce client systems. There is one production order for each client system. Each client system is ordered by one client. The object class diagram also shows that any number of supplier orders can be placed with a supplier. Each supplier order can include any number of parts.

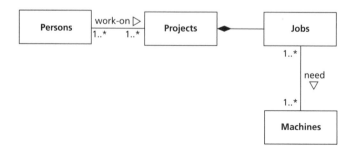

Figure 11.28 *People using machines*

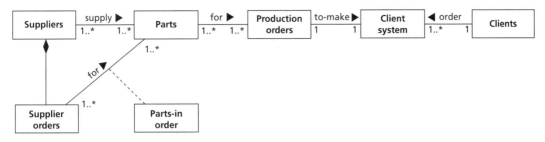

Figure 11.29 *A supply chain*

To improve coordination, the production tasks that make up each production order must be identified. The parts used by each production task are also included, together with the specific supplier order that will deliver the parts for the specific task. How would you therefore change this class diagram to:

- split the production order into a number of production tasks?

- show the parts used by each production task?

- show which of the parts in a supplier order are intended for a particular task?

11.5—Relationships (1)

Figure 11.30 is a detailed object class diagram showing sales made by salespersons. Each sale may be local or global and is made by one salesperson. Each sale is made up of a quantity of items, each of which has a price.

What is the kind of relationship between the following classes?

- SALES and SALESPERSONS?

- SALES and ITEMS-IN-SALE?

- SALES and LOCAL-SALES?

Extend the class diagram to include deliveries of ordered items. External deliveries are arranged with export companies and will need export certificates, whereas local deliveries are arranged with local transport companies.

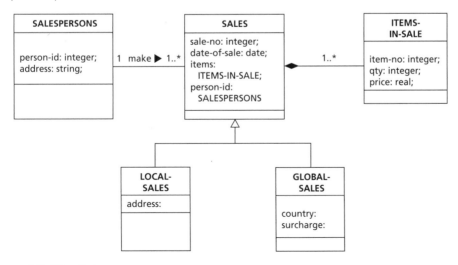

Figure 11.30 *Sales*

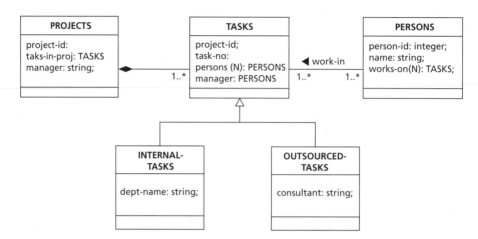

Figure 11.31 *Project tasks*

11.6—Relationships (2)

Figure 11.31 is a detailed object class diagram showing a number of classes. It shows the tasks that make up a project. Each task may be carried out locally or externally and is assigned to a number of persons.

What is the kind of relationship between the following classes?

- PROJECTS and TASKS?

- PERSONS and TASKS?

- TASKS and INTERNAL-TASKS?

Extend the class diagram to include payments made by the tasks. Each payment is made on a given payment-date and is for a given amount. Payments for external tasks are made out as cheques and have a cheque number. Payments for local cheques are made to DEPT-ACCOUNTS and have a transfer-no.

You need to show only the properties and relationships in your class diagram and not the methods.

11.7—Adding Methods

Add methods to the object class diagram shown in Figure 11.32 for the following:

1. Find any persons whose total TIME-ASSIGNED exceeds a given value.

2. Find a person's manager.

3. Assign a person to a project.

4. Find all persons that a manager manages. (How could you change the structure to make this requirement easier to program?)

11.8—More on Methods

What methods would you add to the detailed object class diagram shown in Figure 11.33 to find the total amount of money spent by a given project on parts?

How would you change the object class diagram to allow projects to place orders for parts?

11.9

Figure 11.34 is a use case model for the development process. It defines the kind of activities that take place in system development.

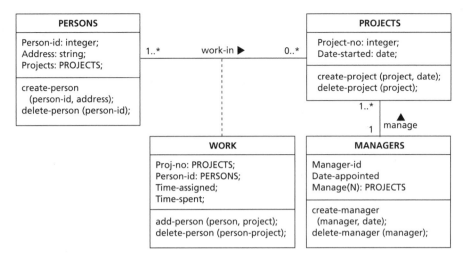

Figure 11.32   *Project work*

Complete the use case by suggesting other actors that may participate in the use cases.

Extend the diagram to include testing, remembering that it is necessary to develop a test strategy separate from development.

11.10

Figure 11.35 is a use case model for semester enrolment. Write use cases of how students proceed through the enrolment process.

11.11—Use Case Extensions

Figure 11.36 is a use case model for customers making requests for either purchases or repairs. Each request requires a search for parts needed to satisfy the request.

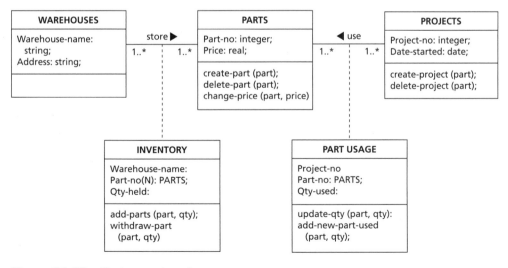

Figure 11.33   *Parts warehousing*

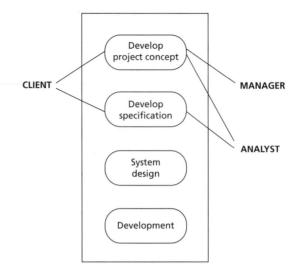

Figure 11.34 *Developing projects*

How would you extend the use case diagram to show that:

- requests for regular customers are processed differently from those made by casual customers?

- the parts can be either purchased or obtained from the organizations's warehouse?

11.12

Try to write some use cases for the interactive marketing exercise.

Example 1: Develop a use case for a producer entering an offer of product into the system.

Example 2: Develop a use case for creating and sending an invoice to the buyer for a sale.

11.13—A Use Case for Requesting an Item

Develop a sequence diagram for the following use case:

1. A customer requests an item from a salesperson.

2. The salesperson looks up the inventory to see whether the item is available.

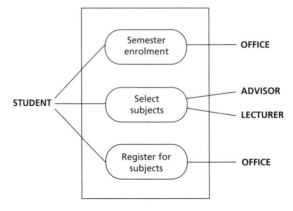

Figure 11.35 *Semester enrolment*

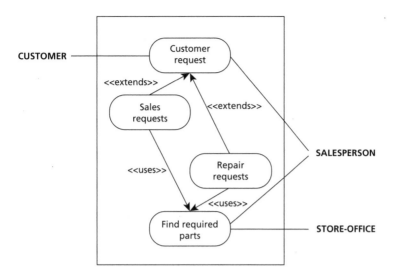

Figure 11.36 *Customer requests*

3.   The salesperson replies to the customer.

4.   The customer places an order with the salesperson.

5.   The salesperson forwards the order to the inventory.

6.   The inventory places the item on hold and informs the salesperson.

7.   The salesperson invoices the customer.

8.   When payment of the invoice is received, the salesperson requests inventory to make the delivery.

9   Inventory sends a delivery note to the customer.

11.14—Investment Queries

Given the following use case, develop a class diagram by first developing a scenario diagram and then reducing it to a class diagram.

1.   A customer arrives at an investment consultant and makes an initial inquiry about their investment requirements.

2.   The consultant records and files the requirements.

3.   The consultant checks the product files.

4.   The consultant proposes a number of financial products to the customer for selection.

5.   The customer then selects one of the products.

6.   The consultant records the decision in the requirements file.

ALTERNATIVES

7.   The customer proposes a variation to one of the alternatives.

8.   The variation is used by the consultant to amend one of the alternatives.

9.   The customer selects one of the alternatives.

11.15

Given the following use case, develop a sequence diagram and use it to develop a detailed class diagram, showing object features and any relationships between the object classes.

1. A customer makes an initial phone inquiry about purchasing tickets for a performance.

2. The reservation officer looks up the reservation file to see whether there are seats available for the performance.

3. The reservation officer responds that seats are available.

4. Once seats are selected, the reservation officer enters the customer details (name, address, credit card number) into an invoice file.

5. The reservation officer updates the reservation file.

6. The invoice is then sent to the customer's credit card company.

ALTERNATIVE 1

7. There are no seats available for the selected performance.

8. The reservation officer suggests alternative times to the customer.

9. The customer selects one of the alternative times.

ALTERNATIVE 2

10. The customer withdraws the request if there are no seats available on the selected date.

11.16

Draw a state transition diagram for the following:

A specialized item is produced for customers. When item production is completed, a completion advice is sent advising that the item is ready for delivery. Following the selection of a transporter, the item becomes ready for delivery. When the transporter arrives, a pickup docket is created. Following delivery, a delivery note is returned. Payment is received following the delivery note, then the item trade is completed.

11.17

Draw a state transition diagram for the following:

Reservations are made for repairs of vehicles. The vehicle is delivered and waits in the dock until repairs can commence. Repairs can commence at any time. During repairs, new parts may be needed, and the vehicle may have to wait for the parts to arrive. When repairs are completed, a test drive is organized. Following the test, more repairs may be needed, or the vehicle will become ready to be picked up. Eventually its owner picks it up.

11.18

Draw a transition state diagram for the following.

An application is received for receipt of a claim. The application is first registered and is then passed for evaluation. During evaluation, expert opinion may be sought any number of times. On completion of evaluation the application can be accepted or rejected. Rejected applications can be appealed. If an appeal is accepted, further information must be submitted and the evaluation is repeated.

BIBLIOGRAPHY

Booch, G. (1994), *Object Oriented Analysis with Applications* (2nd edn), Addison-Wesley, Menlo Park, California.

Booch, G. (ed.) (October 1999), 'UML in action', *Special Issue of the Communications of the ACM*.

Coad, P. and Yordon, E. (1990), *Object Oriented Analysis* (3rd edn), Yordon Press, New Jersey.

Coplien, J.O. and Schmidt, D.C. (eds) (1995), *Pattern Languages of Program Design*, Addison-Wesley, Reading, Massachusetts.

Henderson-Sellers, B. and Edwards, J. (September 1990), 'The object oriented systems life cycle', *Communications of the ACM*, Vol. 33, No. 9, pp. 142–59.

Jacobson, I. (1992), 'Basic use-case modeling', *ROAD*, Vol. 1, No. 2, pp. 15–19 and Vol. 1, No. 3, pp. 7–9.

Jacobson, I., Christerson, P.J. and Overgaard, G. (1992), *Object-Oriented Software Engineering*, Addison-Wesley, Reading, Massachusetts.

McGregor, J.D. and Korson, T. (eds) (September 1990), Special issue on object-oriented design, *Communications of the ACM*, Vol. 33, No. 9.

Rumbaugh, J., Jacobson, I. and Booch, G. (1999), *The Unified Modeling Language Reference Manual*, Addison-Wesley, Reading, Massachusetts.

Sinan Si Albir (1999), *UML in a Nutshell*, O'Reilly, Beijing.

Weiberg, R., Guimares, T. and Heath, R. (Fall 1990), 'Object oriented systems development', *Journal of Information Systems Management*, Vol. 7, No. 4, pp. 18–26.

OBJECT ANALYSIS— IDENTIFYING OBJECTS

CONTENTS

KEY LEARNING OBJECTIVES

How to identify objects

Identifying objects through data analysis

Identifying objects from use cases

INTRODUCTION

The previous chapter introduced object-oriented methods and described object models of data and behaviour. This chapter continues by describing how these models can be used in analysis. The emphasis in analysis is to identify system objects (Figure 12.1).

There are two major approaches in developing an initial object class model:

- *Use a model that has been accepted as identifying the problem situation—for example, the rich picture—and from there define the object classes. This to some extent is similar to E–R modeling, where each set in the E–R model becomes an object class.*
- *Start with use cases and reduce them to an object class diagram using the techniques described in the previous chapter.*

BEGINNING BY DATA ANALYSIS

One of the simplest and earliest object-oriented development processes started with an E–R model. It was widely used when object orientation was first introduced. The process begins by representing each entity as an object. Entity and relationship attributes become object properties. However, it is now necessary to go one step further and add methods, or what some people call services, to each entity and relationship object. These methods can update particular object properties or define local behaviour of objects. They can also be computations that the object can do for other objects.

This process of identifying object classes can sometimes be simplified if we have some higher-level business model to guide us. Rich pictures present one possibility. Although they represent a system more from the perspective of a problem situation, they also give a first indication of object classes. Thus, all roles can become object classes, as can the artifacts produced by the activities. The activities can be described by scenarios leading to the possibility of finding additional object classes using the techniques described in Chapter 11.

In addition, it is necessary to define global behaviour, or how objects interact with each other, much the same as defining a business process object that uses the services of other objects. Methodologies use different techniques for this purpose.

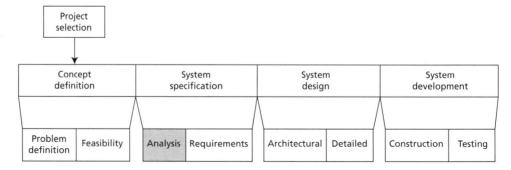

Figure 12.1 *Object analysis in the linear development process*

We now show an example of an object model using the interactive marketing case introduced in Chapter 1.

TEXT CASE A:
Interactive Marketing—Defining the Objects

Readers may recall that in this case producers are marketing their products to consumers. Suppose we now draw an E–R diagram for this system. The diagram is shown in Figure 12.2(a). Each of these sets in Figure 12.1 now becomes an object class.

The first conversion step to an object class diagram is to convert each entity set to an object class, as shown in Figure 12.2(b). The relationship sets convert to object classes only if they have an N:M relationship and carry information. Thus, in Figure 12.2(a), there is an N:M relationship between SALES and PRODUCT-OFFERS, because SALES can include products from more than one PRODUCT-OFFER. In Figure 12.2(b), SALES-LINES is an association class that represent this N:M relationship.

We then identify attributes in the same way as in E–R modeling and place them as object properties. The methods are then defined. Methods selected for interactive trading can include the following:

- method 'initiate-trade' in PRODUCTS to receive a request from a buyer and return the relevant offers and their prices to BUYERS
- method 'request-price' searches PRODUCT-OFFERS, asking them to 'compute-price' for the requested item. It returns the volumes and prices to PRODUCTS
- method 'accept-offer' from BUYERS to PRODUCTS requests PRODUCTS to complete the purchase following acceptance of a quote. This method creates a SALES object and records the PRODUCT-OFFERS that make up the sale.

Other solutions would also be possible here. For example, BUYERS objects could send messages to PRODUCERS objects. In this case, each PRODUCER object manages its own OFFERS and SALES.

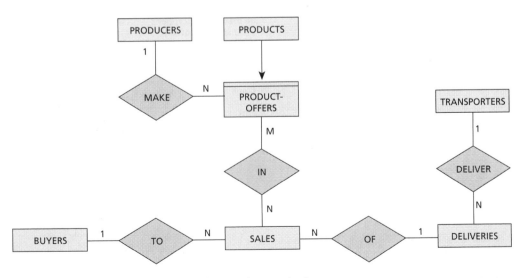

Figure 12.2(a) *E–R diagram for interactive marketing*

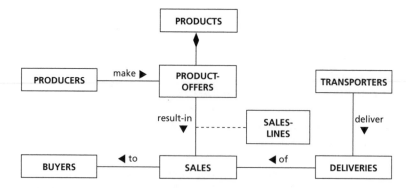

Figure 12.2(b) *Object class diagram for interactive marketing*

BEGINNING WITH USE CASES

As one example of a simple methodology, we show how the techniques described in Chapter 11 can be combined into a methodology. The combination is illustrated in Figure 12.3.

1. Draw a use case model (like Figure 11.8).
2. For each use case, draw a sequence diagram (like Figure 11.14). There can be many sequence diagrams in a design. Remember that there may even be more than one event trace diagram for the one use case, as one diagram is drawn for the general use case process, and separate diagrams may be drawn for exceptions and alternatives.
3. Draw an event flow diagram (like Figure 11.21) that includes flows from all the sequence diagrams.

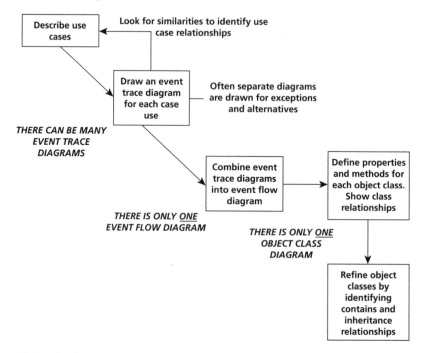

Figure 12.3 *A simple development process*

There should be only one event flow diagram for any problem.
4. Draw an object class diagram (like Figures 11.4 and 11.5) by:
   - providing one method in each object class to cater for each incoming message flow
   - looking at the information carried in or needed by each flow and providing properties to store this information
   - adding association, contains or inheritance links where needed.

   There should be only be one object class model for any problem.
5. Refine the object classes by examining object processes and discovering new object classes in contains and inheritance relationships with initial classes.

VALIDATION AND VERIFICATION

Apart from identifying a set of steps, some additional activities must be added to this process. If you go back to Chapter 5, you might recall that a development process is not only the conversion of techniques from one level to the next. It also requires the model in one step to be checked against the model in the previous step. Good verification usually implies the use of good naming conventions. In that way we can easily check what corresponds to what between models.

Guidelines that can be used in the process shown in Figure 12.3 include the following:

- Each use case and exception must have a unique name. These may be generic. For example, a use case may have a name such as 'Enter product offer'. An exception may take the name 'Enter product offer—product does not exist', or simply 'Enter product offer—exception 1'.
- Each alternative and exception must have its own sequence diagram. The sequence diagram can take the same name as use case (and its alternatives and exceptions).
- An event flow diagram must include all objects and messages in all of the sequence diagrams. The objects should use the same names as in the sequence diagram.
- An object class must have all the objects in the event trace diagram. Each object class must have a method for each incoming message. Each object class must have a property for each parameter of a message. Use the same names for messages and methods to make validation easy. One alternative here is to name each method as 'get-' followed by message name.

TEXT CASE A:
Interactive Marketing—Defining the Object Classes

We now go back to the use cases defined for this system earlier in the chapter and draw sequence diagrams for them. We first construct a sequence diagram for each use case from the use case script. These are illustrated in Figures 12.4(a)–(d). You will note that the sequence diagrams include the messages and their parameters.

Figure 12.4(a) is the sequence diagram for producers registering a product offer. Three object classes have been identified from the use case. These are

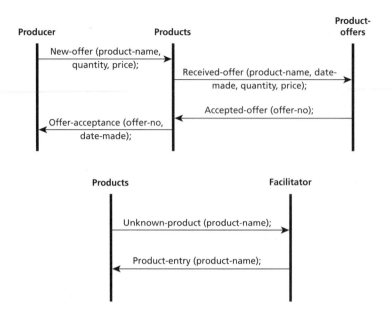

Figure 12.4(a)   *Sequence diagram for entering offers*

Producer, Products, and Product-offers. The messages between them are obtained by following the use case sequence.

Figure 12.4(b) is the sequence diagram for making a sale. Two new object classes are identified, Buyers and Sales-records. Note that each message here has a name, which later becomes or is used to create a method name. It also includes parameters that are carried by the message.

Next we draw the sequence diagram for arranging deliveries. It is shown in Figure 12.4(c).

Finally we draw the sequence diagram for finalizing the sale. It concerns mainly creation and sending of invoices and is shown in Figure 12.4(d).

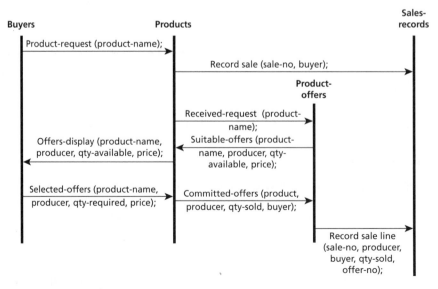

Figure 12.4(b)   *Sequence diagram for making a sale*

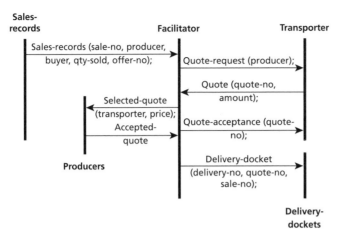

Figure 12.4(c)   *Sequence diagram for arranging deliveries*

CONVERSION TO EVENT FLOW DIAGRAM

We now have a number of alternatives for converting the sequence classes to an object class diagram. The first is to combine the sequence diagrams into the one event flow diagram, as shown in Figure 12.5.

From here we can now get a first cut at the object classes needed in this system. We can go directly to a detailed object class diagram, or we can first try to identify the relationships between the object classes and then create the detailed object class diagram.

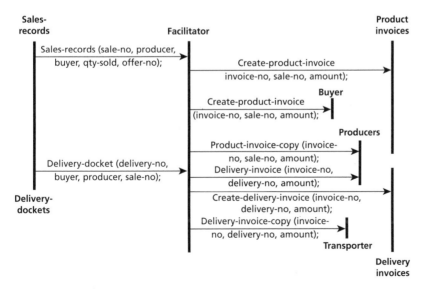

Figure 12.4(d)   *Sequence diagram for finalizing the sale*

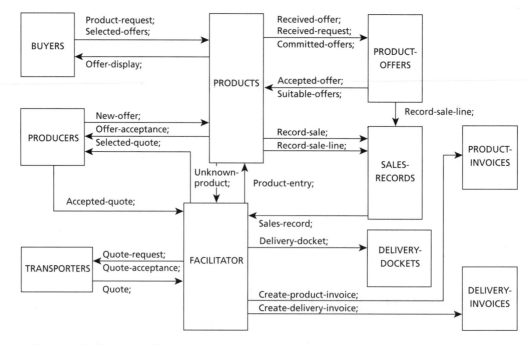

Figure 12.5   *A simplified event flow diagram for interactive marketing*

DEFINING OBJECT CLASSES

Two steps are suggested here to construct an object class diagram—a first cut followed by a refinement.

THE FIRST CUT

The simplest first step is to create an initial set of object classes as follows:

- Each object in the event flow diagram becomes an object class.
- Each exchange of messages becomes an association.

The result of such direct conversion is shown in Figure 12.6.

We then continue by adding features to the object classes:

- Each parameter in an incoming or outgoing method becomes a property of the object.
- Each incoming message becomes a method in the object class.

Applying these rules to Figures 12.5 and 12.6 results in the object class diagram shown in Figure 12.7, which illustrates some of the object classes.

One thing that you might note here is how method names are chosen. We simply take the message name and preface it with 'get-'.

REFINING OBJECT CLASSES

Figure 12.7 represents only a first cut at the object class diagram. Now it becomes necessary to refine it. Refinement includes:

- removing redundant properties from object classes
- identifying additional relationships between object classes
- grouping of properties

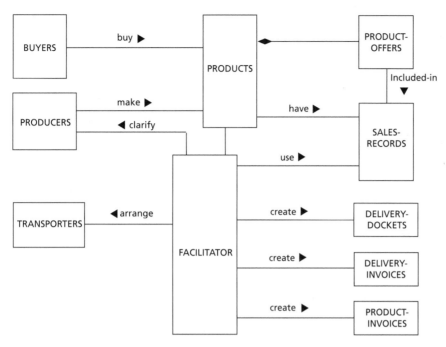

Figure 12.6   *An initial object class diagram*

- discovering any new object classes especially through inheritance and containment.

We now illustrate some of these refinements for selected classes in Figure 12.8.

The first refinement is to eliminate some of the attributes that appear in associated classes, because they have messages between them. These attributes can be replaced by a qualifier that refers to the related object class. Thus, suppose we look at PRODUCT-OFFERS. One thing that we might notice here is that an offer

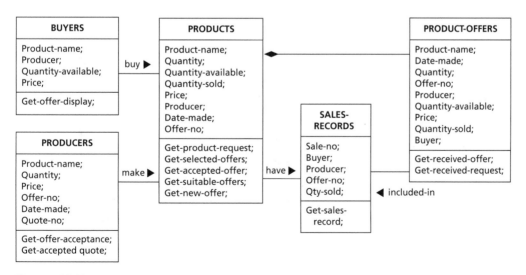

Figure 12.7   *A part of the detailed object class diagram for interactive marketing*

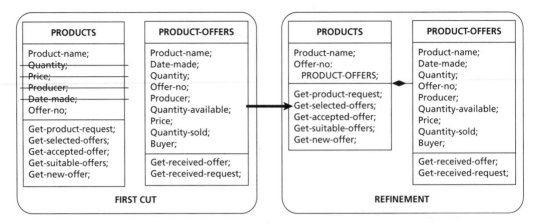

Figure 12.8 *Identifying containment*

for a product can exist only if that product exists and hence there is a containment relationship between them—PRODUCT-OFFERS are contained in PRODUCTS. As a result, we make the change shown in Figure 12.8. In making the change, we make some properties redundant in PRODUCTS, in particular 'price' and 'producer', as these are now contained in PRODUCT-OFFERS, which is now reachable through the qualifier PRODUCT-OFFERS in PRODUCTS.

Going on, we can look at changing the association relationships (Figure 12.9). To do this we look for object classes with similar properties and where the property is an identifier in at least one of the object classes. The most obvious examples are 'producer' and 'buyer'. Thus, the Buyer property in SALES-RECORDS suggests an association with BUYERS. The usual rule here is that we show the association, which defines integrity. Thus, a value of 'producer' can appear in PRODUCT-OFFERS only if there is an instance of PRODUCERS with the same value of 'producer'.

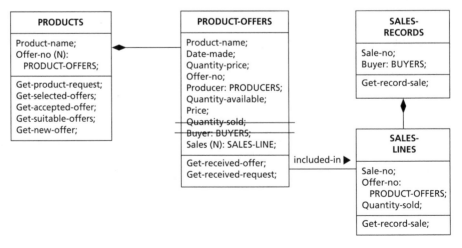

Figure 12.9 *Showing some association relationships*

Figure 12.10 shows how some of these changes are now consolidated into the detailed object class diagram.

The next kind of refinement is to identify inheritance relationships. If you look at Figure 12.6, you will note that it contains DELIVERY-INVOICES and PRODUCT-INVOICES. Many of their attributes will also be common, in particular invoice-no and amount. This suggests that they may be specializations of the object class INVOICE. The conversion to such specialized classes is shown in Figure 12.11.

REFINING METHODS

Methods are also refined during this step. So far the development process required each object class to have a method to receive each incoming message. This of course is essential, as any incoming message must be received and responded to. Now we can have a look at each method and see whether it can be refined, possibly by decomposing it into simpler methods. The kind of refinements can include:

- adding separate methods to create and delete objects
- creating separate methods for sending messages by decomposing methods that read a message into a receiving and sending part
- identifying those components that do purely internal computation and creating internal methods to do this.

The first suggestion is obvious. To see the second, have a look at method 'get-product-request' in object class PRODUCTS in Figure 12.10. This method resulted from a request by a buyer for a product. The method will read the incoming message and send out two messages, 'record-sale' to SALES-RECORDS and 'received-request'

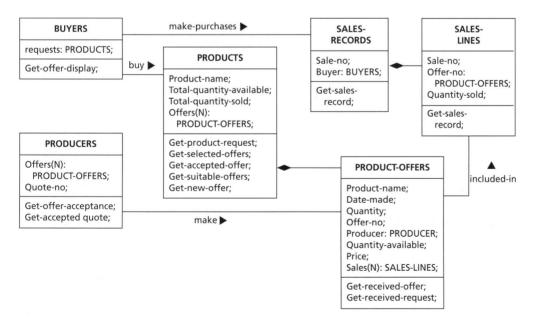

Figure 12.10   *Consolidating the refinements*

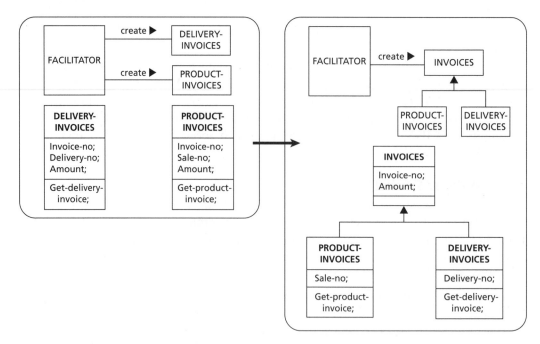

Figure 12.11   *Finding inheritance relationships*

to PRODUCT-OFFERS. Alternatively, we could decompose method 'get-problem-request' into three simpler methods. These methods would be 'get-request' to receive the message, 'send-record-sale' to send a message to SALES-RECORDS, and 'send-request' to send a message to PRODUCT-OFFERS. The method 'get-request' would call the other methods within the PRODUCTS class. The third suggestion is a continuation of such decomposition. Sometimes in decomposition we might identify purely local operations. For example, we might want to include a message that stores references to received requests to make sure that they are eventually dealt with. Typical other examples might be computation of fees, interest, or commissions, all of which are often local to a method.

ALTERNATIVE PROCESSES

An alternative approach is to use collaboration diagrams. Thus, we can start with each use case and its corresponding sequence diagram and construct a collaboration diagram. This can then be converted to an initial object class diagram using the conversion techniques described in the previous section. Then we select the next use case and add to the initial object class diagram.

METHODOLOGIES USED IN PRACTICE

Object modeling methods are now increasingly used in the development process. They provide the ability to commence at the user level, then go directly to a systems specification and implementation using only one modeling method. This is in contrast to structured systems analysis, where a number of modeling methods must be combined with more elaborate conversions from analysis models to implementation. In the next sections the book briefly outlines some of the methodologies used in practice. Each of these requires a text of its own to describe it, and readers should refer to these texts for details about the methods. Readers should consult relevant manuals or texts comparing the techniques they use at different life cycle phases.

Historically, it is perhaps fair to say that object-oriented methods were first used in the implementation phases beginning with languages such as Smalltalk and later C++ and Eiffel. Then object modeling became popular for developing analysis models during the system specification stage. One of the earliest uses was that of Coad and Yordon (1990), which used object analysis in the analysis model, beginning by object identification. This was followed by a large number of proposals, of which three became the most popular. These were Jacobson's OOSE (Object-Oriented Software Engineering), which introduced use cases, Rumbaugh's OMT (Object Modeling Technique), which was oriented toward state transition diagrams, and the methods proposed by Grady Booch. Subsequently, these three methods were combined to form the Unified Modeling Language (UML), which is currently a pseudo-standard. It is developed by Rational, and can be found on the Rational Web site at http://rational.com.au. Each of these methodologies has been described in separate books. Consequently, the descriptions given here are by no means complete, but describe only the general philosophy of each of the methodologies.

RATIONAL ROSE SUITE OF TOOLS

Just like with Structured Systems Analysis, CASE tools are now emerging to support object oriented development methods. Perhaps the best known of these is Rational Rose. Rational, the organization that is supporting the development of the Universal Modeling Language (UML), which was described earlier in Chapter 11, provides a family of tools that support analysis and design based on the object oriented methodology. Details of these tools can be found in manuals that can be located on the Rational Web site at http://www.rational.com. Broadly, the architecture of

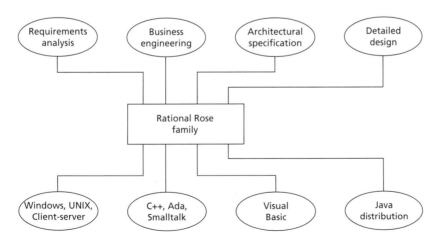

Figure 12.12 *The Rational Rose family of tools*

this family of tools is illustrated in Figure 12.12 and centers on a repository that maintains the data models.

A set of tools is provided for model construction and another set for application generation. Modeling support is also provided for the techniques used in UML and includes:

- use case modeling to create use case models
- scenario modeling
- state modeling, among others.

There are also sets of tools for generating applications on a variety of platforms, including C++, Ada, Smalltalk, Visual basic and Java for Web-based applications. Again, the set of tools provided by Rational Rose is continually increasing and readers are advised to visit its site at http://www.rational.com for the latest versions.

SUMMARY

This chapter describes how to identify the objects in a system. Object identification is central to good design, as finding the correct object classes can mean that the remainder of the design will be conceptually simple. Two ways of identifying object classes are defined. One starts with a data model and converts its sets to object classes. The other, more popular method starts with use cases and converts them to object classes by going through a process that starts with sequence diagrams and results in the identification of object classes, their attributes and methods.

EXERCISES

12.1—Meeting Customer Orders

Following are a number of use cases for meeting customer orders. Develop a use case model for this system showing all the use cases. Then draw sequence diagrams for each use case and combine them.

USE CASE 1—Finding out what the customer wants

1. The salesperson has a list of items for sale.

2. The salesperson goes through the list, retrieving the item name, price, and description and showing them to a client.

3. Whenever a client identifies an item for purchase, the item name and price are recorded in a sales docket, together with the quantity ordered by the client.

Exception

4. The client can request items not on the list. In that case, a record is made of the requested item.

USE CASE 2—Finalizing the sale

1. Once a sales order is completed, the salesperson computes the total value of the order.

2. The salesperson then requests approval from management for a discount to be offered.

3. Following management's decision, a total quote is forwarded to the client; who then makes a decision as to whether to proceed with the purchase.

4. The client decision is recorded on the sales docket.

Exception

5. The client requests a change to the order. In that case, the process is repeated with a new total value computed for the order.

USE CASE 3—Arranging goods for delivery

1. On completion of the sale, the sales docket is retrieved by the expeditor, who begins to look for the ordered items and arranges for their shipment to the customer.

2. The expeditor looks up the availability of each item by examining the warehouse inventory.

3. Whenever the required items are found, they are earmarked for the order, and the item together and location are entered on a delivery docket.

Exceptions

4. If all items are not found in inventory, then a back-order is created, and the customer is informed of possible delay.

5. A customer cancels an ordered item.

USE CASE 4—Arranging deliveries

1. The transport section collects delivery dockets.

2. An invoice is prepared.

3. Arrangements are made with a delivery company to deliver the items in the delivery docket.

4. An invoice is sent once the item is delivered and the client accepts the delivery.

USE CASE 5—Processing payments

1. The customer returns the invoice with the payment.

2. The sale is recorded as completed.

Exceptions

3. The customer does not return a payment within a month, in which case a reminder notice is sent to the customer.

EXTENSION

1. Change the system in Use Case 5 so that a sales order can be delivered as a number of parts. Change the system to allow regular orders to be placed.

2. Draw an event-flow or collaboration diagram for this problem and derive an object class diagram from it.

12.2—Car Hire System

For the following problem:

1. Draw a use case model showing any relationships between the use cases.

2. Develop a class diagram from the use cases using scenario diagrams.

USE CASE 1—Making reservations

1. A customer contacts a reservation officer about a car rental.

2. The customer quotes the start and end dates needed, the preferred vehicle, and the pickup office.

3. The reservation officer looks up a prices file and quotes a price.

4. The customer agrees to the price.

5. The vehicle availability file is checked to see whether an appropriate vehicle is available for the required time at the required office.

6. If the requested vehicle is available at the nominated pickup office, it is reserved for the customer. An entry is made in the vehicle availability registering the reservation.

7. The reservation officer issues the customer with a rental number.

8. A rental agreement is then created on the rental file, including the rental number, the rental period, the vehicle type, and the pickup office.

Exceptions

9. An appropriate vehicle is not available at the pickup office. The customer is offered an alternative vehicle.

10. The customer does not agree to a price and asks for an alternative vehicle or period.

USE CASE 2—Maintaining availability system

1. A vehicle availability file is checked to see whether a vehicle of a given type is available at the requested pickup office for the requested rental period. There is a record for each vehicle, which includes the times when a vehicle is available and when it is rented.

2. If it is, then the vehicle is reserved for the requested period.

Exception

3. If a reservation cannot be made because of lack of vehicles, then a problem report is issued for use in planning vehicle levels.

USE CASE 3—Initiating rental

1. A customer arrives at a pickup office and quotes a rental number to the rental officer.

2. The rental file is checked against the customer's rental number.

3. If it is correct, then the rental agreement is retrieved and discussed with the customer.

4. If the customer accepts, then a rental agreement is printed.

5.   The customer signs the agreement and lodges a credit card number.

6.   The rental officer requests the customer to select one of a number of insurance options.

7.   Following the selection, an insurance policy is filled out and attached to the signed agreement.

Exceptions

8.   A customer does not have a prior reservation. In that case, a vehicle availability check is made. If a vehicle is available, the customer is offered the vehicle and a price is quoted. If the customer accepts, then a rental is initiated.

9.   If the kind of reserved vehicle is not available to a customer with a prior reservation (because of a late return), then an alternative proposal is made to the customer.

USE CASE 4—Processing vehicle returns

1.   The customer records the mileage and fuel level and quotes them to the rental officer.

2.   The amount of fuel to be added to the vehicle is added to the rental account on the rental agreement file.

3.   The rental account is checked by the customer.

4.   The customer pays the rental amount.

Exceptions

5.   The returned vehicle is damaged. This requires a forfeit of the deposit and an insurance claim to be filled in.

6.   The customer disputes the account.

USE CASE—Providing management reports

1.   The reservation statistics file is updated at the time when reservations are made.

2.   The file is also updated when vehicles are rented to customers without prior reservations.

3.   The statistics are updated at the time of return to show the length of reservation and the amount of moneys received.

4.   A rental summary is generated on request from management.

EXTENSION

Extend the design by adding a further feature to support regular rentals to company customers. This will make an agreed number of vehicles available to employees of the company at nominated offices on a daily basis. The company can nominate and provide a list of authorized employees who can make the pickups. The company will be presented with an account on a monthly basis. To do this you may need to first write a new use case or amend existing use cases.

12.3

You should also attempt Case 4 at the end of the book. This case includes a number of use cases that can be used to build up an object model.

BIBLIOGRAPHY

Booch, G. (1994), *Object Oriented Analysis and Design with Applications* (2nd edn), Addison-Wesley, Menlo Park, California.

Coad, P. and Yordon, E. (1990), *Object-Oriented Analysis* (3rd edn), Yordon Press, Englewood Cliffs, New Jersey.

Henderson-Sellers, B. (1992), *A Book of Object-Oriented Knowledge*, Prentice Hall, Englewood Cliffs, New Jersey.

Jacobson, I., Christerson, P.J. and Overgaard, G. (1992), *Object-Oriented Software Engineering*, Addison-Wesley, Harlow, England.

Liang, Y., West, D. and Stowell, F. (April 1998), 'An approach to object identification, selection and specification in object-oriented analysis', *Information Systems Journal*, Vol. 8, No. 2, pp. 163–80.

Rumbaugh, J., Blaha, M., Premerlani, W., Eddy, F. and Lorensen, W. (1991), *Object-Oriented Modeling and Design*, Prentice Hall, Englewood-Cliffs, New Jersey.

Shlaer, S. and Mellor, S.J. (1988), *Object Systems Analysis: Modeling the World in Data*, Yordon Press, Englewood Cliffs, New Jersey.

DEFINING THE
REQUIREMENTS

CONTENTS

KEY LEARNING OBJECTIVES

Formulating the specification

Specification in structured systems analysis

OO specification methods

Rapid application development

Specifications for web designs

Business processes and flowcharts

Job design

INTRODUCTION

This chapter outlines the final step of development of the system specification, which, as shown in Figure 13.1, is the detailed specification of user requirements. We now assume that the conceptual direction for the project has been set in Chapter 7 and an analysis of the existing system is complete. More specific user requirements are now identified to provide an input to system design. Subsequent chapters describe how these user requirements are expanded into detailed design.

Conceptual solutions like those in Chapter 7 are often provided in usage terms but must be converted to system requirements, which must be specified in system terms. Definitions in system terms can then be used by system designers to build the computer system. The methods used to develop the specification may depend on whether we are improving an existing system or building a new one. When an existing process is improved, there is usually an analysis model. The emphasis is on changing this model or rearranging the business process. For a new system there is no analysis model. This often occurs with strategic development, especially when building electronic commerce applications, where analysts have to create a new model. This chapter covers both of these approaches. The main difference is that prototyping and rapid application development are often found with strategic developments, especially where they concern building innovative business systems, the Worldwide Web, or electronic commerce. Linear development processes are more likely to be used when improving an existing process, as we can predefine what is needed.

CREATING THE SPECIFICATION

Requirements specification is a design activity that defines how the new system will work. Its goal is to propose a way of doing things, which requires analysts to creatively develop new solutions. Again, as discussed in Chapter 4, of all the material in any text, the hardest to describe and structure is the way new designs emerge. Chapter 4 identified a number of characteristics of the design process. It stressed that design is a creative process and that it is about identifying possible solutions, selecting the best of these and putting them into practice. It described how information systems development must fit with the strategic and organizational objectives. These objectives in turn set the objectives for the information system.

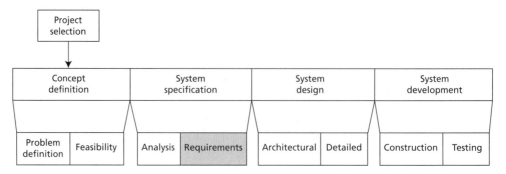

Figure 13.1 *Requirements Analysis*

Chapter 4 also pointed out that it is not possible to define a process that guarantees a good design. One can only give guidelines and suggestions and suggest techniques that improve all of these activities.

A systematic way of identifying what needs to be done is illustrated in Figure 13.2:

1. List the statement of requirements and the expected business improvements and identify the detailed objectives.
2. From the detailed objectives, identify affected components of the existing system.
3. Propose how each affected component will be changed.
4. Define new business processes.
5. Describe (or demonstrate) how the benefits will be obtained.
6. Develop a specification for the changes, or build a prototype.

We examine the statement of requirements together with the analysis model to identify detailed objectives of the design. We then look at which components of the existing model will be affected by the detailed objectives. These affected components need not necessarily be just computer systems but can be business units. The latter is particularly the case in strategic initiatives. We then define a set of new components and how they will be put together into a system. It is also useful to carry out some prototyping here to elicit specific user requirements and to validate that expected improvements are feasible.

The process shown in Figure 13.2 is integrated into the development process in the sense that it elaborates on the conceptual solutions produced early in the process. It takes the statement requirements and shows what must be done in detail to satisfy them. The specification should also recommend the technology to be used to

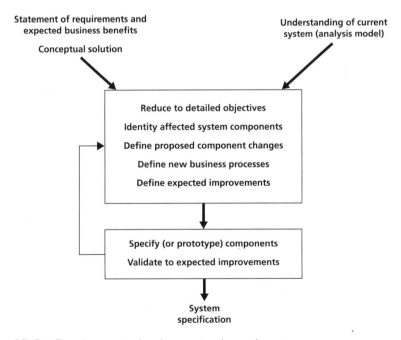

Figure 13.2   *Requirements development—the major steps*

implement the system. In many cases this may be to use the equipment and software already available in an organization. In some cases, additional software must be acquired. This is particularly the case when we are designing interactive electronic commerce sites that provide large number of alternatives for implementing interactive WWW pages.

IDENTIFYING DETAILED OBJECTIVES

Developing the system specification often benefits us by elaborating the conceptual solution in terms of more detailed objectives. Such objectives can be specified in terms of improvements to the organization's processes and functions and what is to be done to realize these improvements. It is therefore important to state the objectives in a way useful to design. What is needed is precision rather than generality, to give designers precise goals. For example, we would prefer to:

'ensure that errors during input are less than 1%' instead of to 'improve the data capture process', or

'ensure that all reports on stock movements are produced by the end of the month' rather than to 'improve timeliness of the stock reporting process'.

KINDS OF OBJECTIVES

There are many kinds of objectives. Common types are listed here:

1. *Functional objectives* state new or amended functional system requirements. For example:
   - new or changed output reports or displays
   - new services to be provided by the system functions
   - revised security and access controls.
2. *Process improvements* include:
   - changes to the way data is accessed
   - changes of the sequence in which things are done
   - changes to the process steps
   - changes to input and output methods; for example, regular instead of *ad hoc* reporting.
3. *Operational objectives* specify performance standards to be attained by the new system. These define system accuracy and various timing requirements.
4. *Personal and job satisfaction needs*. Important objectives include designing systems that are easy to use and that allow users to be creative rather than simply responding to computer outputs.

Different ways may be used to satisfy each of these objectives. Functional objectives may require changes to the kinds of computations in the system—for example, new accounting reports may be needed. It may also be necessary for new data to be stored in the system. Process objectives may require rearranging the way a service is provided or some internal objective to be carried out.

The operational objectives, on the other hand, require different solutions. They may, for example, require a different physical implementation of some of the system processes to improve performance or to improve flexibility, such as changing from a batch- to a transaction-oriented environment. Another typical operational change may be to propose an *ad hoc* method of data access instead of regular reporting. Operational objectives may also call for a rearrangement of procedures used for some tasks, such as including more checks to improve accuracy, rearrangement of functions to reduce the number of steps in a procedure, or greater use of computers to increase throughput. Operational objectives will also identify the training needed to enable users to use the new system.

Personal and job satisfaction objectives may call for changes to the user interface, new report layouts, or changes to the data flows. Presenting delivery advices sorted by area may improve the satisfaction of schedulers, who do not have to perform this routine task any more. Instead, schedulers can concentrate on the important problem of selecting routes for each vehicle.

Specifications must address all these issues. They must specify the layout of any input screens or forms. The outputs are also designed, showing the layout of reports or screens. In addition, the files, structures, and programs are also specified, showing the different record types and program modules needed to make the system work. They must also specify what people in the system will have to do to make it work.

Another way is to identify key performance factors. These key factors are then analyzed. To do this, analysts talk to the users to define precisely what each key goal entails. Usually it is necessary to ascertain things like:

- the information requirements to satisfy the key goals
- performance targets (such as time, volume, sample size)
- comparison to existing systems
- personnel issues such as skill levels.

New and specific values can then be set against each key goal. These target values become the system objectives. Often the process can be repeated through a number of levels. The objectives are defined, then they are broken down into more detailed objectives. Specific target values are assigned to these detailed objectives, and so on.

The specification development is integrated within development processes.

SYSTEM SPECIFICATION USING STRUCTURED SYSTEMS ANALYSIS

Proponents of structured systems analysis have suggested a way to fit system specification into the development process. Their process is illustrated in Figure 13.3 and fits into the linear development process described in Chapter 5. Figure 13.3 also illustrates the next phase, system design, and the two models developed in that phase. System requirements definition follows the following four steps:

1. Develop an analysis model to describe how the system works now.
2. Develop an analysis model to describe what the system does now.
3. Develop a requirements model to describe what the new system will do.
4. Develop a requirements model to describe how the new system will work.

Of course, the first two steps are unnecessary if you are building a completely new system. All the steps, however, apply if you wish to re-engineer an existing system. Each of the models uses DFDs, E–R diagrams, and process descriptions, as

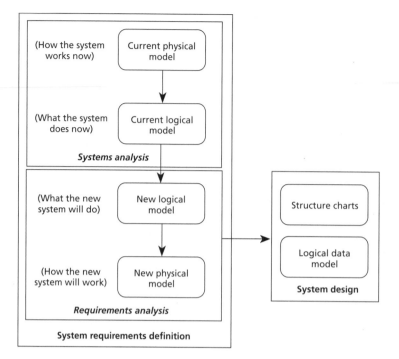

Figure 13.3 *Problem-solving steps in structured systems analysis*

described in Chapters 8–10. The model of the new system can become part of the system specification and be converted to an implementation model during system design. Thus, the DFD part of the model is converted to structure charts, as described in Chapter 16, and the E–R diagram is converted to a logical data model, as described in Chapter 15.

THE FIRST STEPS—SYSTEMS ANALYSIS

The first two steps in Figure 13.3 develop the analysis model. As described in Chapter 8, they begin by analyzing the current operation, developing the current physical model, and identifying system problems. The physical model often uses user terms and is thus easier to develop as a first step. Then the logical model, which describes what the system does, is developed. The logical model closely approximates the subject world model, as described in Chapter 4.

Although we are after the logical, or subject world, model in systems analysis, in practice it is easier to build this model in two steps. The first step is to construct a physical model using user terms and then convert the physical model to a logical model using the methods suggested in Chapter 8.

A physical model is built first because it is closer to the user world. Most analysts find it easier to start analysis with physical objects. The analyst gathers the information needed about the system from interviews with system users and by examining documents, procedure manuals, or existing computer programs. During this search, the emphasis is on the physical system components, and the tendency is to develop an initial data flow diagram that includes these physical components. Once a physical model exists, it is elaborated in terms of its logical components in the way described in Chapter 8. To do this, analysts often start on defining system-level terms and use them to name the logical model processes. These lower-level logical processes can then be recombined into a higher-level logical model in the manner described in Chapter 8.

THE NEXT TWO STEPS—SPECIFYING THE NEW SYSTEM

The requirements analysis steps in Figure 13.3 commence once the logical model of the existing system is completed. Their goal is to develop a model of the new system. Development of the new system model is driven by system objectives for the new system. The new model defines the requirements and creates a system specification that is later used in the system design stage.

Just like analysis, this development proceeds in two steps. First, the new logical model is developed primarily in subject-level terms. It includes any new processes or changes to existing processes necessary to meet system objectives. This step calls for considerable creativity. There are many ways to meet system objectives, and some are better than others. The designer must examine as many solutions as possible to make sure that a good solution is found.

Then the new physical model is developed. The development of the new physical model includes many activities. The physical model corresponds closely to the idea of a user-level model and describes how the new system will work. Here decisions are made on which processes will be manual and which are to be computerized. User processes and interfaces with computers are defined in broad-level terms. Physical devices to store any data are chosen, and methods for carrying out system functions are defined. The interface between system users and computers in the new system is designed. A number of physical alternatives are usually produced during physical design, and one of these is selected on economic and social criteria. Again, considerable creativity is needed to propose possible solutions.

A number of DFD models are produced as we proceed through the cycle shown in Figure 13.4. These stages, shown in Figure 13.4, are:

- a current physical model
- a current logical model
- a new logical model
- a new physical model.

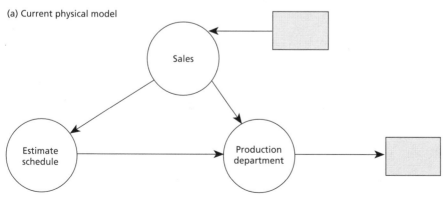

(a) Current physical model

Figure 13.4 *The structured systems analysis problem-solving cycle* *(continues . . .)*

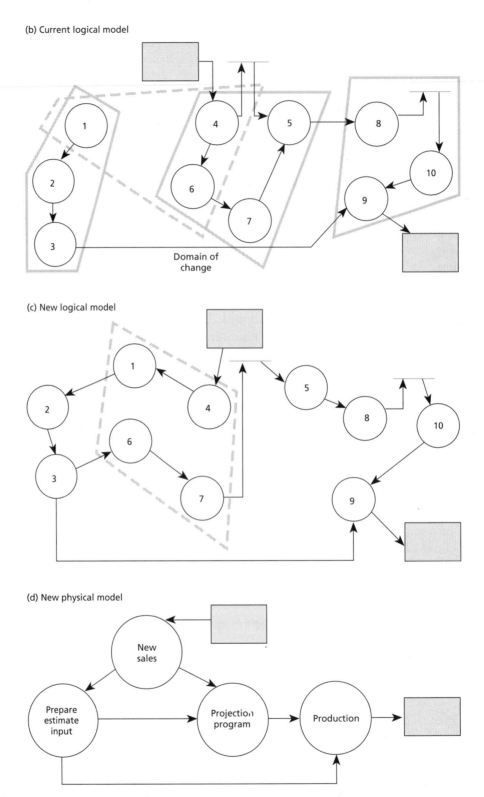

(b) Current logical model

Domain of change

(c) New logical model

(d) New physical model

Figure 13.4 *(Continued)*

This does not mean that four separate models have to be maintained at all times. Rather, they gradually evolve into a model of the new system. Thus, we may begin with a physical model and then convert it to a logical model using the techniques described in Chapter 8. Then we change the logical model of the existing system into a new logical model that satisfies the objectives of the new system. Then a physical model is created to show how the new system is to work. It is of course possible, particularly if one is using computer support tools, to retain snapshots of the models as one proceeds through this cycle.

DESIGNING THE NEW LOGICAL MODEL

DeMarco (1978) has proposed a method for creating a new logical model from the current logical model of the existing system. This method is shown in Figure 13.5. The first step is to see which processes are affected by the objectives for the new system. These processes are included in what is called the **domain of change**, which looks something like Figure 13.6. It includes all the processes that are affected by the system objectives, together with the interface between these processes and the remainder of the system. The affected processes may be adjacent to each other, as shown in Figure 13.6(a), or they may be made up of disconnected or disjointed data flow diagrams, as in Figure 13.6(b).

As an example, let us turn back to Figure 8.16. Suppose we decide to include an inventory system to contain frequently used parts. A new process will be needed to check whether a project request can be met from inventory or whether it must

Domain of change That part of a logical DFD of an existing system that will be changed in the new system.

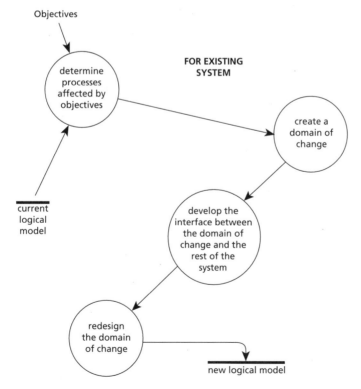

Figure 13.5 *Steps for developing the new logical problem*

(a) A single domain of change

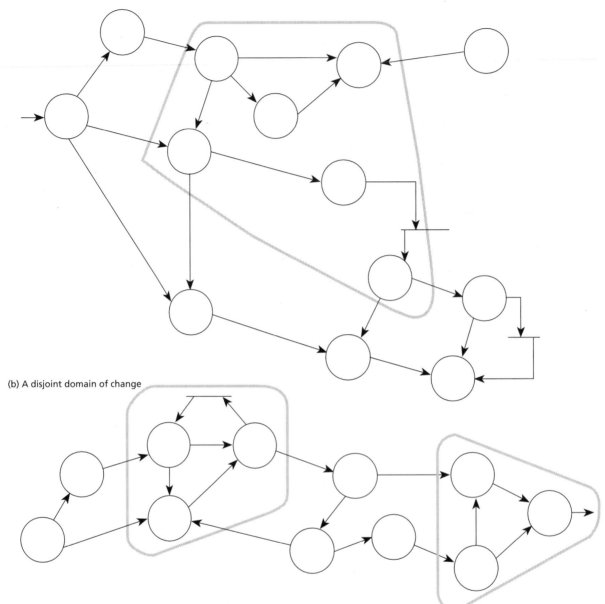

(b) A disjoint domain of change

Figure 13.6 *Domains of change*

be met by purchasing parts from a supplier. Only project requests for parts not in inventory will go to Process 1 to select suppliers. Any such request will be held in data store PO-REQUESTS until the parts are received from suppliers. In this case, the domain of change includes Process 1, which must be changed to get its input from the newly created process.

Note that designers may see different ways of meeting system objectives and thus define alternative domains of change. One of these alternatives must be selected during design.

The domain's processes, data flows, and data stores are designed once a domain of change is selected. Design cannot be described as a prescriptive process. It is possible to suggest guidelines, but it is up to designers to use their knowledge of the system to suggest ways to satisfy system objectives. Two approaches can be identified. One is to completely redesign the domain of change, and the other is to amend it. A mix is also possible, where a part of the system is amended and another part is redesigned.

REDESIGNING THE DOMAIN OF CHANGE

We can use the methods described in Chapter 8 and illustrated in Figures 8.27 and 8.28 to redesign the domain of change.

DATA ANALYSIS IN DESIGN

One part of the new logical model is the data structures. As data stores are defined, the system data model is amended to include any new data requirements. The existing E–R diagram is amended to include any new entity sets, relationship sets, and attributes, and an E–R diagram for the new system is produced. The newly designed E–R model should then be discussed with system users to gain their approval.

During detailed design, the new E–R model is converted to a set of relations, normalized to eliminate data redundancies, and then converted into a database. Techniques used to do this are described in Chapter 15.

THE REQUIREMENTS MODEL IN STRUCTURED SYSTEMS ANALYSIS

The output of this phase consists of the new DFD and the new E–R diagram.

TEXT CASE D:
Construction Company—The New Logical Diagram

We now develop the new logical model for the construction company. The logical model for the existing system in this case was illustrated in Figure 8.16. The system goal is to reduce the time between the project request and goods delivery. Feasibility analysis (see solution 5 in Figure 7.7) indicated that this is to be realized by adding an inventory function and integrating the GRS and POS to eliminate manual checking of deliveries and invoices against purchase orders.

These specifications thus construct the new logical model now shown in Figure 13.7 and specify the changes to be made to the existing systems shown in Figure 8.16.

CHANGES

1. Add the new data store INVENTORY to represent the inventory.
2. Add Process 8 to check project requests against inventory. Project requests now go to Process 8 in the new system rather than to Process 1, as in the existing system.
3. A project request will be forwarded to Process 1 only if it cannot be met from current inventory. If a project request can be met from inventory, then it is sent directly to Process 5 as a 'store request' requesting the items to be withdrawn from store and sent to the project site.

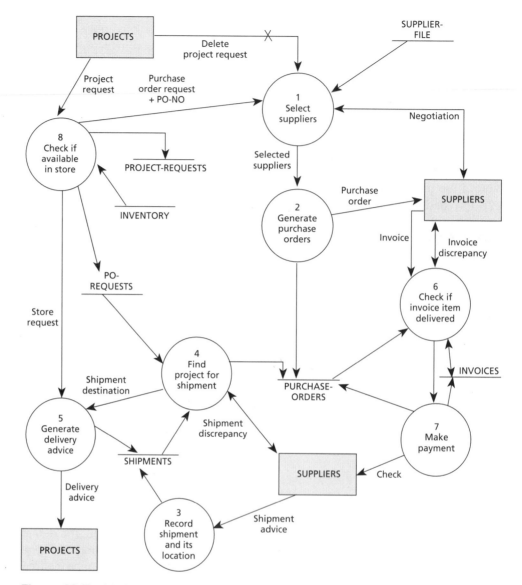

Figure 13.7 *Logical model*

4. Change Process 5 to receive data flows 'store request'. These data flows are used to withdraw parts from inventory and a generate data flow 'delivery advice'.

5. Add data store PROJECT-REQUESTS to store all project requests. In Figure 8.16, all project requests appeared in data store PO-REQUESTS. In the new system, PO-REQUESTS contains only those requests whose item requirements cannot be met from INVENTORY. Thus, PROJECT-REQUESTS is added to the system to store all project requests. Note also that Process 4 would be simplified in the new system because of the more simplified matching.

Note that in the new system, only project requests, which require purchase orders to be sent to suppliers, are stored in data store PO-REQUESTS and sent to Process 1. A project request directed to Process 1 has a purchase order number assigned to it by Process 8. This purchase order number is sent to Process 1 together with the project request. Data store PO-REQUESTS contains the requests sent to Process 1 together with the purchase order numbers allocated to them. Process 1 will select a supplier to supply the requested items, and Process 2 sends the purchase order to that supplier. A record of all purchase orders sent out by Process 2 is held in data store PURCHASE-ORDERS.

THE OO REQUIREMENTS MODEL

There are a variety of ways to specify requirements in object-oriented development processes. These often depend on the kind of development process that is used. For traditional process specification, models are often similar to analysis models, although they may include some development aspects. Rapid application development on the other hand might use models that combine modeling constructs from a number of linear phases. One early example is the requirements model in OOSE. It is made up of the components shown in Figure 13.8—behavior, presentation, and information. The OOSE method also has a number of specific ways to design interfaces and support processes. Each use case is seen as a process, and a special process object is created to model the process. This object initiates all the process steps by calling other objects as needed. There are also interface objects created for each user in a use case.

In Jacobson's method these components become the entity, interface, and control objects. These correspond to use cases and model the interaction of objects within the one use case.

TEXT CASE A:
Developing an OO Specification
An example is shown in Figure 13.9. This shows the analysis model that is part of the requirements specification for interactive marketing. It shows the major objects that are identified from the use cases as ellipses. They include the object classes that were illustrated in Figure 12.7 and relationships between them. Figure 13.9 also shows a number of interface objects.

Start with each use case and determine the control object for the use case. A rough rule is that there should be one control object for each case. Figure 13.9

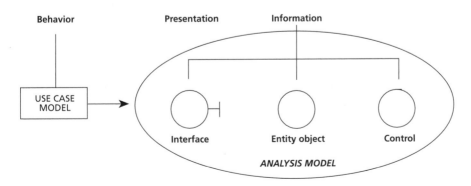

Figure 13.8   *The requirements model*

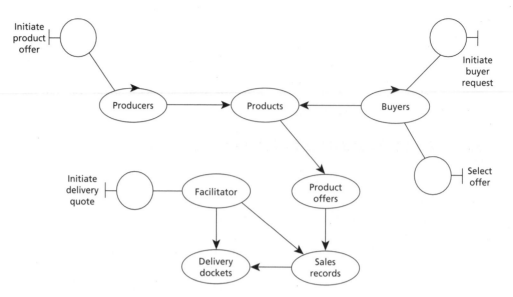

Figure 13.9   *A requirements model*

illustrates one such object, PRODUCERS, that initiates the registration of new offers. Similarly, BUYERS initiates purchase of products for a buyer, commencing with finding a suitable offer, getting the buyer to agree, and then confirming the sale with the producer and the buyer. A system may be made up of a number of control objects—for example, there may be another control object to arrange deliveries. The actions taken by that control object will be defined by its use case and include the definition of interface objects.

Each interface object will be eventually implemented, as a user interface, as program modules or perhaps WWW pages.

USING UML IN SPECIFICATIONS

UML provides additional modeling constructs to specify systems through stereotypes and the realization association. One intention of stereotypes is to tailor the modeling method to a particular application domain. Thus, classes can be stereotyped to imply additional meaning within an application domain and often imply new constraints. For example, we could stereotype buyers, producers, and facilitators as <<people>> in the analysis model. We could then have a constraint that no aggregation relationships exist between people. For a specification model, we can stereotype an implementation class that uses the elements of some existing element. A realization association denotes the objects the properties of which are used by the implementation stereotype.

Figure 13.10 illustrates the use of stereotypes in the specification model based on UML. Here we have the stereotype <<interface>>. The realization association is modelled by the dotted line. Thus, for example, SELECT-OFFER is an interface used by a buyer to select a product offer. The dotted line from PRODUCTS to SELECT-OFFER means that the interface gets the values that it presents to users values from PRODUCTS.

One question that may come up here is what detail one should go to in a specification. For example, should interfaces be specified precisely at this stage, or

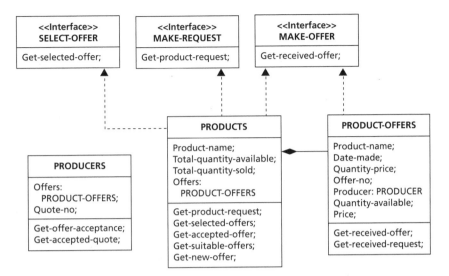

Figure 13.10 *Requirements model using UML stereotypes*

should this be left later in design? Thus, we can have a broad specification, which is elaborated into detail in design. The answer often depends on the type of development process used. In a strict linear cycle, detail can be left to later. During prototyping or rapid application development, such detail is captured earlier.

TEXT CASE C:
Universal Securities—Realizing the Strategy

Chapter 3 defined the broad requirements for Universal Securities, described project objectives in Figure 3.15, and described the following feasibility study, which outlined the conceptual solution shown in Figure 7.5. We now begin to expand on the strategy by more clearly showing the way the components will be realized.

One strategic objective is to 'outsource some specialist production and form necessary alliances for the production of parts'. The question now is how to realize this objective. If you go back to Figure 3.15, you will see that the conceptual solution suggested a WWW interface that can be used by potential suppliers to respond to supply required parts. However, the system must go beyond simply allowing suppliers to respond to requirements to allowing them to bid for orders and keep track of delivery schedules using the WWW. Thus, suppliers must be integrated into the system to the extent that they are notified when required parts are needed and, if necessary, get involved in their installation. They can then access information about these parts.

Object modeling can be used to specify requirements. A way to start is to develop some initial object classes. To develop the initial object classes, we might look at what people do. Thus, there might be scenarios or use cases for creating a contract or for getting a customer order. There would also be scenarios for suppliers responding to parts requirements and delivering parts. We find, for example, that orders are arranged by local marketing managers and are broken up into parts. Facilitators then place supplier orders with suppliers for the required parts.

We can, for example, develop a use case that describes how the facilitator would place a supplier order on the Internet and follow up the take-up of this order by suppliers. Here, we do not have a use case to start with, so specifications will have to be used to define use requirements. A simple use case can take the following form:

1. Facilitator is notified that a new set of parts is required.
2. The facilitator checks the required parts order and enters the required parts as requests for quotes.
3. Suppliers that have the capability to supply the parts are notified.
4. Suppliers place their responses to the requests for quotes.
5. The facilitator selects from these responses and places supplier orders quoting required delivery times.
6. The supplier delivers the parts when needed.

Some such initial object classes are shown in Figure 13.11. This figure shows contracts arranged by regional managers with suppliers. It also shows the parts required in each order as the association object class 'Required parts'. Facilitators then create requested quotes, which are extensions of required parts. Suppliers respond to these quotes, and these responses are used to create supplier orders.

Developers may now define some prototype WWW pages that show how the system can be implemented on the WWW. These can be shown to the users (in this case the facilitator and suppliers) or tested by making them publicly available. User response is then evaluated, and changes are made if necessary.

The problem now is to implement the proposal. Use of WWW interfaces is not simply a technical problem but requires social acceptance by all parties. For this

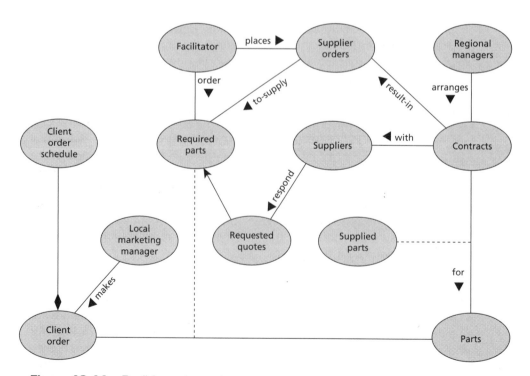

Figure 13.11 *Realizing outsourcing*

reason, any system will be implemented gradually, and the design itself is gradual. Thus, one page (or set of pages) might be implemented and trialed with users. This may be amended and tried again or integrated with a new page.

INTEGRATION WITH DESIGN

Many Web-based applications are developed using rapid application development. In this case, there is a strong link between requirements modeling and architectural and detailed design. This can then lead to more detailed specifications than those shown in Figure 13.11. As rapid application development combines specification with design, it can benefit from the greater use of stereotypes to specify some implementation details.

RAPID APPLICATION DEVELOPMENT

Rapid application development is often used to realize a strategic development that must support new ways of doing business. It is very often used in developing electronic commerce applications on the WWW. Broad objectives for the system are identified, although development usually proceeds in an evolutionary manner. Often what happens is that strategic directions are defined for the information system, but development proceeds in an evolutionary manner or using a rapid development cycle.

Little is known about processes based on the WWW, as they depend heavily on the social acceptance of new ways of doing business. A trial and error process, inherent in prototyping, combined with rapid application development is often suitable under these circumstances. As shown in Figure 13.12, this combines some analysis, some requirements specification, and some design. Often the overall strategy is to build a system in stages. Each stage defines some detailed requirements from the conceptual design, then that part of the system is designed, prototyped, and put out to use. The acceptance of the system is then evaluated, the system is improved, and the process is repeated. Pages are defined one at a time, checked

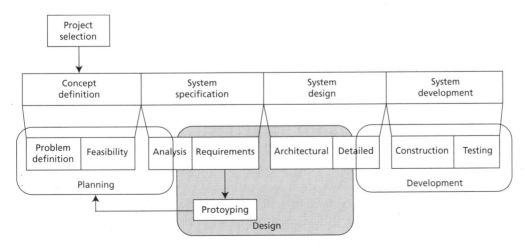

Figure 13.12 *Prototyping*

with users, refined, and so on. Thus, there is a close link with detailed design, with the architectural and detailed aspects going on almost concurrently. On the other hand, the linear process is often useful in process improvement, as there the process is often well understood, and objectives are quite clear and precise.

Requirements analysis is now strongly integrated with analysis, prototyping, and design. Thus, an analyst may be proposing a WWW solution and prototyping them with users very early in the development process. Consequently, models that combine aspects of requirements and design would be useful here.

STEREOTYPES FOR WEB DESIGN

Web design itself has a number of alternative implementations, which can be specified using stereotypes. Conallen (1999) describes a set of stereotypes for defining Web-based systems. Primarily, these are used to define how particular classes will be implemented. Thus, in rapid prototyping, the model must include elements of specification and architectural design. The model must include some objects that specify the architectural design components and provide an initial specification for a prototype.

TEXT CASE C:
Universal Securities—Initial Design

Universal Securities has decided to use the rapid application development process to implement the link with outsourced manufacturers. The first step is to define a scenario or use case for the process to be followed in placing an order.

The proposed scenario to develop the initial prototype is as follows:

1. A facilitator looks up a client order.
2. The facilitator looks up contracts that include these or similar parts.
3. The facilitator places supplier orders for these contracts.

The resulting object class diagram is shown in Figure 13.13.

Figure 13.13 contains a number of stereotypes that indicate access to data through the Web. One stereotype is <<server databases>>, which indicates object classes that are stored on a server. There is a <<control page>> called PLACE-ORDER, which corresponds to the scenario and presents Web pages to users in the sequence that corresponds to the scenario. The <<form>> stereotype indicates Web pages that allow users to input information. These are <<submitted>> to objects residing on the <<server database>>. The facilitator can thus select a number of pages that include forms to access or update databases.

Figure 13.13 is an initial specification for a prototype. Once some experience is gained, the model can be expanded to include more options and alternates.

DESIGNING PROCESSES

Part of the specification must describe what people will do in the new system and the processes they will follow. It is not sufficient to do this simply by defining methods or programs. Processes that define what people must do to make the system work must also be included. One may at this stage argue that perhaps the process should be the first thing that appears in the specification of the new system, particularly as it is increasingly realized that good business processes are essential in

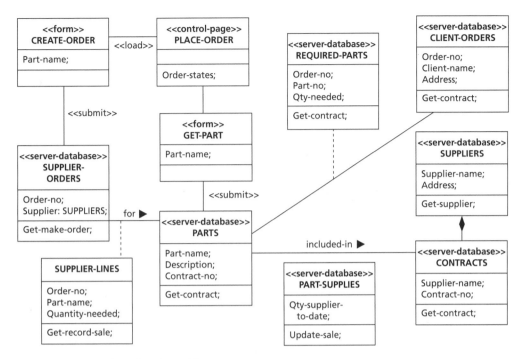

Figure 13.13 *Specification with stereotypes*

organizations. Processes define the way the organization works and must therefore be properly designed if we are to have a high-quality organization. Analysts will increasingly be required to define the process first, including the jobs in the process, and then choose the technology to implement it.

A number of things are important in process design. One is to correctly define all the tasks in the processes and make sure they fit together. These tasks must transform and move the data in the way specified by logical design. Another important part of detailed design is job design. Job design specifies what people must do to carry out the tasks. Jobs must then be combined into the process. Processes, jobs, and tasks are thus related, and tasks must be designed so that they can be carried out by people and put together into an effective process.

PROCESS MODELS

Business processes are often represented by flowcharts or by what are now more commonly known as workflows or process models. Flowcharting is one of the earliest representation tools for physical systems and is still frequently used to describe physical procedures. This is especially the case where specifications are made to select reference processes. A variety of flowcharting techniques have been used in practice. A flowcharting technique uses a finite set of symbols to represent system components, which may be physical hardware devices, information stores, flows, or processes. Figure 13.14, for example, shows symbols used to represent processes by the language EPC (Event-driven Process Chain), which is used by SAP to model processes.

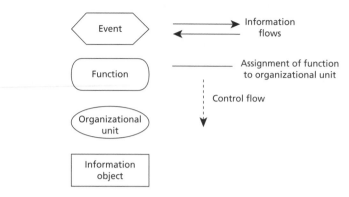

Figure 13.14 *Symbols used in EPC*

When drawing process models, we represent each process component by one of the model symbols. Then we look at information flows between these components and join the corresponding representations on to the flow chart.

Figure 13.15 illustrates a typical process model. It illustrates how loan applications are processed. First, the application is received and loan conditions are determined by the loans department. This department uses guidelines and creates records of approved loans. The customer is informed of the decision and decides whether to accept the conditions or request an alternative loan. If the conditions are accepted, then a loan account is set up by the accounts department.

TEXT CASE D:
Construction Company—A Flowchart
Figure 13.16 is a slightly more complex EPC model that illustrates the process of meeting project requests for parts. Project requests are received from PROJECTS

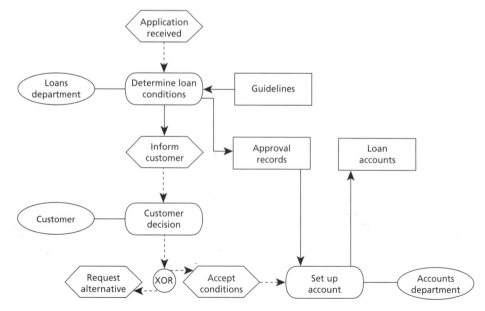

Figure 13.15 *An EPC model*

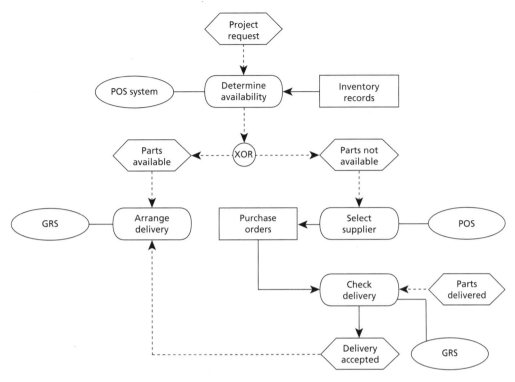

Figure 13.16 *A process model for the construction company*

and checked manually for parts availability in the POS. If parts are available in-store, then a delivery is arranged; otherwise the part must be purchased from a supplier. In the latter case, a supplier is selected manually and a purchase order is issued. Once the parts are received from a supplier, a check is made to see whether the parts on the invoice have been received. If so, the purchase order file is updated and delivery to the project is arranged.

JOB DESIGN

It is now generally accepted that systems will work much better if they are properly designed for users. Job design must encourage user involvement with the system. People will be more likely to accept the system if they are satisfied with it, and their skills will improve while using the system.

Good job design involves a number of things. One is to ensure that people are not in the service of computers. This can be done in a number of ways. First, the interface to the computer should be made as easy and pleasant as possible. The interface should use dialog that approaches natural language rather than using computer jargon. Dependence on computers is also reduced by creating jobs that contain a variety of work, some dealing with computers and some independent of computing. Designing jobs with such variety is often called 'job enrichment'.

Another important consideration in job design is the skills and skill levels required for jobs. A skill profile can be developed for each job, and the skill level needed can be specified. Skills include dealing with people, various technical skills, and decision or supervision abilities.

Finally, once the skills needed for the job are defined, it is important to precisely define the duties required of the job and make sure that these duties can be carried out within the allocated times. Precise specifications are needed to avoid uncertainties with consequent indecision and loss of respect for the system. Procedures must clearly specify what is needed at each step and what is to be done at each step. They must specify any checks to be made on incoming data and what to do in case of errors. Such precise descriptions make the system easy to use and consequently increase user satisfaction with the system. Users who know precisely what to do under all conditions will avoid making mistakes and thus accomplish their work as quickly as possible.

Job descriptions are usually documented in a user manual, which is produced in conjunction with users during system design. It includes an entry for each task and the jobs necessary to accomplish that task.

JOB ENRICHMENT

Job enrichment, now a common term in practice, is used to make jobs more interesting and rewarding to system users. It gives individuals greater initiative and responsibility, enabling people in the organization to improve themselves through involvement in a variety of activities. They also get more satisfaction from their work because they are contributing to the development of the organization rather than doing some minor and repetitive task.

An important part of job enrichment is to eliminate routine and repetitive tasks. For example, suppose we have a procedure where orders are received and checked for accuracy, then entered into the computer for processing. If the procedure calls for one person to receive the orders, another to check the orders, and a third one to enter the orders into the machine, there is little scope for people to improve themselves, as they are restricted to one small task and have no opportunity to do other work. They do not see the final outcome of their work, the completed order, because they are only a small cog in the ordering process. Consequently, there can be considerable dissatisfaction and lack of commitment to the work.

A better way would be for each person to follow the order through the whole ordering process—that is, one person takes an order, checks it, enters it into the machine, and follows it through from there. Satisfaction is generally increased, because people feel responsible for a larger task. In this method, a number of people are doing the same kind of work, and hence there are greater possibilities of team work, and people can learn from each other.

Jobs can also be enriched by providing a good interface with the computer, ensuring that people do not feel they are in the service of computers rather than the reverse.

SUMMARY

This chapter describes the development of the system specification. Some design methodologies use objectives to focus the design whereas others allow objectives to evolve as design proceeds. The waterfall development cycle focuses design by defining objectives. The chapter describes ways of specifying the objectives. Broad design in the linear cycle is made up of two steps—producing a new logical model and following this with a new physical model. The chapter defines some techniques used to define these new models. It then outlines how the physical model is used to derive processes within the new system. A similar approach can be followed in object modeling, although the object approach tends to favor prototyping or evolutionary development processes. The chapter shows how these processes are defined, and emphasizes the importance of defining jobs to be carried out by people within the process.

DISCUSSION QUESTIONS

13.1   What is the role of objectives in system specification?

13.2   Describe the four problem-solving steps suggested by DeMarco. Do you think they are natural to the way analysts proceed in analysis?

13.3   What are the advantages of developing an E–R diagram separately from a data flow diagram? Consider the advantages in cases where you are developing a database intended primarily for on-line inquiry or where you are designing a system from scratch.

13.4   What is the domain of change and how would you create it?

13.5   What are the advantages of using object modeling in a prototyping development process?

13.6   Do you agree that using stereotypes is useful in rapid application development?

13.7   Suggest some typical flowcharting symbols you would be likely to use.

13.8   Do you think that physical flowcharts are useful, and could they be integrated somehow into structured systems analysis?

EXERCISES

13.1   Develop a specification using UML constructs for Case 2. Assume that a WWW implementation is proposed.

13.2   Suppose someone has proposed an additional extension to the project support system whose logical model is given in Figure 13.7. It has been suggested that projects be allowed to initiate their own orders for urgent parts requirements without going through the system. However, before doing so, they should check the inventory holdings. How would you amend the logical model to accommodate this suggestion, and what physical implementations would you suggest?

13.3   Draw a physical flowchart for the following process. An organization services a variety of equipment. Following a fault report, which may be by letter or phone, a fault report form is filled in and sent to the dispatch center. A repairperson is selected at the dispatch center, and a partially filled in repair report is prepared. This report includes the person's name, the fault description, and the promised date or time of repair. Following completion of repair work, details of the work are entered onto the repair report. These details include the time spent and parts used. After the repair form is received, it is costed, and an invoice is sent to the customer. Use the symbols shown in Figure 13.14 to describe this system.

BIBLIOGRAPHY

Butler, K.A., Esposito, C. and Hebron, R. (January 1999), 'Connecting the design of software to the design of work', *Communications of the ACM*, Vol. 42, No. 1, pp. 38–46.

Conallen, J. (October 1999), 'Modeling Web application architectures with UML', *Communications of the ACM*, Vol. 42, No. 10, pp. 63–70.

DeMarco, T. (1978), *Structured Analysis and System Specification*, Yourdon Press, New York.

Gane, C. and Sarson, T. (1979), *Structured Systems Analysis*, Prentice Hall, Sydney.

Jacobson, I., Christerson, P.J. and Overgaard, G. (1992), *Object-Oriented Software Engineering*, Addison-Wesley, Harlow, England.

USER INTERFACE DESIGN

CONTENTS

KEY LEARNING OBJECTIVES

How to judge a good interface
What is a user workspace?
Usability and how to measure it
Interface presentation and dialog
Windows and design of window systems
Designing multi-user interfaces
When should offline processing be used?

CHAPTER 14

INTRODUCTION

This and the next three chapters cover system design and describe activities that are carried out during the system design phase. As shown in Figure 14.1, system design follows system specification and includes interface design, database design, and program design. There is also a distinction between architectural and detailed design. Architectural design defines the overall system and its logical parts, whereas detailed design describes its detailed physical implementation. Often the step between architectural and physical design is blurred, because it is useful to know the physical methods used to build the system when designing the system architecture. This chapter covers interface design.

The goal of user interface design is to provide the best way for people to interact with computers, or what is commonly known as human–computer interaction (HCI). *Provision of good interfaces is important, because most people in organizations are spending more and more time interacting with computers—they enter transactions, retrieve data, design artifacts, and do the other myriad things that need to be done in organizations. It is important that interaction with computers be as simple as possible so it helps rather than hinders people in their everyday work. Better interfaces can lead to greater satisfaction and an improvement in the quality of work and the effectiveness of the organization.*

The ideal interface would be one where the user can interact with the computer using natural language. This is especially the case where computers are used in everyday work, where they should be ubiquitous, almost like a telephone. The user types in a sentence on the input device (or speaks into a speech recognition device) and the computer analyzes this sentence and responds to it. The user could then follow up with another sentence to which the computer would respond, and so on. However, technology has not yet advanced to the stage where it can support such natural interfaces, and most dialogs between users and machines take a much more structured form. It is, however, important to ensure that interaction support problem-solving methods that are natural to the user.

Many people believe that improving interaction is one of the most important activities in system design. One reason for paying more attention to HCI is that, nowadays, computers are used by nearly everyone, not only people closely associated with computers. People are no longer interested in the technology behind the computer, they simply want

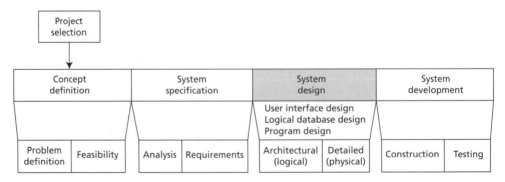

Figure 14.1 *Interface design within the development process*

a tool that is easy to use and can help them with their problems. They don't want to spend a lot of time learning about computer software, they just want computers to make their own work easier. A good interface certainly helps satisfy this goal.

The interface requirements are defined in broad terms during architectural design, and specific layouts are defined in detailed design. Architectural design usually defines how interfaces fit into the new business processes and the kinds of input and output that they should provide. Detailed field positions are defined in detailed design but very frequently are part of prototyping during requirements analysis. This chapter begins by describing what makes a good interface and then describes some typical interfaces and the techniques used to develop them.

DEFINING INTERFACES

An interface realizes the way the user sees the information in the computer systems and how the user can interact with the system. There are many reasons for people to use computers. The most common are to:

- capture information for storage and later use by system users
- retrieve information needed in work
- support everyday operations, such as keeping appointment diaries or writing reports
- solve a problem or make decisions, such as designing a system
- control remote facilities, such as a machine tool
- carry out a task in a transaction
- develop an artifact, such as writing a report or proposing a budget
- support workgroup interactions, often when jointly developing an artifact.

Interfaces must make it easy for users to carry out these tasks. How then do we design good interfaces? A good interface must first of all provide the information needed by users in a process, usually as a series of transactions, and present these transactions in an easy-to-use way. The most important consideration is how close the presentation approximates the way the user sees their problem—what we call the presentation metaphor. Then we check on the actual way the user interacts with the screen.

THE PRESENTATION MODEL

Systems users usually have a view of the world in which they operate. They are likely to find interfaces that closely model the user view or their **mental model** easier to use. The gap between this user mental view and what is presented on a screen is usually called a cognitive gap (Figure 14.2). This gap defines the amount of thinking that must take place for a user to make sense of what they see on the screen. Designers must provide a metaphor that will help users solve their problems in terms of objects that they can readily understand in terms of their everyday work.

The difficulty of reducing the cognitive gap depends on the kind of work process supported by the interface. Often, effectiveness is improved by providing an interface that closely parallels the user's mental model of their problem—that is, the way a user perceives the problem. What we need to do is design interfaces that

Mental model The way a user sees a problem.

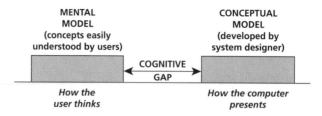

Figure 14.2 *The cognitive gap*

support the user's mental model so that the user can easily recognize the issues and work on the problem in a natural way. Mental models, especially for dialog, are different for different kinds of problems. Thus, a model for information input and retrieval is different from one used in decision making. A large number of presentation and dialog techniques have been developed for use in interfaces and most people are now familiar with them. One way to view interfaces in a general way is to consider them as workspaces.

WORKSPACES—A GENERAL VIEW OF THE INTERFACE AS A WORK SUPPORT SYSTEM

Workspace The space and facilities provided for a user on a screen.

A **workspace** defines all the information we need for our work and the presentation of this information. When we work, we usually lay out information before us in a way that will simplify our tasks. A good workspace will support the mental model of our task and will provide facilities for carrying out the actions associated with the task. The workspace will also define the actions that users can take within the workspace. The computer screen is then seen as part of a user's workspace. It must become part of the user's information and be integrated with other media that form the workspace, including documents, books, and tables.

This wide variety of interfaces and users means that workspaces must support a wide variety of users. Different principles and methods are often needed to support the different kinds of uses. The main distinction is between:

- personal systems, which keep records about a person's activities and must present information to users in an easily understood way
- transaction interfaces, such as those based on processing forms, which must be precise and capture the information needed by the transaction, and
- workspaces to support collaboration in groups, which must keep users aware of what others are doing.

According to some, the ultimate goal is for information technology to eventually make up most, if not all, of a user's workspace. We now have multimedia systems that include text, video, and voice. Thus, we can support communication in all these media. Why then do we need phones, faxes, and other devices when all of these can be integrated on the computer screen? All we need to do is link the computer to the phone system and dial through the computer, or have the computer dial numbers for us. Representing the entire workspace using information technology may be the way of the future, but at the moment, most computer interfaces are restricted to only a part of the user's workspace. Designers must identify the part of a user's workspace to be represented on the screen, at the same time presenting information in a way that is natural to use and complements the remainder of the workspace. This information can include reports, tables, and documents and should be provided in a way that can easily be manipulated in natural chunks rather than through computer-specific commands.

MULTIMEDIA DISPLAYS WITHIN WORKSPACES

Screen presentations are now becoming richer, because it is possible to use more than one medium in a display. A screen may now contain some text, an image, a video, and even speech. Each of these may use their own window. This provides more options to the screen designer. For example, a user may be controlling tools in a machine shop. There may be a video that monitors the tools and displays them on a multimedia display. The user can then activate the tools through the screen.

INTERFACES FOR PERSONAL SUPPORT

Personal work may include work used solely to build up a person's expertise, or work involved as part of some wider process. The kinds of personal support needed include document processing software, such as word processors and spreadsheets. Personal support systems pay special attention to the presentation and layout of artifact information so that a user can select and view parts of the artifact and apply various transformations to them. The most common example is working on a report, a budget, or a diagram. Parts of these artifacts may then be used in a business process or be produced for personal use. The computer can be used in the development of the artifact and in transferring parts of it into the business process.

There is increasing emphasis on using a variety of media in preparing artifacts and in combining information from more than one artifact. This includes combining a document from many parts, such as moving a drawing into a text, and using reference material as background.

The artifact itself may be presented by using different media such as text, graphs, images, sound, or video. The usual approach is to use windows to allow users to lay out their work in front of them and menus to select transformation commands to apply to the artifacts. An accurate representation of the user's mental model on the computer screen is again important for the user to quickly recognize the problem and thus simplify the decision to reach a solution. This requires special attention to the layout of the screen. Important features of such interfaces are:

- definition of mental domain concepts and their representation by *icons*
- identifying the actions to be taken by users
- subdividing the problem space and representing each part by a separate window.

Similar features are needed to move information from personal artifacts onto screens provided by a business process. What we need is a multi-window display, where one window is part of a business process and another is a personal document. We should be able to select part of a personal document and move it into the business process window.

WINDOWS AS A WORKSPACE

One of the most common presentations is windows. A window is an enclosed area of a computer screen that provides a well-defined object. The window will contain all the objects needed by a user and present the user with a variety of commands to operate on these objects. Figure 14.3 shows a window with which most people are familiar—the window for Microsoft Word. It displays the current content, which is some text, and a menu bar that provides the user with a number of actions.

Dropdown menu A menu selection that provides more detailed selections.

Figure 14.3 also illustrates some other common features now generally found in interfaces. One is the idea of a **dropdown menu**, where selection of one of the entries on a menu bar provides the user with further options. In Figure 14.3, the selection of 'File' gives the user a number of options of what to do with the current file. Another is an icon bar, which again provides a user with a variety of options. The user selects the option by clicking on an icon.

There are other ways of interacting apart from menu commands:

Templates are equivalent to forms on a computer. A form is presented on the screen and the user is requested to fill it in (Figure 14.4(c)). Usually several labeled fields are provided, and users enter data into the blank spaces. Fields in the template can be highlighted or blink to attract attention. The advantage that templates have over menus or commands is that the data is entered with fewer screens.

Commands and prompts: In this case the computer asks the user for specific inputs. On getting the input, the computer may respond with some information or ask for more information. This process continues until all the data has been

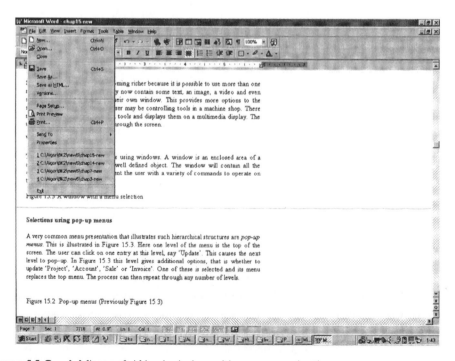

Figure 14.3   *A Microsoft Word window with a menu selection*

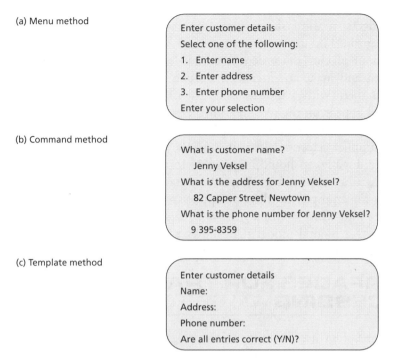

Figure 14.4   *Methods used in dialog design*

entered. In Figure 14.4(b), the system prompts the user for successive pieces of data. It first asks the user for the person's name. When the user enters the name, the computer asks for the address. Then it asks the user to enter the phone number.

MULTI-WINDOWS WITHIN A WORKSPACE

Often users must consider relationships between different artifact objects in their work. A multi-window presentation is often used for this purpose. The presentation in Figure 14.5 includes four windows. One window shows orders, another shows

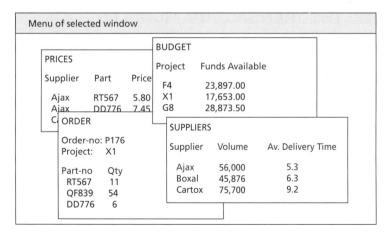

Figure 14.5   *A multi-window presentation*

project budgets, a third shows prices quoted by suppliers, and a fourth shows supplier records. The user is presented with all the information needed to make a decision about placing the order. They can look up the prices for the parts, select the supplier, and make an entry in the order. Windows can be selected, usually by clicking on them with a mouse, one at a time, when the menu for the selected window will appear on the screen. A further extension is to include a work window used for simple computations, such as computing the total value of an order following supplier selection. In that case, the screen would be the total workspace for the user. Finally, it should be possible to move information from one window to another, thus supporting simple integration of information from different functions—for example, to move a supplier price to an order.

With a multi-window display, users have all the information in front of them. This supports a mental model that allows all the factors to be jointly considered when making a decision.

INTERFACES FOR TRANSACTION PROCESSING

Windows are often used in transaction processing. Design for transaction systems must encourage rapid interaction between the user and machine. Interactions in transaction processing are usually simple. Each interaction deals with one user, who usually enters one or a very few related facts. This may be an entry of one record, such as an appointment, or an update to a record, such as a payment. Transactions are usually made through a workstation, and messages are interchanged in a relatively short space of time. One important property is that an output is obtained from the computer very soon after the input, allowing the user to make additional inputs if necessary. Transaction outputs need to produce the minimum information for a particular purpose. If the information is insufficient, the user can ask for more information. Thus, the computer can be selective in the information it outputs. Interactive interface design calls for user–machine dialogs that permit rapid interchange of information between computers and their human users.

CHOOSING THE TRANSACTION MODULES

Defining the information presented in a transaction is often part of the system specification. The requirements model produced in Jacobson's method, for example, includes the interface objects and what information they should present. Each such object defines one interface module that will interact with the user in some way. Each such interaction results in one transaction with the system. For example, if we return to the requirements model in Figure 13.10, we see that it defines four transactions—initiate product offer, initiate delivery quote, initiate buyer request, and select offer. Each of these can be implemented as one or more screens.

COMPARING DIALOG METHODS FOR TRANSACTION PROCESSING

Different methods may be appropriate for different purposes. The template is probably the best way to enter information about an entity, like a person, because all the data is on the screen at the time of input. The user can view it, check it, and decide to input it in one screen. However, a template may not be the best way to change certain details about a person, such as their address or telephone number. A menu system may be more useful in this case, because it displays the changes that can be made and asks the user to select that change.

Most dialogs, however, use combinations of all these methods. A dialog usually goes through a number of screens, with the response to one screen usually resulting in a new screen being displayed. A response to that screen leads to yet another screen, and so on. Often dialogs are described by screen hierarchies.

One such screen hierarchy is shown in Figure 14.6. It starts with a menu that asks the user to select one of three actions—enter a new customer, delete a customer, or change customer details. A new screen is displayed, depending on the choice. If a new customer is to be entered, then a template asking for customer details may be displayed. If deletion is specified, there may be a command asking for the name of the customer to be deleted. If details are to be changed, then another menu may be displayed asking the user to define the kind of change.

The dialog hierarchy may go down any number of levels. In Figure 14.6, the hierarchy goes down four levels to enter a new customer. First it gets the customer details, then it asks for customer status. If the customer is to be an account customer, the hierarchy goes down one more level to set up an account for the customer.

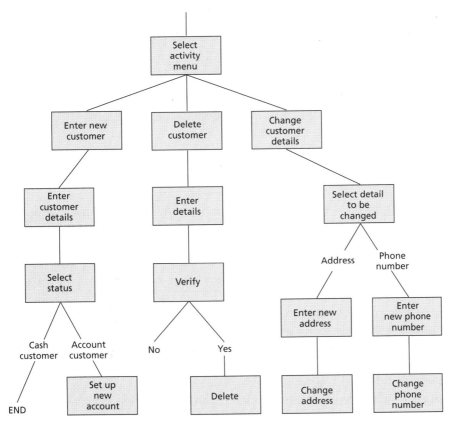

Figure 14.6 *Dialog hierarchy*

CONTROLS FOR INTERACTIVE TRANSACTION INPUT

Most transaction input systems include controls to prevent erroneous data from entering the system. Such controls are made by edit programs, which check every field of data entry at the time of input and inform the user as soon as there is an error. The user is usually informed by a loud beep together with a message that describes the error.

Some techniques are also commonly used to help users avoid errors. One method is to highlight a blank field for data entry. The size of the field ensures that a user does not enter more characters than can be accepted.

Another error-reducing technique is to reduce keystrokes required by users. Responses should preferably use only one keystroke and should require users to make a yes/no choice. When a yes/no choice is made, the system design should permit entries such as Y, y, N, n, Yes, YES, No, NO, and so on.

DESIGNING WORKSPACES FOR COLLABORATION

In collaborative workspaces users in a workgroup interact continuously over a period of time on one problem. The range of problems here is quite large. It may be specifically aimed at one particular activity such as:

- designing an artifact such as a document or a report, or the screen layout itself
- making a decision about a course of action, such as what route to take to make a set of deliveries
- communicating and coordinating with other group members about the progress of an activity.

Within the emerging global enterprise, collaboration may concern support for a knowledge-intensive process that may require expertise located at a number of locations. Typical examples here may be:

- preparing a contract or a response to a contract
- identifying new strategic directions
- developing new products or services.

Collaboration on a specific artifact often requires one workspace, although the kind of workspace may be different for each kind of activity. Such workspaces have some additional features compared with those used for personal support—for example, showing interaction between users. The term multi-user domain (MUD) is often used to imply that the same screen must support a number of different users. One overriding requirement is that the workspace itself should evolve as new ideas come up. Furthermore, the idea of *awareness* must be included in the interface. This allows each user to be aware of what other users in the group are doing. Responsibility for artifact manipulation may be distributed: the user may be working on one part of a document but at the same time must be aware of what other users are doing to other document parts.

Support for a wider community often requires more workspaces for each activity. It also requires awareness to be maintained between the workspaces. Generally, support must not only be provided for tasks but also provide ways for people to familiarize themselves with the work supported by the workspace, its background, and important documents. This is often known as the work context. Support also includes finding out about the other people in the workspace and being able to do more than just communicate with them. It should be possible to make commitments across a distance to work together to an agreed-upon goal. This often requires not only appropriate technology but also changes to people's ways of working.

Figure 14.7 describes an interface that includes the kinds of objects found in asynchronous collaboration. The documents define the context. The roles specify the responsibilities of people and who occupies the roles. A discussion database allows asynchronous interaction. Users can select any objects in these windows to work on in any sequence they like. Awareness is maintained through easy access to goals, news items, terminology, milestones, and any surprises that may occur in the workspace. Workspaces like that shown in Figure 14.7 include other objects such as news items, statements of goals, and ways to send messages following the occurrence of events within the workspace.

MUD WORKSPACE REQUIREMENTS

It is often necessary to support synchronous collaboration. In that case, MUD workspaces must provide facilities for users to carry out their personal work and interact synchronously though the interface. For example, suppose a group of people are working at the same time but at different places, and all group members need

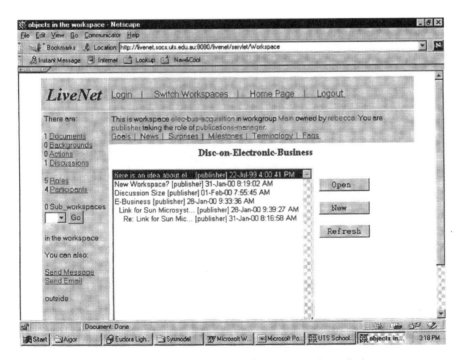

Figure 14.7 *Representing a collaborative environment using windows*

WISIWYS 'What I
see is what you see'
interface.

to share the information. What is needed is an interface known as a **WISIWYS** (what I see is what you see) interface, where any change made by one user appears on the screens of all the other users, as shown in Figure 14.8. Each user has a window or windows for a personal workspace and another window for a group workspace. The group workspace also has a set of identifiers, which may be faces, of all the group members. One face is shaded to show who currently has control of the workspace. The entry made by that member appears in the workgroup window of all the other users. Control from one user to another may be negotiated through a shared cursor. A controlling member must 'park' the cursor at some spot and another member can then grab it. Alternatively, control can rotate from one member to another, perhaps with a predetermined time limit.

The kinds of additional windows needed to support coordination vary with the kind of group. Figure 14.8 assumed a simple interchange of messages, or perhaps changes made on an artifact in a sequence. These windows may show other group members or they may contain information on who is working on particular parts of a jointly owned artifact. WYSIWYS is only one possible kind of group support.

DESIGNING WORLDWIDE WEB INTERFACES

It is now more and more common to design interfaces for the WWW. The design shown in Figure 13.10 can be implemented on the WWW, with each of the transactions being implemented as a set of WWW pages. Thus, for example, we can have a home page that allows users to select any of the transactions through a menu. Selecting a transaction will open up new pages.

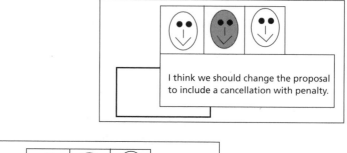

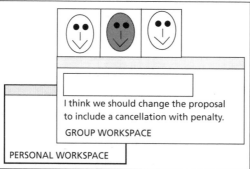

Figure 14.8 *Additional windows for collaboration*

It is often a good idea to build a directory of pages for the site. Figure 14.9 illustrates an initial directory of pages to initiate a trade. It is assumed that there is a home page for the interactive trading system. The home page may include a description of the system together with a menu (often a set of graphic images on the WWW) that allows a user to select any transaction available to that user. The menu selections are shown on the lines emanating from the page. When the Initiate Trade transaction is selected, new options are provided to the user on a TRADE PAGE. One of these is to search through the products. A particular product may be selected through an entry in a form field, and this in turn will allow the users to view details about the product in a PRODUCT PAGE. Alternatively, the user may select a particular producer. If the user searches a PRODUCT PAGE and becomes interested in purchasing that product, they may open up a page to set up a trader for the product, or alternatively look at the product's sales history, before initiating a trade. Often scenarios of such sequences are defined, and storyboards are built with skeleton pages.

TEXT CASE B:
Managing an Agency

The manager of the agency defined rough layouts of workspaces needed to support the task teams and committee members. One of these is shown in Figure 2.12. It outlines access to documents and ways to comment on them. The next step is to implement them using Internet technologies. The workspace in this figure is predominantly asynchronous and provides people with access to the latest information. A workspace like that shown in Figure 14.7 could be useful here. The documents window would contain the latest documents, and a number of discussions could be set up to exchange comments about them. Furthermore, the idea of roles is also useful, as it makes it possible to identify the various people associated with the agency project.

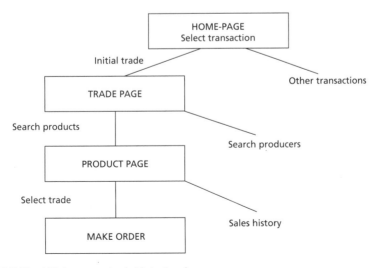

Figure 14.9 *Web pages to initiate trade*

OFFLINE TRANSACTIONS

This chapter would not be complete without referring to offline interfaces as well as interactive interfaces. In offline interfaces there is no direct response to transactions. Instead, a number of transactions may be collected and input as a batch. This batch is processed later, sometimes overnight, and outputs are distributed later to users. Thus, considerable time can elapse from when transactions are input to when a response is obtained. A typical batch run is shown in Figure 14.10. It begins by collecting a set of transactions that are entered onto a form. The form is passed to data entry operators, who enter the data through a computer terminal.

The transactions are stored in an input file. The input file goes through an edit run, which outputs any errors found in the transactions. The error transactions can be corrected and input again on a subsequent batch run. The correct transactions are passed to a FILE UPDATE program, which updates existing files using transaction data. The file is then processed by a report program to produce a set of reports.

OFFLINE INPUT INTERFACE

Data entry Transaction input for later batch processing.

The term **data entry** is often used to describe transaction input. Transactions are collected onto forms at the point of capture. A whole batch of these forms may then be given to a data entry operator, who will input them into the computer. The form that is used to capture the transactions can be an important component of batch processing. Form design is important. Forms must be easy to fill in and should not lead to unnecessary delays while users decide how to enter some unusual

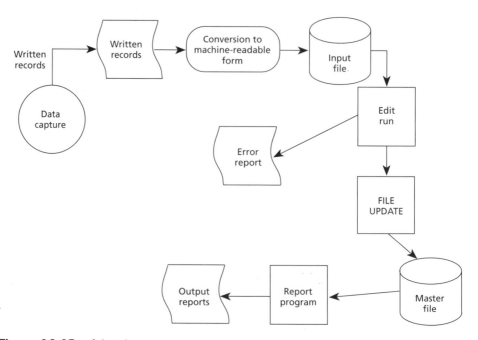

Figure 14.10 *A batch run*

transaction. Form layout must be clear and must capture the complete data necessary for the transaction. Sometimes forms allow codes to be used to minimize the time needed to fill them out.

Figure 14.11 illustrates a typical form, called 'onsite sales'. This form is used to collect information about sales made by salespersons in an organization. One form is filled out following each sale, at the time and place of the sale. The items sold are also recorded. The form is then sent to a central computer site for data entry.

The 'onsite sales' form is divided into a number of parts. One part records information about the customer, another contains information about the salesperson, a third part contains the items sold to the customer, and finally the method of payment is recorded.

Some efforts have been made to make it easier to fill in the form. The salespersons have a code and must enter that code. This reduces the number of keystrokes later, because a short code rather than a long name must be entered into the computer. Similarly, only a circle is needed to select an area for a user, so one keystroke is needed to enter the area code into the computer. Only a tick is needed to enter the method of payment.

CONTROLS WITH OFFLINE INPUT

One important offline input requirement is to ensure that all data is correct and no errors are entered into the computer. If erroneous data is entered, then incorrect data will be stored in the computer. This in turn will lead to errors in the system when the data is used.

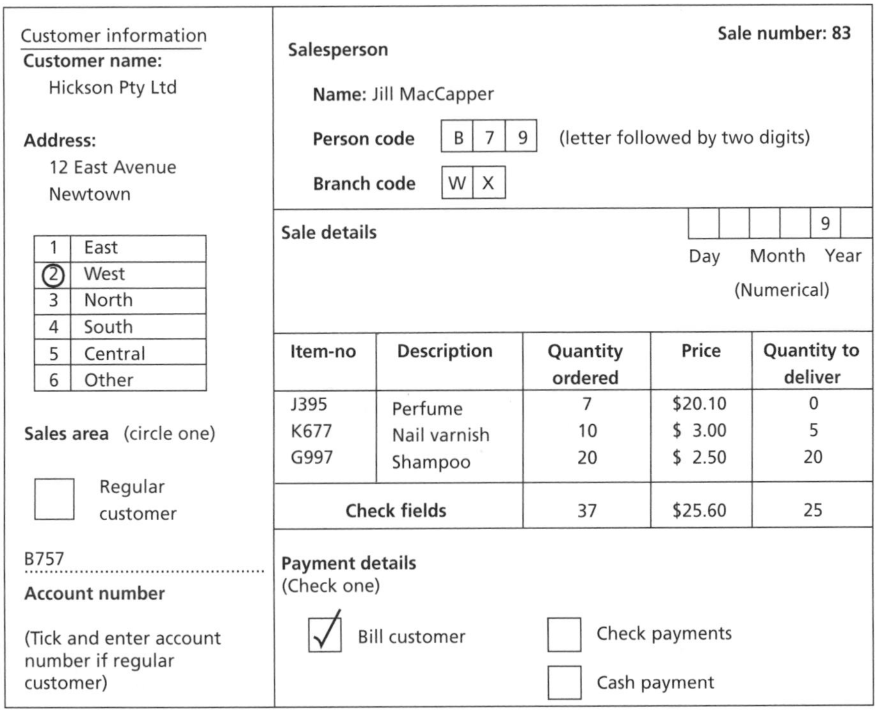

Customer information
Customer name:
Hickson Pty Ltd

Address:
12 East Avenue
Newtown

1	East
(2)	West
3	North
4	South
5	Central
6	Other

Sales area (circle one)

☐ Regular customer

B757
Account number

(Tick and enter account number if regular customer)

Sale number: 83

Salesperson

Name: Jill MacCapper

Person code B 7 9 (letter followed by two digits)

Branch code W X

Sale details

_ _ _ _ 9 _
Day Month Year
(Numerical)

Item-no	Description	Quantity ordered	Price	Quantity to deliver
J395	Perfume	7	$20.10	0
K677	Nail varnish	10	$ 3.00	5
G997	Shampoo	20	$ 2.50	20
Check fields		37	$25.60	25

Payment details
(Check one)

☑ Bill customer
☐ Check payments
☐ Cash payment

Figure 14.11 *Onsite sales form*

Two types of errors commonly appear in inputs. One type occurs when the transaction data itself is incorrect—for example, incorrect dates and errors in data formatting. The other type occurs when there is an error in entering the transaction onto the form or transcribing the data from the form into the computer. This is called a transcription error. Finally, there is the possibility of someone forgetting to input a transaction into the computer.

Different methods are used to protect the system against these kinds of errors. First, guidelines are included on the input form to guard against erroneous entries being made by the user. Thus, on the 'onsite sales' form are simple instructions such as telling the user how to enter a date. The day comes first, then the month, then the year. Computer edit programs are used to check data input to the computer. The computer system contains information about expected data formats and ranges and checks the incoming data to see if it satisfies these requirements.

Transcription errors are detected by including check fields with the transaction data. The 'onsite sales' form includes a number of check fields. There is a check field for QTY-ORDERED. Thus, all entries in the QTY-ORDERED column are summed, and the sum is entered into the check field. The computer checks whether the input check field equals the sum of the fields. If the operator makes an error entering a field, this computer check will fail. When an error is detected, someone must match the input against the form to find and correct the error.

Another method to reduce errors is to reduce the number of keystrokes needed to enter data—the fewer key strokes, the less the possibility of error. This is done by using codes, circling precoded fields, and using short names whenever possible.

Finally, input forms are usually numbered to check whether any forms have been lost. The 'onsite sales' form has a sale number. The salesperson will fill out these forms in sequential order. The computer will keep track of sale numbers received from each salesperson and report any missing numbers.

OFFLINE OUTPUT

Offline output is usually produced as paper listings by a line printer. Considerable care must be taken to present the output in an easily understood way. Titles must be provided for rows and columns, and important data must be highlighted or otherwise made to stand out clearly. Any column or report headings should be repeated on each new page, and pages should be numbered.

SOME GENERAL ASPECTS

Irrespective of the presentation, there are relatively general aspects of presentation and dialog criteria that are recognized as essential to making interfaces easy to use. Some of these are described below.

MEANINGFUL MESSAGES

There are many ways of evaluating interfaces. One is to determine how **user-friendly** the interface is. User-friendliness means that the interface should be helpful, tolerant, and adaptable, and the user should be happy and confident to use it. Thus, outputs or messages such as:

User-friendly Describes a helpful interface.

SYNTAX ERROR, or
IMPROPER DATA—TRY AGAIN

do not satisfy the criterion of friendliness. They simply say 'you made a mistake', and it's up to you to find out what it is and fix it. Instead, the interface should be explanatory and helpful, and say:

AT THIS POINT YOU SHOULD PUT IN THE DATE IN THE FORMAT DAY/MONTH/YEAR.
YOU SEEM TO HAVE A MONTH WHOSE VALUE IS HIGHER THAN 12 BUT I EXPECT A VALUE IN THE RANGE 1 TO 12.

These simple things must be kept in mind when designing interfaces. Paying attention to friendly interactions results in better interfaces, which not only make users more productive but also make their work easier and more pleasant. The terms 'effectiveness' and 'efficiency' are also often used to describe interfaces. An interface is effective when it results in a user finding the best solution to a problem, and it is efficient when it results in this solution being found in the shortest time with least error.

ROBUSTNESS

Another important feature of the interface is its **robustness**. This means that the interface should not fail because of some action taken by the user, or indeed that a user error leads to a system breakdown. This in turn requires checks that prevent users from making incorrect entries.

A good interface can deal with incorrect inputs and prevent errors from entering the system. It also helps the user correct any errors. There are usually **controls** associated with the input to ensure that no erroneous data enters the system. Similarly, output design must ensure that all data needed by users is provided by the output and that this output is laid out in an easy-to-read way. Furthermore, it must capture this data without introducing any errors.

USABILITY

It is important for a designer to know how to evaluate interfaces. Although it may be easy to devise an interaction, it is often harder to tell how good it will be in practice. Effectiveness and efficiency are ways to evaluate interfaces. For example, a simple criterion for data capture is to minimize the number of keystrokes required of the user. Inputs must be well laid out and easy to understand and use. They must use precise names and allow abbreviations where necessary to speed up input. Repetitive inputs should be avoided.

The term **usability** is often used when evaluating interfaces. The computer interface defines how users interact with a computer and has an important bearing on users' acceptance of a system. There is considerable work in progress on how to define and measure the usability of an interface, or, to use more contemporary terminology, to define the **usability metrics**. Metrics are the things about usability that can be measured. What we need to do next is to choose their method of measurement. Metrics cover both objective factors and subjective factors. The MUSiC project in Europe identifies four kinds of metrics.

Robustness Ability to prevent interface errors from corrupting the system.

Controls Checks to ensure that system inputs are correct and do not destroy system integrity.

Usability A term that defines how easy it is to use an interface.

Usability metrics The things that can be measured to describe usability.

- *Analytical metrics*, which can be directly described—for example, whether all the information needed by a user appears on the screen.
- *Performance metrics*, which include the time used to perform a task, and system robustness, or how easy it is to make the system fail.
- *Cognitive workload metrics*, or the mental effort required of the user to use the system. These cover how closely the interface approximates the user's mental model or reactions to the screen.
- *User satisfaction metrics*, which include how helpful the system is and how easy it is to learn.

The next question is how to measure these factors.

MEASURING USABILITY

Usability is becoming more important, and consequently many software providers emphasize this issue and often require interfaces to be evaluated for their usability. However, no standard usability measures exist at present. Usability measurement depends on the kind of metric being measured. It is usually measured by questionnaire or observation. Thus, analytical metrics, such as 'Is all the needed information on the screen?', can often be observed. Performance metrics, or how long it takes to carry out some task, can often be measured by observing how long users take to carry out a task. Other metrics may be harder to measure, because they must consider the mental effort that is required of a user. Metrics on cognitive overload often center on the stress level of the user, which can be measured by a user's heart rate or blood pressure, and it is important that these measurements be made under actual working conditions. User satisfaction metrics can usually be evaluated through questionnaires or by interviewing users to find out their general attitude to the system.

Useability itself depends on many factors, sometimes called the context, which include:

- the kind of user, ranging from the expert to the novice user
- the kind of task carried out by the user
- the organizational environment
- the workplace conditions
- the technical system.

The whole area of selecting usability metrics and their measurement is outside the scope of this text. It may be worthwhile to say that most metrics have concentrated on transaction interfaces, and very little is known about what to measure in collaborative interfaces. Nevertheless, considerable work is in progress to define both measures and standards of usability.

DESIGNING FOR USABILITY

The next question is how to design for usability. Do we simply build a system and then test and adjust interfaces, or do we progressively check for usability as the system design proceeds? The second approach is obviously preferable as it minimizes the amount of rework. To make it work, however, we have to integrate usability considerations into the system development cycle. Design for usability must be user-centered and must ensure user participation. It is also experimental and iterative. This means that we must get the users involved early in system design and thus integrate usability design into the development cycle.

INTERFACE DESIGN TOOLS

Developing interfaces is a relatively tedious process. Programs must be written for menus, there must be provision for all selections, and it should be easy to move from one screen to another. Interface design can thus be quite a complex and laborious process unless we have tools that enable us to easily construct trial interfaces and quickly change them. The kinds of things are common to all applications. It is now very common to have development environments that provide high-level tools to define interface components. Some such tools are defined in Chapter 18.

SUMMARY

This chapter describes design of the computer interface. This defines how users in the system use computers. The chapter introduces the idea of the computer workspace and how this should be designed to make the work of users effective. It makes a distinction between workspaces for personal support, those for transaction processing, and those that support collaboration. It stresses that interfaces must be designed to capture all the information in a meaningful user interface and describes some criteria that interfaces should possess to facilitate computer use.

DISCUSSION QUESTIONS

14.1   What do you understand by the term *cognitive gap*?

14.2   What do you understand by the term *transaction*?

14.3   How would you design a transaction?

14.4   What is a user-friendly interface?

14.5   Why is usability measurement important?

14.6   What would you measure in a collaborative interface?

14.7   What factors are important when measuring the usability of an interface?

14.8   What kinds of interactions are commonly found in organizations?

14.9   What do you understand by the term *workspace*?

14.10   Do you think the computer should encompass all of a user's workspace?

14.11   Compare different types of online user dialog for transaction processing and describe the advantages of each.

14.12   What features distinguish group support workspaces from workspaces that support one user?

14.13   Why are user interface tools built up in layers?

14.14   What are the better known layers?

14.15 Why is it important to link user interface design tools with other software systems used in application design?

14.16 How would an interaction for computer entry differ from that used in decision making?

14.17 List some criteria for good form design.

EXERCISES

14.1 Design a transaction dialog for entering a project request for the construction company.

14.2 Suggest useful windows for a multi-window display for the schedulers in Universal Securities.

BIBLIOGRAPHY

Barfield, L. (1993), *The User Interface: Concepts and Design*, Addison-Wesley, Wokingham, England.

Bass, L. and Dewan, P. (eds) (1993), *User Interface Software*, Wiley, Chichester, England.

Brave, S., Ishii, H. and Dahley, A. (November 1998), 'Tangible interfaces for remote collaboration and communication', *CSCW98*, Seattle, Washington.

Bullinger, H.-J. and Fahnrich, K.-P. (1991) 'User interface management—the strategic view', in Bullinger, H.-J. (ed), *Human Aspects in Computing: Design and Use of Interactive Systems and Work with Terminals*, Elsevier Science Publishers, The Hague.

Dix, A., Findlay, J., Abowd, G. and Beale, R. (1993), *Human–Computer Interaction*, Prentice Hall, New York.

Hurley, W.D. (1992), 'Integrating user interface development and modern software development', *International Journal on Software Engineering and Knowledge Engineering*, Vol. 2, No. 2, pp. 227–50.

Mayhew, D.J. (1992), *Principles and Guidelines in Software User Interface Design*, Prentice Hall, Englewood Cliffs, New Jersey.

Molich, R. and Nielsen, J. (March 1990), 'Improving a human–computer dialog', *Communications of the ACM*, Vol. 33, No. 3, pp. 338–48.

Nielsen, J. (1993), *Usability Engineering*, Academic Press, San Diego.

Nielsen, J. (January 1999), 'User interface directions for the Web', *Communications of the ACM*, pp. 65–72.

Shafran, A. (1996), *Creating and Enhancing Netscape Web Pages*, Que, Indianapolis.

Shneiderman, B. (1992), *Designing the User Interface* (2nd edn), Addison-Wesley, Reading, Massachusetts.

LOGICAL DATABASE DESIGN

CONTENTS

KEY LEARNING OBJECTIVES

Defining logical data structures
Converting E–R diagrams to logical record structures
How to draw access paths
How to satisfy access requirements
Defining a relation
Why we should avoid redundancy
Normal relational forms

Functional dependencies
How to normalize relations
Multivalued dependencies

INTRODUCTION

This chapter continues with system design by describing how to design the logical database structure, which is part of architectural design, shown in Figure 15.1. The logical database is an important part of the design specification. The logical database is used to create the detailed physical design, which is described in Chapter 17.

Figure 15.2 illustrates the steps used in database design. One is the common path followed in structured systems analysis. Here the E–R model created in the system specification is first converted to a set of logical record types, with each record made up of a number of fields. The set of record types is called the logical record structure (LRS). The LRS is then converted to a relational model. Later in physical design, the relational model is converted to a definition for a database management system (DBMS). The database definition, which is the implementation model, depends on the DBMS.

Such conversion is not needed with object models, which are converted to object-oriented implementations. In that case, the system specification is converted directly to object classes in the implementation. There is also the alternative where the user requirements are specified as object classes but the implementation uses a record-based DBMS. In that case, the object model must be converted to an LRS, which is then converted to a database implementation in the same way as occurred with structured systems analysis.

This chapter first describes logical record structures and how to derive them. It then continues with the second goal, to create a design that is in some way a good design. Such criteria for good design are provided by relational theory, which has been used for this purpose for a number of years. The way this is done is to convert the LRS to a set of relations. The relations are then examined for redundancy and, if necessary, changed into a non-redundant form. The output of this step is a non-redundant relational model of the user system.

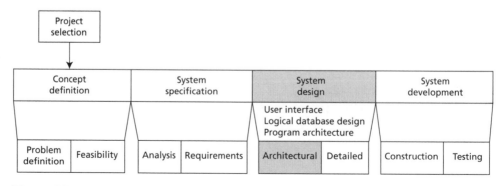

Figure 15.1 *Logical database design*

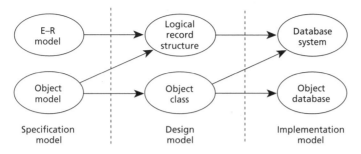

Figure 15.2 *Database design steps*

WHAT DO LOGICAL RECORD STRUCTURES LOOK LIKE?

There is no standard for **logical record structures** at the system level. A typical LRS is shown in Figure 15.3. The LRS is made up of a number of record types. Each record type is represented by a rectangular box and has a unique name. The three record types in Figure 15.3 are named PROJECTS, USE, and PARTS. Each record type is made up of a number of fields. In Figure 15.3, record type PROJECTS is made up of three fields, PROJECT-NO, START-DATE, and BUDGET. To distinguish the logical structure from the E–R diagram, the record type name appears outside the box, inside which the record type fields are placed.

The logical record structure also contains links between record types. Each logical record link is labeled by the fields that appear in both linked record types and is directed from one record to another. Thus, the link between PROJECTS and USE is labeled PROJECT-NO, because PROJECT-NO is a field in both the PROJECTS and USE record types.

The following convention is used to determine link direction. The link originates in the record type that contains only one record with a given value of the link label. Thus, in Figure 15.3, the link labeled PROJECT-NO originates from the PROJECTS record type, as there will be only one PROJECTS record with a given value of PROJECT-NO. The link terminates on a record type that may have one or many USE records with that value of label. Thus, in Figure 15.3, there may be many records with a given value of PROJECT-NO, because a project may use many parts. Another interpretation of the link semantic is that it originates in a record that must exist before a record type with the same value of the link label can be created. Thus, a record with a given value of PROJECT-NO in PROJECTS must exist before a USE record with the same value of PROJECT-NO can be created. Otherwise, no existent projects would be using parts.

Logical record structure A way of describing records at the system level.

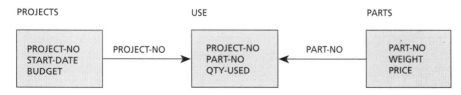

Figure 15.3 *Logical record structure*

There is an additional semantic interpretation for such links. Links can also be seen to define owner records of other records. In Figure 15.3, PROJECTS records own USE records. The common field, PROJECT-NO, identifies the particular records owned. Thus, a PROJECTS record with a given PROJECT-NO value will own all USE records with that PROJECT-NO value. There is also an application of this link. Suppose we want to find all the parts used in a project. We would first find the PROJECTS record with the required PROJECT-NO. The links are then used to find all the USE records owned by that PROJECTS record. These USE records contain the PART-NO of parts used by a given project.

It is, of course, possible to use other conventions for linking records, and different conventions can be found in different methodologies.

CONVERSION TO AN LRS SYSTEM MODEL

Now we have described what an LRS looks like, we can begin to describe how to derive it from the E–R model. Both E–R diagrams and object models can be converted to system designs that use logical record structures.

STRUCTURED SYSTEMS ANALYSIS—CONVERTING E–R MODELS TO LRSs

The first database design step in structured systems analysis converts the E–R analysis model to LRSs and specifies how these records are to be accessed. These access requirements are later used to choose keys that facilitate data access. Quantitative data such as item sizes, numbers of records, and access frequency are often also added at this step. Quantitative data is needed to compute storage requirements and transaction volumes to be supported by the computer system.

The combination of logical record structure, access specifications, and quantitative data is sometimes known as the system-level database specification. This specification is used at the implementation level to choose a record structure supported by a DBMS.

The simplest conversion is to make each set of E–R diagram into a record type. However, there is one small variation to this where we combine object sets. One such conversion is illustrated in Figure 15.4. Here, sets PROJECTS, FOR, and PARTS are each converted to a logical record type. The entity set ORDERS and the relationship set MAKE are combined into the one logical record type. You will find that such combinations are always possible in 1:N relationships. In a 1:N relationship, entities in one of the entity sets appear in one relationship only. In Figure 15.4, each order in the ORDERS set appears in one MAKE relationship only and hence they are combined.

Links are then added. Links start in logical record types that represent entities and terminate in logical records that represent relationships. There is a link from PARTS, which represents an entity set, to FOR, which represents a relationship set. The same rule applies where sets have been combined. There is a link from PROJECTS, which represents an entity set, to ORDERS, which contains the relationship MAKE.

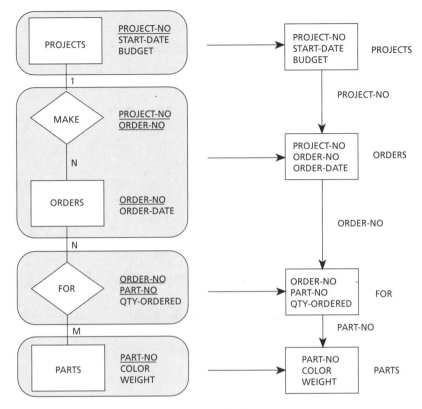

Figure 15.4 *Converting an E–R diagram to a logical structure*

Each dependent entity set is also converted to one logical record type. A link is then added from the entity set to its dependent entity sets. The label on the link will be the entity set identifier.

Subsets and dependent entity sets can also be converted to logical record structures. One example of this conversion is given in Figure 15.5. This shows the E–R diagram previously illustrated in Figure 9.15 but changed to model course offerings as a dependent entity set. COURSE-OFFERINGS is dependent on COURSES. Thus, a course must exist on the university's statutes before it can be offered. It can then be offered in many semesters, in a different room for each semester. Each of the sets is now converted to a record type in an LRS, with links labeled by the common attribute names. The links point from the owner record to the subset, as the owner must exist before it can become part of a subset—teachers must be hired before they can teach. Similarly, there is an arrow from the set to the dependent entity set—a course must exist before it can be taught.

TEXT CASE D:
Construction Company—Developing the LRS

We now illustrate the application of these techniques to the data of the construction company. The logical structure is shown in Figure 15.6. You may wish to go back to Chapter 9 and see if you can derive the logical structure in Figure 15.6 from the E–R diagram in Figure 9.5.

ENTITY-RELATIONSHIP DIAGRAM

LOGICAL RECORD STRUCTURE

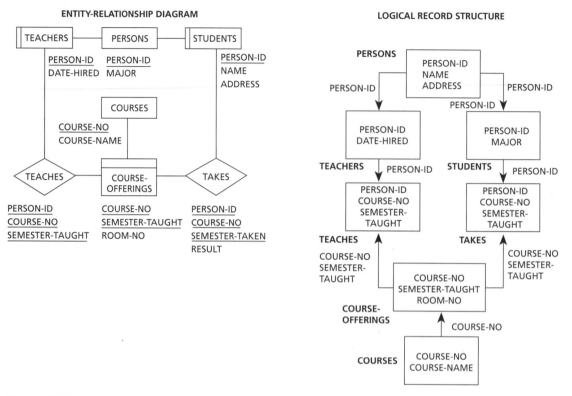

Figure 15.5 *Converting subsets and dependent entity sets*

The method illustrated in Figure 15.4 is used to convert the E–R model in Figure 9.5 to the logical record structure in Figure 15.6. Each entity set or dependent entity set is combined with any 1:N relationships in the manner shown in Figure 15.4 to form one logical record type. For example, dependent entity set INVOICE-LINES can be combined with relationship sets ABOUT, ON, and FOR to construct the logical record INVOICE-LINES. Such combinations are made for all other entity sets and dependent entity sets.

The next step is to add links to the logical record structure. Again, the method used in Figure 15.4 is used. The links always terminate in logical records that represent or contain relationship sets. Logical record type INVOICE-LINES contains relationship sets ABOUT, ON, and FOR. There are three links terminating on INVOICE-LINES for these relationship sets. There is also a fourth link from logical record type INVOICES to show the dependence of INVOICE-LINES on INVOICES. You may like to go through Figure 8.11 and see how the rest of the logical record structure in Figure 15.6 is constructed.

CONVERTING OBJECT MODELS TO LRSs

Object models must be converted to LRSs when a logical analysis model is implemented by a conventional DBMS. The simplest conversion is for each object class to become a logical record, with each class attribute converted to a field. A structured attribute may become a separate record. Different methodologies give their logical records different names—Jacobson calls them blocks, Henderson-Sellers names them schemas, and Booch calls them modules. The ways of specifying links

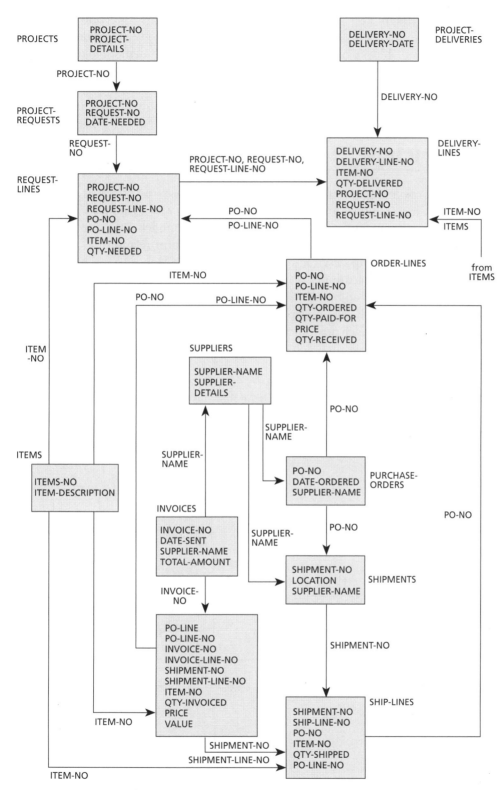

Figure 15.6 *A logical record structure for Text Case B*

and directing them also varies. According to Booch, the direction is to the dependent module. Figure 15.7 illustrates the first step of conversion from an object analysis model to blocks or modules based on Jacobson's methodology. It shows the analysis model shown earlier in Figure 12.6 converted to module blocks, with arrows labelled in the direction of dependency. You will note that the interface objects have also been converted to blocks, as they will eventually be implemented as program modules.

RELATIONS

Once we have a logical record structure, the next step is to convert it to a set of relations and use relational analysis. Relational analysis was introduced in 1970 by Codd, and is now a standard technique for reducing models into non-redundant forms by reducing the data model to a set of normal form relations.

There are two ways to describe the relational model. One is to use more formal terms such as relation, attribute, or tuple. The other is to use more familiar terms such as table, column, and row. The relational model describes data as a set of **relations** or tables. Each relation or table has a table name. Each such relation has a set of **attributes**, and each attribute has a unique name in the relation. The relation attribute is the same as a table column. Each relation has a set of tuples, or rows in a table.

Relation A table or list of values.
Attribute in a relational model A column of the table or list of values.

There is a close correspondence between the relational model and E–R diagrams. Figure 15.8 shows how the E–R diagram of Figure 9.2 has been converted to a set of tables or relations. Each entity set is converted to a table and each entity set attribute is converted to a column. Individual entities in an entity set become rows

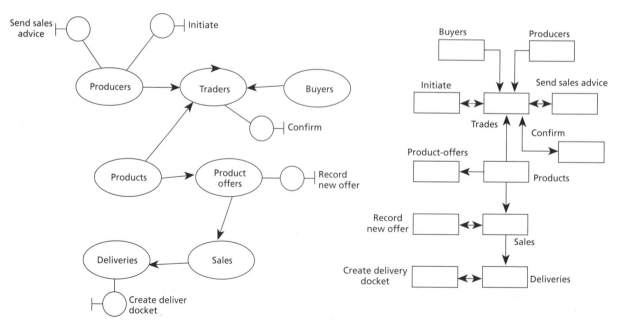

Figure 15.7 *Conversion of object modules*

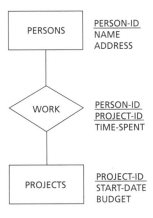

PERSONS

PERSON-ID	NAME	ADDRESS
PX1	Jackson	London
PX2	Maine	Liverpool
PZ5	Oldham	London

WORK

PERSON-ID	PROJECT-ID	TIME-SPENT
PX2	Proj3	30
PX1	Proj2	15
PZ5	Proj2	40
PX2	Proj5	30
PZ5	Proj3	75

PROJECTS

PROJECT-ID	START-DATE	BUDGET
Proj3	1 Mar 98	50
Proj2	15 Feb 98	30
Proj5	1 Nov 98	60

Figure 15.8 *E–R diagram and relations*

in the relation that represents the entity set. Similarly, relationship sets are also converted to relations.

A tabular representation of data can be useful in explaining the data structure to users. However, representing a system by tables can become clumsy, especially if the system is large. Besides having to supply a lot of space and drawing effort, we are always faced with providing columns for data and either leaving these columns blank or filling them with example values. To avoid such drawing effort, it is convenient to use a relational notation, which shows only the relation name and its attributes. The relational notation for the tables shown in Figure 15.8 is:

PERSONS (<u>PERSON-ID</u>, NAME, ADDRESS);
WORK (<u>PERSON-ID, PROJECT-ID</u>, TIME-SPENT);
PROJECTS (<u>PROJECT-ID</u>, START-DATE, BUDGET).

In this notation, each relation is represented by one line. Each line starts with the relation name and is followed by the names of the relation attributes in parentheses. The underlined attributes are the relation keys. Values of the relation keys identify unique rows in the relation. You will recall that when 1:N relationship sets are converted to relations, the relationship set relation has the same relation key as one of the entity set relations. The relations with the same relation key are then combined into one relation. Thus, if a person worked on one project, then the key of relation WORK would be PERSON-ID. Relations WORK and PERSONS would then be combined into the one relation:

PERSONS (<u>PERSON-ID</u>, PROJECT-ID, NAME, ADDRESS, TIME-SPENT).

There are two reasons for converting the E–R model to a set of relations. First, such a conversion is a convenient step for going from an E–R model to a set of files. Each table eventually becomes a database file. Each attribute becomes a field in the file, and each row becomes a record occurrence. The other reason is more important. A theory based on sound mathematics specifies how we should construct tables to avoid data redundancies. What we are after is a set of 'normal' relations. If we organize the data into a set of such relations, we are guaranteed a good data design.

A number of normal forms have been defined for relations. They are commonly known as first normal form (1NF), second normal form (2NF), third normal form (3NF), and Boyce-Codd normal form (BCNF). Two other normal forms are known as the fourth normal form (4NF) and fifth normal form (5NF). Relations in 1NF, 2NF, 3NF, and BCNF must satisfy a different set of constraints from those in 4NF and 5NF. Database designers must ensure that their relations are in the highest normal form.

It should also be pointed out that, although much of the work on relations has been of a mathematical nature, the beauty of the **relational model** is that the mathematical results can be explained in terms of constraints relevant to practical database design. Two main constraints must be met by normal relations. First, relations must be flat—that is, all the column values must have simple values and cannot be groups of values. Flat file structures are easier to access and change in order to meet new user requirements. Normal relations must also not contain redundancy—normal relations store each fact once only. To explain how to obtain a set of normal relations, we must first define some terminology.

Relational model A set of tables that describe the data in a system.

TERMINOLOGY AND PROPERTIES

Relations can be thought of as tables made up of columns and rows. In relational work, columns are often called attributes, and rows are called **tuples**. Each relation has a unique name within the system, and each column or attribute has a unique name in the relation. In Figure 15.8, the relations were given the names PERSONS, WORK, and PROJECTS. Individual rows describe objects modeled by the relation. Thus, in relation PERSONS (see Figure 15.8), each row (or tuple) describes one person, whereas each column is one property of the person. The column NAME contains the person's name, the column PERSON-ID contains the person's identifier, and the column ADDRESS contains the person's address.

Tuple A row in a relation.

Another property of relations is that the order of columns and rows is insignificant. Furthermore, a relation cannot contain two identical rows.

NORMAL-FORM AND NON-NORMAL-FORM RELATIONS

The difference between relations in normal form and those in non-normal form is illustrated in Figure 15.9. The relation ORDERS in Figure 15.9(a) is not in any normal form. Each row in relation ORDERS represents one order. This order is identified by the value of ORDER-NO and has one value of ORDER-DATE. Each order is for any QTY-ORDERED of parts with a given PART-NO. The parts ordered in the order are themselves modeled by a relation in column ORDER-LINES in relation ORDERS. Thus, an order identified by a value of 'ord1' of

(a) Non-normal-form relations

ORDERS

ORDER-NO	ORDER-DATE	ORDER-LINES	
Ord1	6 June 1998	PART-NO	QTY-ORDERED
		P1	10
		P6	30
Ord2	3 May 1998	PART-NO	QTY-ORDERED
		P5	10
		P6	50
		P2	30

(b) First normal-form relation

ORDERS

ORDER-NO	ORDER-DATE	PART-NO	QTY-ORDERED
Ord1	6 June 1998	P1	10
Ord1	6 June 1998	P6	30
Ord2	3 May 1998	P5	10
Ord2	3 May 1998	P6	50
Ord2	3 May 1998	P2	30

Figure 15.9 *First normal-form and non-normal-form relations*

ORDER-NO is for 10 parts identified by 'P1' and 30 parts identified by 'P6'. The important thing to notice for relation ORDERS is that one of the attributes (ORDER-LINES) does not have simple values. Attributes ORDER-NO and ORDER-DATE have only one single value for one row and thus are said to have simple values. Attribute ORDER-LINES has a value made up of a number of rows in another relation, and this value is not simple. Normal-form relations can have only simple values, and thus relation ORDERS in Figure 15.9(a) is not in normal form.

Figure 15.9(b) illustrates a normalized relation that contains the same data as the relation in Figure 15.9(a). Each value in each column of this relation is a simple value. Relations that are not in normal form can always be normalized. To do this, take each row in the unnormalized relation and look at the relation in the non-simple column of the unnormalized relation. Now combine each row of the relation in the non-simple column with the values of other columns in the unnormalized relation to make a row in the normalized relation. For example, take the first row in the unnormalized relation ORDERS. Now take the first row in the relation in column ORDER-LINES. In this row, PART-NO = 'P1' and QTY-ORDERED = 10. Now combine these values with the values in the other columns in ORDERS—that is, ORDER-NO = 'ord1' and ORDER-DATE = '6 June 1998'. All of these values now become the first row in the normalized relation ORDERS in Figure 15.9(b). This combination is now repeated for every row of the relations in column ORDER-LINES of the unnormalized relation ORDERS.

Relations with only simple attribute values, such as that shown in Figure 15.9(b), are in the first normal form (1NF) or are normalized. The important advantage of normalized relations is that each attribute has the same importance. This is useful when we convert the relation to computer software. It then becomes possible to add an index on any attribute field and retrieve data using any attribute as a key.

The relation ORDERS in Figure 15.9(b) is in 1NF. Relations in 1NF can still contain redundancy. For example, in relation ORDERS, the ORDER-DATE for a given order can be stored more than once for the same order. To have no redundancy, relations must satisfy additional constraints. Relations that satisfy such constraints and contain no redundancy are in higher normal forms. Constraints satisfied by relations in higher normal forms are defined in terms of functional dependencies and relation keys.

FUNCTIONAL DEPENDENCIES

Functional dependencies describe some of the rules that hold between attributes in a system. In particular, they state whether a particular value of one attribute (X) in a relation determines a particular value of another attribute (Y) for that relation. This is a way of saying that if we know the value of X, then we can determine a unique value of Y. For example, if we know the value of PERSON-ID for a person, then we can determine the value of NAME for that person. A functional dependency is often expressed as:

PERSON-ID → NAME

Functional dependency Where one value of an attribute determines a single value of another attribute.

We can now say that attribute NAME is **functionally dependent** on attribute PERSON-ID. Alternatively, we sometimes use the terminology that PERSON-ID *determines* a unique value of NAME or that NAME is determined by PERSON-ID. The important thing to remember is that for each value of PERSON-ID there is *one* value of NAME.

Functional dependencies often involve more than one attribute on the left-hand side. Thus, to know how much time a person spent on a given project, we must know both the value of the project identifier, PROJECT-ID, and the value of the person identifier, PERSON-ID. The functional dependency then becomes:

PERSON-ID, PROJECT-ID → TIME SPENT

Again, there will be only one value of TIME-SPENT for a given combination of PERSON-ID and PROJECT-ID.

DERIVED FUNCTIONAL DEPENDENCIES

Sometimes, one functional dependency can be derived from other functional dependencies. For example, a project identifier PROJECT-ID may determine one value for attribute BUDGET. Now suppose each manager manages one project. Given a manager, we at once know the project managed by that manager, and from the project we can derive the budget managed by that manager. Mathematically, we would say that if:

MANAGER →   PROJECT-ID and PROJECT-ID   → BUDGET
then
MANAGER →   BUDGET

This simply says that if you know the one project managed by the manager, and the one budget of this project, then you know the one budget for which the manager is responsible.

It is up to the designer and analyst to ensure that there are no derived functional dependencies in their model.

HOW TO FIND FUNCTIONAL DEPENDENCIES

Functional dependencies are not just mathematical relationships. They arise from the nature of the information system and must be found by careful and detailed analysis. They usually arise from rules that hold in a system. For example, a rule that a person can work on only one project will lead to the functional dependency

PERSON-ID → PROJECT

However, if a person could work on more than one project, this dependency would no longer be true. The functional dependencies require careful and detailed examination of the system rules. This detailed analysis is carried out during design. Analysts must question the relationships between attributes and the detailed rules that determine these relationships to find the functional dependencies. Collections of the dependencies are usually documented as a functional dependency diagram.

FUNCTIONAL DEPENDENCY DIAGRAMS

Functional dependency diagrams show all the functional dependencies between a set of attributes in a pictorial way. One example of such a functional dependency diagram is shown in Figure 15.10, where there are eight attributes, each enclosed by a rectangle. Other enclosures could of course be used—for example, an oval-shaped ellipse. Directed arrows are drawn from attributes that determine other attributes. Thus, if you find that each project has a single budget, there is a directed arrow from PROJ-NO to BUDGET. Similarly, if you discover during analysis that each supplier has one address, you should represent it by a directed arrow from SUPPLIER-NAME to ADDRESS on the functional dependency diagram. Also, there is a directed arrow from PART-NO to WEIGHT because each part has one weight. Whenever two attributes determine a single value of a third attribute, then these two attributes are enclosed, and a directed arrow is drawn from the new

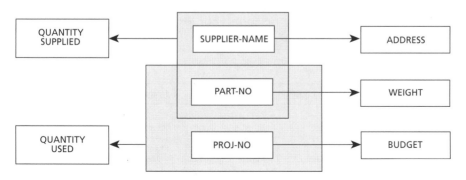

Figure 15.10   *A functional dependency diagram*

enclosure to that third attribute. Because there is a single value of QUANTITY-USED for each combination of PART-NO and PROJ-NO, then a directed arrow originates from an enclosure that contains PART-NO to PROJ-NO and terminates at QUANTITY-USED. There is also a directed arrow from the enclosure of SUPPLIER-NAME and PART-NO to QUANTITY-SUPPLIED. This states that there is one value of QUANTITY-SUPPLIED for a given combination of SUPPLIER-NAME and PART-NO.

You should note two things about functional dependency diagrams. First, each attribute appears once only on the functional dependency diagram. Second, all the attributes of interest to the system appear on the one diagram.

Let us return to the discussion about relations in higher normal forms. Such relations must satisfy additional constraints to having only simple attribute values. To describe such constraints, it is necessary to define a relation key.

RELATION KEYS

Relation key A set of attributes whose values identify a unique row in a relation.

A **relation key** is a set of columns whose values select unique relation rows. Thus, PERSON-ID is the relation key of relation PERSONS in Figure 15.8. Once we are given a value of PERSON-ID, we can immediately select one unique row.

Relation keys can be made up of more than one column. For example, the relation ORDERS in Figure 15.9(b) stores all the facts about orders (i.e. ORDER-DATE) and the facts about the parts (i.e. QTY-ORDERED) in those orders. The relation key of ORDERS is the two columns ORDER-NO and PART-NO. ORDER-NO is not sufficient on its own, as there can also be more than one row with one value of PART-NO. There is, however, only one row for each order–part combination. We use the notation {ORDER-NO, PART-NO} to describe relation keys made up of more than one column.

Relations can have more than one relation key, as illustrated in Figure 15.11. This relation stores records of patient consultations with doctors. It is assumed that there is only one patient in each consultation and only one doctor participates at each consultation. This relation has two relation keys. One is {DOCTOR-NO, TIME-OF-VISIT}, because a doctor can be in only one place at a time. The other is {PATIENT-NAME, TIME-OF-VISIT} because each patient can be in only one place at a time.

One important thing to remember about relation keys is that they must apply for all possible contents of the relation. For example, if you look at relation PROJECTS in Figure 15.8, you may think that BUDGET is also a relation key,

CONSULTATIONS

DOCTOR-NAME	PATIENT-NAME	TIME-OF-VISIT
DR. SMYTHE	A. BLAND	10am 15/1/94
DR. TAN	B. JACKO	10am 15/1/94
DR. MACK	A. BLAND	10am 17/1/94
DR. TAN	J. COPES	11am 15/1/94
DR. SMYTHE	K. BLISS	11am 15/1/94

Figure 15.11 *A relation with two relation keys*

because each row has a different value of budget. However, the uniqueness property of BUDGET does not hold for all possible relation contents. As soon as there are two projects with the same budget value (which is quite possible), a value of BUDGET no longer identifies one row only. PROJECT-ID, however, is a relation key, because for any possible contents of the relation PROJECTS, there can never be two rows in the relation with the same value of PROJECT-ID.

Two other definitions are related to relation keys. These definitions are needed to define normal relations. An attribute is a *key attribute* if it is at least part of one relation key. It is a *non-key attribute* if it is not part of any relation key. Thus, in relation CONSULTATIONS, each attribute is a key attribute because each attribute is part of the relation key. In relation ORDERS in Figure 15.12, only ORDER-NO is a key attribute; the other attributes are non-key. You should bear in mind the distinction between a relation key and a key attribute. An attribute may be a key attribute but not a relation key. For example, attribute DOCTOR-NAME is a key attribute in relation CONSULTATIONS because it is part of the relation key {DOCTOR-NO, TIME-OF-VISIT}. It is not a relation key in its own right because, on its own, a value of DOCTOR-NAME does not identify a unique row.

NORMAL-FORM RELATIONS

The concept of **normal-form relations** is based on functional dependencies and relation keys. Informally, we would like relation keys to determine unique values of the non-key attributes. For example, look at the PERSONS relation in Figure 15.8. The relation key here is PERSON-ID. The other columns—the person's NAME and ADDRESS—are functionally dependent on the value of PERSON-ID. Note that there is no redundancy in relation PERSONS, because a person's NAME and ADDRESS are stored once for each person.

Normal-form relations A set of relations that describes the data in a system, but where each data component is a simple value.

DATA REDUNDANCIES

Now let us consider a relation that stores some facts more than once. The relation ORDERS in Figure 15.9(b) is such a relation.

As explained before, the relation key of relation ORDERS is made up of two columns, ORDER-NO and PART-NO. You should note that in relation ORDERS, the value of ORDER-DATE is determined by part of the relation key only. It is determined by ORDER-NO only and not by the combination of ORDER-NO and PART-NO. You will also note that the ORDER-DATE for the same order can appear more than once in relation ORDERS. It is stored as many times as there are parts in the order. When facts are stored more than once, a relation is no longer in the highest normal form.

The design goal is to eliminate such redundancies. Relations that do not have such redundancies satisfy a number of constraints. The constraints are defined in terms of functional dependencies and relation keys.

SECOND NORMAL FORM

Relations in second normal form (2NF) must satisfy the additional constraint that all non-key attributes must be functionally dependent on the whole of each relation

ORDERS

ORDER-NO	ORDER-DATE
Ord1	6 June 1998
Ord2	3 May 1998

ORDER-CONTENTS

ORDER-NO	PART-NO	QTY-ORDERED
Ord1	P1	10
Ord1	P6	30
Ord2	P5	10
Ord2	P6	50
Ord2	P2	30

Figure 15.12 *Relations in second normal form*

key. Relations in 2NF cannot have non-key attributes that depend on only part of a relation key. To show how this constraint applies, let us turn to relation ORDERS in Figure 15.9(b). Here the relation key is {ORDER-NO, PART-ID}. The key attributes are ORDER-NO and PART-ID. The non-key attributes are ORDER-DATE and QTY-ORDERED. The value of ORDER-DATE, however, depends only on ORDER-NO, because all you have to know to determine ORDER-DATE is ORDER-NO. The relation is therefore not in 2NF.

Relations that are not in second normal form can always be decomposed into second-normal-form relations. To do this we can use a very simple rule:

Remove the offending functional dependency.

This means that we take the functional dependency that violated the 2NF constraint and make a new relation out of all the attributes in this functional dependency. We then remove the attributes on the right-hand side of this functional dependency from the original relation to make another, new, second relation.

For example, the relation ORDERS in Figure 15.9(b) is not in 2NF. The offending functional dependency is:

ORDER-NO → ORDER-DATE.

As shown in Figure 15.12, we make a new relation ORDERS from this functional dependency and remove ORDER-DATE from the original relation. Relation ORDERS now stores facts about orders only. The original relation is now replaced by relation ORDERS-CONTENTS, which stores facts about order lines (or the parts ordered in orders). The relation key of relation ORDERS is ORDER-NO, and the relation key of ORDER-CONTENTS is {ORDER-NO, PART-NO}. You will note that in these relations, non-key fields are determined by the whole (and not part of) the relation key, and hence these relations are in second normal form.

THIRD NORMAL FORM

Relations in 2NF can still contain redundancies, and additional constraints must be satisfied to eliminate such redundancies. Relations in third normal form (3NF) must satisfy yet another constraint. In such relations there must be no dependencies between non-key attributes.

An example of a relation that is in 2NF but not 3NF is relation VEHICLES in Figure 15.13. Relation VEHICLES stores information about cars. Each car is uniquely identified by its REGISTRATION-NO and has one owner. Values in all the other columns—that is, its OWNER, MODEL, MANUFACTURER, and NO-CYLINDERS—are determined by the value of REGISTRATION-NO. The

(a) Relations not in third normal form

RELATION KEY

VEHICLES

REGISTRATION-NO	OWNER	MODEL	MANUFACTURER	NO-CYLINDERS
YX-01	George	Laser	Ford	4
YJ-77	Mary	Falcon	Ford	6
YW-30	George	Corolla	Toyota	4
YJ-37	Mary	Laser	Ford	4
YJ-83	Andrew	Corolla	Toyota	4

(b) Relations in third normal key

RELATION KEY

REGISTRATION

REGISTRATION-NO	OWNER	MODEL	MANUFACTURER
YX-01	George	Laser	Ford
YJ-77	Mary	Falcon	Ford
YW-30	George	Corolla	Toyota
YJ-37	Mary	Laser	Ford
YJ-83	Andrew	Corolla	Toyota

RELATION KEY

VEHICLE 1

MODEL	MANUFACTURER	NO-CYLINDERS
Laser	Ford	4
Falcon	Ford	6
Corolla	Toyota	4

Figure 15.13 *Decomposing a relation to third normal form*

column REGISTRATION-NO is the whole relation key. The values of the non-key attributes are determined by the whole relation key and therefore the relation is in second normal form.

However, you may note that there are functional dependencies between the non-key attributes of relation VEHICLES. For example, NO-CYLINDERS is functionally dependent on a combination of MODEL and MANUFACTURER. If there are more than two cars of the same MODEL and MANUFACTURER in relation VEHICLES, then the NO-CYLINDERS of this model and manufacturer will be stored twice—again an undesirable characteristic. Here facts are stored more than once because NO-CYLINDERS (which is a non-key column) is functionally dependent on other non-key attributes, MODEL and MANUFACTURER.

Relation VEHICLES is in second but not third normal form. To be in third normal form, a relation must first be in second normal form. In addition, it should not have any functional dependencies between non-key attributes.

Relations in second but not third normal form can always be decomposed into third-normal-form relations. Again, we can use our simple rule 'remove the offending functional dependency'. In this case the offending functional dependency is:

MODEL, MANUFACTURER → NO-CYLINDERS.

Thus, relation VEHICLES can be decomposed into the two relations shown in Figure 15.13(b). The relation key of relation REGISTRATION is REGISTRATION-NO, and the relation key of relation VEHICLE1 is {MODEL, MANUFACTURER}. You will note that in Figure 15.13(b) each relation contains facts about the relation key only and has no facts between non-key columns.

Each relation in Figure 15.13(b) is therefore in third normal form.

OPTIMAL NORMAL FORM

All the relations discussed so far had one relation key. Normal forms must also cover relations that have more than one relation key. The particular problem with relations that have more than one key is that part of one key may be functionally dependent on part of another key. For example, look at relation WORK in Figure 15.14(a). This relation describes the TIME-SPENT by persons working on projects. It also includes the MANAGER of each project. In this system, each project has one manager, and a manager can manage at most one project. Relation WORK in Figure 15.14(a) has two overlapping keys. One relation key is (PROJECT-ID, PERSON-ID) and the other is (MANAGER, PERSON-ID). These keys overlap because they have a common attribute, PERSON-ID. The only non-key attribute in relation WORK is TIME-SPENT.

In relation WORK, part of relation Key 2, namely MANAGER, is functionally dependent on part of Key 1, namely PROJECT-ID (i.e. the manager of the project). Similarly, PROJECT-ID is functionally dependent on MANAGER, as it gives the

(a) A non-normal-form relation with overlapping keys

PROJECT-ID	PERSON-ID	MANAGER	TIME-SPENT
Proj1	J1	Vicki	30
Proj2	J1	Joe	12
Proj1	J2	Vicki	11
Proj2	J2	Joe	79
Proj3	J2	Belinda	17
Proj2	J3	Joe	3

WORK — RELATION KEY 1 = (PROJECT-ID, PERSON-ID); RELATION KEY 2 = (MANAGER, PERSON-ID)

(b) Decomposing to normal-form relation

PROJECTS — RELATION KEY

PROJECT	MANAGER
Proj1	Vicki
Proj2	Joe
Proj3	Belinda

WORK — RELATION KEY

PROJECT-ID	PERSON-ID	TIME-SPENT
Proj1	J1	30
Proj2	J1	12
Proj1	J2	11
Proj2	J2	79
Proj3	J2	17
Proj2	J3	3

Figure 15.14 *Non-normal-form relations with overlapping keys*

project managed by the manager. Note also that the MANAGER for a project identified by a given PROJECT-ID can be stored more than once. However, using the strict third normal definition, relation WORK in Figure 15.14(a) is in third normal form. The only non-key attribute in relation WORK is TIME-SPENT. This attribute depends on the whole of each relation key. There are no dependencies between non-key attributes (as there is only one of them).

What is needed is a higher normal form—one that takes overlapping keys into account. Such a higher normal form is the Boyce–Codd normal form (BCNF). A relation R is in BCNF if, for every functional dependency $(X \rightarrow Y)$ between any relation attributes, the attributes on the left-hand side (that is, X) are relation keys.

Relations where this property holds are in BCNF, which is sometimes called **optimal normal form**. The relation WORK in Figure 15.14(a) is not optimal. There is a functional dependency:

PROJECT $\rightarrow$ MANAGER,

but PROJECT is not a relation key of WORK.

Again, a non-BCNF relation can be decomposed into a set of optimal relations by removing the offending functional dependency. The result of decomposing Figure 15.14(a) is shown in Figure 15.14(b).

> **Optimal normal form** Set of relations that have no single-value redundancy.

> ### TEXT CASE D:
> ### Construction Company—Constructing Relations
>
> Let us now see how we convert the E–R diagram for our construction company into relations. Figure 15.15 shows the relational model of the system shown in Figure 7.6 and modeled by the E–R diagram in Figure 9.5. The conversion proceeds in two steps. First, each set in Figure 9.5 is converted to a relation. Then, relations with the same relation key are combined. This means that on the E–R diagram, a relation for a 1:N relationship set is combined with a relation for one of the entity sets that participate in that relationship set. For example, the relation for entity set PURCHASE-ORDERS is combined with the relation for relationship set TO into

---

PROJECTS (<u>PROJECT-NO</u>, PROJECT-DETAILS)
PROJECT-REQUESTS (<u>PROJECT-NO, REQ-NO</u>) DATE-NEEDED)
REQUEST-LINES (<u>PROJECT-NO, REQ-NO, REQ-LINE-NO</u>, ITEM-NO, QTY-NEEDED, PO-NO, PO-LINE-NO)
PURCHASE ORDERS (<u>PO-NO</u>, DATE-ORDERED, SUPPLIER-NAME)
ORDER LINES (<u>PO-NO, PO-LINE-NO</u>, ITEM-NO, QTY-ORDERED, QTY-PAID-FOR, PRICE, QTY-RECEIVED)
ITEMS (<u>ITEM-NO</u> ITEM-DESCRIPTION)
SUPPLIERS (<u>SUPPLIER-NAME</u>, SUPPLIER-DETAILS)
SHIPMENTS (<u>SHIPMENT-NO</u>, LOCATION, SUPPLIER-NAME)
SHIPLINES (<u>SHIPMENT-NO, SHIP-LINE</u>, PO-NO, PO-LINE-NO, ITEM-NO, QTY-SHIPPED)
INVOICES (<u>INVOICE-NO</u>, DATE-SENT, SUPPLIER-NAME)
INVOICE-LINES (<u>INVOICE-NO, INVOICE-LINE-NO</u>, SHIPMENT-NO, SHIP-LINE-NO, ITEM-NO, QTY-INVOICED, PRICE, VALUE, PO-NO, PO-LINE-NO)
DELIVERIES (<u>DELIVERY-NO</u>, DELIVERY-DATE)
DELIVERY-LINES (<u>DELIVERY-NO, DELIVERY-LINE-NO</u>, ITEM-NO, QTY-DELIVERED, PROJECT-NO, REQ-NO, REQ-LINE-NO)

Figure 15.15 *Relations for Text Case D*

one relation, PURCHASE-ORDERS. The combination is made to reduce the number of relations. The combination is possible because each purchase order is sent to one supplier, and the relation key of relation TO is PO-NO. Hence, both PURCHASE-ORDERS and TO have the same relation key, and combining them does not destroy normal form.

FINDING THE HIGHEST NORMAL FORM OF A RELATION

It is perhaps worthwhile here to outline a procedure for finding the highest normal form of a relation. This procedure is illustrated in Figure 15.16. It follows the arguments described in the previous paragraphs. We begin by finding the relation keys and then listing the key and non-key attributes. We then check whether each non-key attribute depends on the whole key. Relations that satisfy this constraint are in at least second normal form. Then we check whether there are any dependencies between non-key attributes. Relations that have no such dependencies are in at least third normal form. We then check whether the relation has more than one relation key. If not, then it is also in BCNF. If the relation has more than one relation key, then one more check is needed. It is necessary to check whether

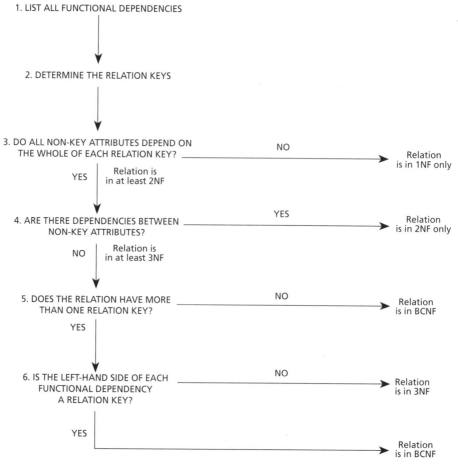

Figure 15.16 *Finding the highest normal form of a relation*

the left-hand side of each functional dependency is a relation key; if so, then the relation is in BCNF.

Of course, in practice it is not always necessary to follow such a procedure. What we are really interested in is whether a relation is in optimal form. To find this out we must first find all the functional dependencies and all the relation keys. All we need to do then is see whether all the left-hand sides of each functional dependency are a relation key. If so, then the relation is in optimal normal form. If not, we may then try to determine the highest normal form to determine how to decompose the relation into optimal normal form.

NORMAL-FORM RELATIONS AND MULTIVALUED DEPENDENCIES

Normal forms up to and including the optimal form had to satisfy constraints defined in terms of functional dependencies. A number of problems can also arise with multivalued dependencies, and higher normal forms have been developed to deal with them. A **multivalued dependency** exists in a relation when a value of one column or set of columns, X, determines a set of values in another column, Y. Multivalued dependencies are often expressed by the notation:

Multivalued dependency Where one value of an attribute determines a set of values of another attribute.

PERSON-ID ↠ SKILL.

This means that a value of PERSON-ID always determines a set of values of SKILL.

Some problems that are caused by multivalued dependencies are illustrated in Figure 15.17. A person identified by PERSON-ID has a number of SKILLs and works on a number of projects, identified by PROJECT-ID. Thus, both SKILL and PROJECT-ID are multivalued dependencies of PERSON-ID (because a person can have many skills and work on many projects). How do we store SKILLs and PROJECT-ID for persons in the one relation? Some possibilities are shown in Figure 15.17. In Figure 15.17(a), each row contains a value from either one SKILL or one PROJECT-ID only, and the other column is blank. The problem with this method is that there are excessive null fields (resulting in wasted storage), and system programs would have to handle these null fields.

Figure 15.17(b) minimizes the number of rows by storing values from both SKILL and PROJECT-ID in the same row. If the number of SKILL and PROJECT-ID values for a given person is the same, then there are no blank values (as for 'Jill'). If the number of values is different, however (as is more likely), then there will be blank values. System programs in this case can become difficult, for two reasons:

- Insertions of a new value may have different effects. For example, adding a new project for 'Jill' adds a new row (with a NULL skill value). Adding a new project for 'Arnold' changes a NULL value to a project identifier.
- Deletions may also have different effects. Deleting 'Economics' for 'Arnold' would delete a row. Deleting 'French' for 'Arnold' would mean that 'Proj1' would have to be moved to another row and then a row would have to be deleted. Deleting 'French' for 'Jill' would replace a value by a blank value.

(a) Storing different multi-fact values in different rows

PERSONS

PERSON-ID	SKILL	PROJECT-ID
Jill	Computing	—
Jill	French	—
Jill	—	Proj1
Jill	—	Proj3
Arnold	French	—
Arnold	Economics	—
Arnold	—	Proj1
George	Computing	—
George	—	Proj2
George	—	Proj1

(b) Minimizing number of rows

PERSONS

PERSON-ID	SKILL	PROJECT-ID
Jill	Computing	Proj1
Jill	French	Proj3
Arnold	French	Proj1
Arnold	Economics	—
George	Computing	Proj1
George	—	Proj2

(c) Cross product

PERSONS

PERSON-ID	SKILL	PROJECT-ID
Jill	Computing	Proj1
Jill	French	Proj1
Jill	Computing	Proj3
Jill	French	Proj3
Arnold	French	Proj1
Arnold	Economics	Proj1
George	Computing	Proj1
George	Computing	Proj2

Figure 15.17 *Storing multiple facts*

All these effects would make the processing program complex and difficult to write and test.

Figure 15.17(c) is an alternative representation that does not make use of NULL values. We simply have one row for each possible SKILL-PROJECT-ID combination for each person. Thus, all skills possessed by 'Jill' appear in combination with all projects that 'Jill' works on. In this case, to insert or delete a new value we must insert or delete more than one row. Thus, if 'George' gained a new skill 'Accounting', we would have to add two rows to relation PERSONS in Figure 15.17(c). One row would contain <'George', 'Accounting', 'Proj1'> and the other <'George', 'Accounting', 'Proj2'>. You will now note, however, that there is redundancy in Figure 15.17(c). For example, we store all of 'Jill's' SKILLs once for each project on which 'Jill' works. Relations must satisfy additional constraints to avoid such redundancies. These constraints are expressed in terms of multivalued dependencies.

FOURTH NORMAL FORM

All the relations in Figure 15.17 are in 3NF or optimal form because there are no functional dependencies between their attributes. However, they are not in fourth normal form (4NF). Relations in 4NF must satisfy a constraint in terms of multivalued dependencies. A relation in 4NF must not contain more than one independent multivalued dependency or one independent multivalued dependency together with a functional dependency.

In Figure 15.17, SKILL and PROJECT-ID are independent multivalued dependencies of PERSON-ID. A person's skills are independent of the projects they work on, and the projects a person works on are independent of their skill. Therefore, a person's skills and a person's projects should be stored in separate relations, as shown in Figure 15.18. A relation is in 4NF if it does not contain independent multivalued dependencies.

It can be shown that any relation in 4NF is automatically in BCNF. This arises because each functional dependency is also a multivalued dependency.

FIFTH NORMAL FORM

Fifth normal form (5NF) can be viewed as an extension of 4NF in the sense that now the multivalued dependencies are no longer independent. For example, let us return to Figure 15.17(c). Suppose that instead of containing information about a person's SKILLs and PROJECT-ID, the relation also stores information about what skills the person uses in a given project. It is assumed that if a person possesses a skill, then that person will apply that skill to a project if the project needs it. Thus, suppose 'Proj1' needs only 'Computing' and 'Economics' skills but not 'French'. The relation would now be as shown in Figure 15.19(a).

Fifth normal form A set of relations that have no multivalued redundancy.

The relation in Figure 15.19(a) differs from the relation in Figure 15.17(c) in that it does not contain rows that include both 'French' and 'Proj1'. This relation is in 4NF but still has undesirable properties in that:

- some facts are stored twice (e.g. 'Jill' and 'George' possess the skill 'Computing')
- there are blank fields (e.g. 'Arnold' has the skill 'French' that he is not currently applying to any project).

These undesirable properties arise because APPLYING-SKILLS contains dependent multivalued dependencies—that is, the value of SKILL that is associated with PROJECT-ID depends on the SKILLs needed by the project. A property of a relation that is in 4NF but not in 5NF is that it cannot be decomposed into two relations but must be decomposed into three relations. Thus, to remove the undesirable properties in relation APPLYING-SKILLS in Figure 15.19(a), we must decompose it into the three relations shown in Figure 15.19(b).

The relations in Figure 15.19(b) differ from the decomposed relations in Figure 15.18 because they contain the additional relations NEEDS-SKILL. The NEEDS-SKILL relation eliminates some of the rows in relation PERSON in Figure 15.17(c).

KNOWLEDGE

PERSON-ID	SKILL
Jill	Computing
Jill	French
Arnold	French
Arnold	Economics
George	Computing

ASSIGNMENTS

PERSON-ID	PROJECT-ID
Jill	Proj1
Jill	Proj3
Arnold	Proj1
George	Proj1
George	Proj2

Figure 15.18 *Relations in 4NF*

(a) Relation with dependent multivalued facts

APPLYING-SKILLS

PERSON-ID	SKILL	PROJECT-ID
Jill	Computing	Proj1
Jill	Computing	Proj3
Jill	French	Proj3
Arnold	French	—
Arnold	Economics	Proj1
George	Computing	Proj1
George	Computing	Proj2

(b) Decomposing into 5NF relations

KNOWLEDGE

PERSON-ID	SKILL
Jill	Computing
Jill	French
Arnold	French
Arnold	Economics
George	Computing

ASSIGNMENTS

PERSON-ID	PROJECT-ID
Jill	Proj1
Jill	Proj3
Arnold	Proj1
George	Proj1
George	Proj2

NEEDS-SKILL

PROJECT-ID	SKILL
Proj1	Computing
Proj1	Economics
Proj2	Computing
Proj3	Computing
Proj3	French

Figure 15.19 *Fifth normal form*

There is yet a third possibility, where a person may not apply all their skills to a project even if they are needed. Thus, for example, 'Jill' may not use her 'computing' skill in project 'Proj3' even if 'Proj3' requires a 'computing' skill. For that reason, relation APPLYING-SKILLS would not contain the row <'Jill', 'Computing', 'Proj3'>. In that case, relation APPLYING-SKILLS could not be decomposed and would be in 5NF. However, the other three relations in Figure 15.19 must be in the database to avoid null fields in relation APPLYING-SKILLS. Thus, an informal definition of a 5NF relation is a relation that cannot be decomposed without losing information.

CONSTRUCTING RELATIONS FROM FUNCTIONAL DEPENDENCY DIAGRAMS

So far we have assumed that relations are first constructed. They are then checked to see whether they are in the highest normal form and, if not, then they are decomposed to higher normal forms. Sometimes, however, an alternative approach works. Why not start with functional dependencies and construct normal-form relations from them?

A simple way to convert functional dependencies to relations is illustrated in Figure 15.20. This is to convert each set of dependencies with the same determinant

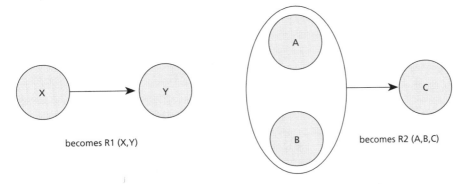

Figure 15.20 *Converting functional dependencies to relations*

to a relation. In this case, the left-hand side of the dependency will become the relation key. All the other non-prime attributes will be dependent on that key, and thus the relation will be in at least 2NF. To ensure that such relations are also in optimal form, it is necessary to remove any redundant functional dependencies before conversion takes place. The functional dependencies in Figure 15.20 would be converted to the following relations:

SUPPLIERS (SUPPLIER-NAME)
PARTS (PART-NO, WEIGHT)
PROJECTS (PROJ-NO, BUDGET)
SUPPLY (SUPPLIER-NAME, PART-NO, QUANTITY-SUPPLIED)
USE (PROJ-NO, PART-NO, QUANTITY-USED).

SUMMARY

This chapter describes the first database design step: converting the E–R model to a relational model, which is then checked to see whether it contains any redundancies. A number of criteria for such tests are described. These criteria are defined in terms of functional dependencies and relation keys. Relations that satisfy the criteria are known as normal relations. A number of normal forms are defined in this chapter. Designers should ensure that their data model contains only relations in the highest normal form.

EXERCISES

15.1   You are given the logical record structure shown in Figure 15.21. The LRS describes a delivery system. Each trip is made by one driver using one vehicle. A number of deliveries can be made on each trip. Any number of part kinds can be delivered in each delivery. Draw access paths on the logical record structure for the following access requirements:

1.   List the NAME of drivers who made deliveries to a customer with a given CUSTOMER-NAME.

2.   List the NAME of drivers who used a vehicle with a given REGISTRATION-NO.

3.   List the REGISTRATION-NO and MAKE of vehicles used on a given TRIP-DATE. Assume that a trip never takes more than one day.

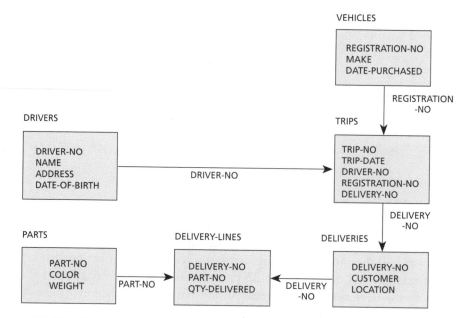

Figure 15.21 *A logical record structure*

4. List QTY-DELIVERED of a given PART-NO on a given TRIP-DATE.

15.2 Construct relations for the following. Give each relation a name and specify all its attributes.

1. Deliveries (identified by DELIVERY-NO) of parts (identified by PART-NO) are made. The QTY-DELIVERED of each part is stored.

2. The distance that cars (identified by REGISTRATION-NO) are driven on trips (identified by TRIP-NO) is stored together with the driver (identified by DRIVER-NO) and TRIP-DATE.

3. A person's weight is recorded each day. The person is identified by PERSON-ID.

4. The amount of rainfall is recorded each day for a number of locations. Locations are identified by LOCATION-NAME.

5. The quantities of parts (identified by PART-NO) withdrawn by projects (identified by PROJECT-NO) from warehouses (identified by WAREHOUSE-NO) are recorded. The date of the withdrawal is also recorded.

6. The quantities of items (identified by ITEM-NO) purchased by customers (identified by CUSTOMER-ID) from stores (identified by STORE-NO) are recorded together with the PURCHASE-DATE.

7. The ADDRESS, DATE-OF-BIRTH, and SURNAME are recorded for persons identified by PERSON-ID.

15.3 Draw a functional dependency diagram showing functional dependencies between the capitalized attributes in the following problem:

Policies identified by a POLICY-NO can be established in an organization. Each policy has one DATE-SET-UP and is set up for one customer. Each customer has a CUSTOMER-ID. The customer also has an ADDRESS, but addresses can change. The START-DATE is kept for each address. There is one RISK-LOCATION for each policy.

A policy can include any number of special items. Each special item has a unique SPECIAL-ITEM-NAME within the policy. The VALUE of each special item is recorded.

Claims can be made against policies. Each claim has a unique CLAIM-NO and is made on a given CLAIM-DATE and a CLAIM-AMOUNT. Any special items included in the claim are recorded together with the ADDITIONAL-AMOUNT-CLAIMED for each special item.

15.4   Convert the E–R diagrams shown in Figure 15.22 to relations.

15.5   Examine the four relations in Figure 15.23. The rules for these relations are given below. Find the relation key for each relation and state the highest normal form for each relation.

RULES FOR LOAN-APPLICATIONS-NO:

1.   Each loan application has one APPLICANT and is identified by a unique value of LOAN-APPLICATION-NO.

2.   Each application has one LOAN-TYPE.

3.   Each application has one APPLICANT-ADDRESS.

4.   An applicant can make many applications.

RULES FOR CUSTOMER-INVOICES:

1.   Each invoice is sent to one customer and is identified by a unique value of INVOICE-NO.

2.   A number of CUSTOMER-SERVICES can be included on one invoice.

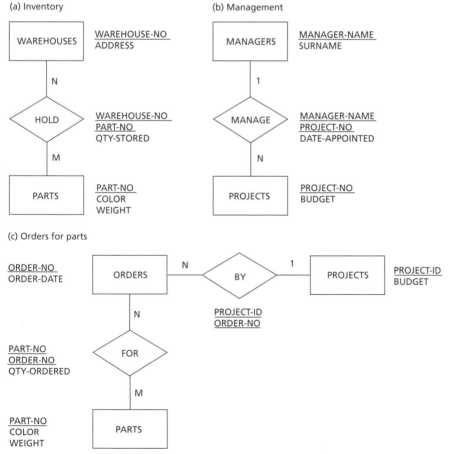

Figure 15.22   *E–R diagrams for Exercise 15.4*   (continues . . .)

(d) Vehicle registration

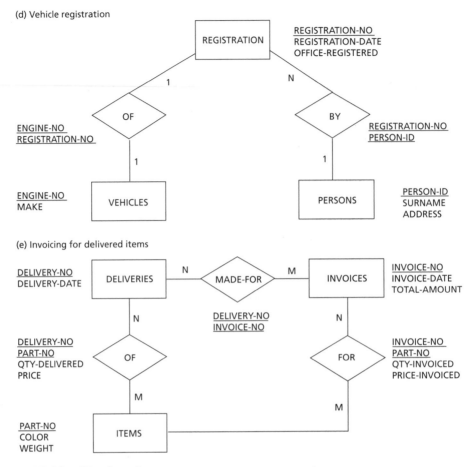

(e) Invoicing for delivered items

Figs 15.22 *(Continued)*

3. There is a separate SERVICE-COST for each CUSTOMER-SERVICE on an invoice.

4. Each customer is identified by a unique CUSTOMER-NO and has one CUSTOMER-ADDRESS.

RULES FOR MACHINE-USE:

1. A machine can be used by only one operator on a given date.

2. An operator can use any number of machines on the one day.

3. TIME-SPENT-ON-MACHINE is the time spent by an operator on the machine on a given DATE. QTY-PARTS-PRODUCED is the quantity of parts produced on that machine by the operator on the given date.

RULES FOR PART-USE:

1. Each PROJECT-ID has one MANAGER.

2. Each MANAGER manages one PROJECT.

3. Each TASK belongs to one PROJECT.

4. Each task uses a given QTY of a PART and may use any number of parts.

LOAN APPLICATIONS

LOAN-APPLICATION -NO	APPLICANT	APPLICANT- ADDRESS	LOAN-TYPE
1	Jill	Canberra	Home
2	Joe	Sydney	Mortgage
3	Jill	Canberra	Personal
4	Max	Melbourne	Home

CUSTOMER-INVOICES

CUSTOMER- NAME	INVOICE-NO	CUSTOMER- ADDRESS	CUSTOMER- SERVICES	SERVICE- COST	SERVICE- DATE
Joe	6	Sydney	Repair	120	June 1998
Joe	6	Sydney	Course	320	July 1998
Jill	3	Canberra	Repair	80	Aug 1998
Joe	6	Sydney	Repair	150	Oct 1998

MACHINE-USE

OPERATOR	MACHINE	DATE	QTY-PARTS- PRODUCED	TIME-SPENT-ON- MACHINE
Joe	Mach1	1 June 1998	15	10
Joe	Mach2	1 June 1998	20	12
Bill	Mach3	1 June 1998	12	6
Bill	Mach2	2 June 1998	20	14

PART-USE

PROJECT- ID	MANAGER	TASK	PART	QTY
Proj1	Jenny	3	Hammer	7
Proj2	Henry	9	Drill	9
Proj1	Jenny	4	Saw	11
Proj2	Henry	9	Hammer	6

Figure 15.23 *Relation for Exercise 15.5*

15.6 Construct a set of normal relations described by the following statements.

1. Persons (identified by PERSON-ID and a SURNAME) are given AUTHORITY-NOs by PROJECTS (identified by PROJECT-NO and a given BUDGET). Each such AUTHORITY-NO authorizes the person to place orders for a project. A person can have many authority numbers, and many persons can have an authority for a given project.

2. The AUTHORITY-NO is assigned on a given AUTHORITY-DATE for a given MAX-AMOUNT for one project only.

3. The order is made to one supplier on a given ORDER-DATE. It can be for any number of different PART-KINDS. The order includes a QTY-ORDERED for each PART-KIND. Each order is placed under one authority number.

4. Each PART-KIND has a PRICE. The PRICE depends on the supplier.

5. The supplier is identified by SUPPLIER-ID and can have any number of ADDRESSes. An order is placed at one ADDRESS.

15.7 Continue the analysis of Cases 1 and 2 begun at the end of Chapter 8 by developing E–R and relational models for the data.

15.8 What is the highest normal form of the following relations?

1. | JOB-NO | PLACE | PERSON-ID | DATE STARTED |

   A job exists in one place, and a number of people are working on a job. Each person starts once only on each job.

2. | ROOM | COMMITTEE | CHAIRPERSON | TIME | DATE |

A committee can meet any number of times in different rooms. The committee has the same chairperson at each meeting and can meet more than once on the same day, possibly in different rooms.

3. | LOT-NO | DESCRIPTION | PURCHASER | DATE | PURCHASER-ADD |

Lots are sold to any purchaser at auction. A lot can be sold only once on any given day. A lot has one description, and the purchaser has one address. There can at most be one purchase with a given LOT-NO on any day, but the same LOT-NO can be used on different days.

4. | ACCOUNT-NO | NAME | DATE | BALANCE | ADDRESS |

Account balances change over time. ADDRESS is for account. Each account has one address only.

5. | PROGRAM | DAY | TIME | CHANNEL |

A television program guide.

BIBLIOGRAPHY

Aho, A.V. et al. (1979), 'Theory of joins in relational databases', *ACM Transactions on Database Systems* Vol. 4, No. 3, pp. 297–314.

Armstrong, W.W. (1980), 'Decompositions and functional dependencies in relations', *ACM Transactions on Database Systems* Vol. 5, No. 4, pp. 404–30.

Beeri, C. (August 1977), 'A complete axiomatization for functional and multivalued dependencies', *SIGMOD International Conference on Management of Data*, Toronto, Canada.

Codd, E.F. (1971), 'A relational model of data for large shared data banks', *Communications of the ACM*, 13 (6), pp. 377–87.

Date, C.J. (2000), 'An introduction to database systems' (7th edn), Addison-Wesley, Reading, Massachusetts.

Davenport, R.A. (1979), 'Logical database design—from entity model to DBMS structure', *Australian Computer Journal*, 11(3), pp. 82–97.

Hawryszkiewycz, I.T. (1991), *Relational Database Design: An Introduction*, Prentice Hall, Sydney.

Kent, W. (1983), 'A simple guide to five normal forms in relational database theory', *Communications of the ACM*, 26(2), pp. 120–5.

McFadden, F.R. and Hoffer, J.A. (1999), *Modern Data Management*, Addison-Wesley, Reading, Massachusetts.

Sen, A., Jacob, V.S. (eds) (September 1998), 'Industrial strength data warehousing', *Special Issue of the Communications of the ACM*.

Silberschatz, A., Korth, H.F. and Sudarshan, S. (1997), *Database System Concepts* (3rd edn), McGraw-Hill, New York.

PROGRAM DESIGN

CONTENTS

KEY LEARNING OBJECTIVES

Identifying physical modules

Steps to be followed in program design

Structure charts

Good characteristics of structure charts

Converting data flow diagrams to structure charts

Object component diagrams

CHAPTER 16

INTRODUCTION

This chapter continues architectural design by describing program structure design for both structured systems analysis and object modeling. Program design is the third part of system design (Figure 16.1). The other parts, interface and database design, were described in previous chapters.

Program design in structured systems analysis begins with the DFD system specifications derived in Chapter 13. The objective of architectural design is to take the logical DFD specification and convert it to a program module design. The module design must satisfy a variety of good design criteria, which later result in modular programs that are easy to develop and, more important, change. Modular program design localizes each well-defined user function to one program module or object class. Modularity is consistent with the criteria for good data flow diagrams (DFDs). In a well-designed DFD, each process represents one well-defined function. When the DFD process converts to a program module, the DFD process specification can be converted to program code.

Object modeling methods also provide ways to model object architecture designs. The most common way is to use the component diagrams proposed in UML. The components can correspond to object classes produced during system specification. However, object modeling is often used with rapid application development, where logical and physical design are integrated into the one step. This may then extend component diagrams to include physical design.

FROM SPECIFICATION TO SYSTEM MODELS

Figure 16.2 shows the steps usually followed in going from system specifications to defining an architectural program design and then to an implementation. Architectural design starts with two steps:

1.   An initial broad-level design to identify those processes that will be automated.
2.   An architectural phase that describes the module structure.

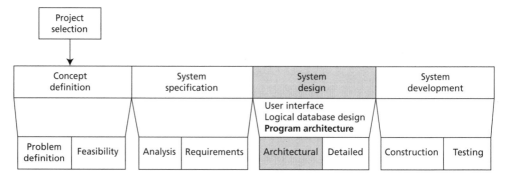

Figure 16.1 *Physical procedure design*

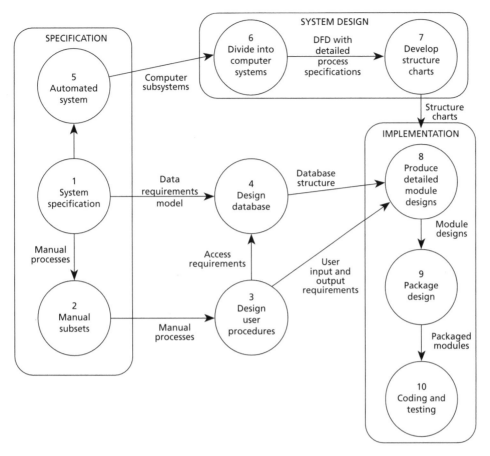

Figure 16.2 *Program design*

Process 1 in Figure 16.2 is the actual receipt of the specification, which in the case of structured systems analysis is a logical DFD. Then we define what is to be automated and what is to remain manual. Processes 5, 6, 7, 8, 9, and 10 in Figure 16.2 then describe the steps used to implement the systems. Processes 6 and 7 develop the structure charts and are described in this chapter. Processes 3 and 4 are the database and business process design that must be supported by the program modules. Processes 8, 9, and 10 are then described in the following chapters.

NEW LOGICAL TO NEW PHYSICAL

System design investigates alternative ways to implement the system by drawing a boundary to identify those processes that are to be automated. This boundary is represented by a dashed line and is often called the automation boundary. The automation boundary can be drawn on high- and low-level DFDs or on a class diagram. An example is shown in Figures 16.3 and 16.4. The automation boundary at the high-level DFD in Figure 16.3 shows that Processes 2 and 3 are to be automated. Figure 16.4 illustrates the same boundary for the leveled Process 2 of Figure 16.3. The leveled diagram shows that the transaction is entered in two messages. Message 1 is entered first and examined for errors. A request is made for

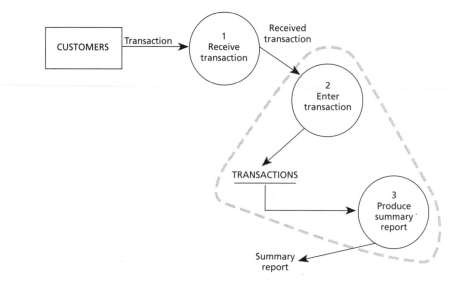

Figure 16.3 *High-level automation boundary*

Message 2 of the transaction if no errors are found in Message 1. Otherwise, a request is made for a correction.

Figure 16.4 shows both the processes and the interfaces provided for users.

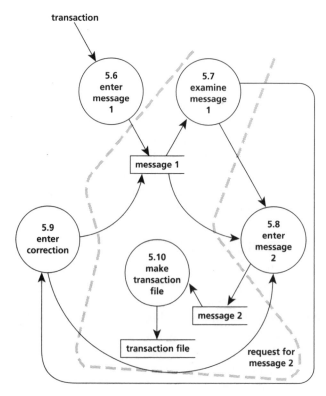

Figure 16.4 *Low-level automation boundary*

It is common to define alternative automation boundaries during system design. Usually, one of the options is a 'total' option—everything is automated. Another option is a minimum automation option, which automates the processes necessary to realize only the most important objectives while leaving a considerable portion of the system non-automated. Finally, an option that includes no additional automation but uses the revised logical flows should be considered.

The operational, technical and economic feasibility of the proposed designs is then evaluated. A preferred alternative may be somewhere in between the three you started with. The evaluation uses the methods described in Chapter 7 but now carried out in greater detail.

TEXT CASE D:
Construction Company—Broad Design

The broad-level automation boundary for the new parts distribution system in the construction company is shown in Figure 16.5. Two boundaries are drawn on this figure. One is for the new project ordering system (POS) and the other for the new goods received system (GRS). You should note that the boundaries between the two systems are now defined in more detail than shown in Figure 7.6, which was only a very rough proposal. Now you can see that one flow from the new POS to the new GRS is the purchase order request. This request is generated when the requested items are not found in store. The purchase order request contains the purchase order number. A copy of the purchase order number and its matching project request are kept in data store REQUESTS in the new POS.

The new POS now checks incoming shipments against the data in data store PO-REQUESTS to find project requests that match the shipment and sends 'delivery advices' to PROJECTS. Thus, manual matching is no longer necessary. Advices about 'received shipments' are sent by the new POS to the new GRS, where they are used to update the PURCHASE-ORDERS data store by the received goods. Process 6 in the new GRS then checks any invoices against PURCHASE-ORDERS to see whether the invoiced goods have been received.

DIVIDING INTO COMPUTER SUBSYSTEMS

Architectural design then continues with DFDs in the automation boundary and uses outputs from database design and user procedure design to produce a structure chart. In structured systems analysis, the first system design step (Process 6 in Figure 16.2) is to divide the DFD into computer subsystems. Logically connected processes are grouped into computer subsystems. These subsystems usually involve one transaction or some connected transactions and become transaction programs or batch suites of programs.

The most common method of grouping DFD processes is to follow through one kind of input, which usually identifies a part of a business process. Figure 16.6 shows how DFD processes can be broken up into computer subsystems. The DFD in Figure 16.6 describes a business process that takes a customer order and delivers the parts requested in the order. In Figure 16.6, this business process is subdivided

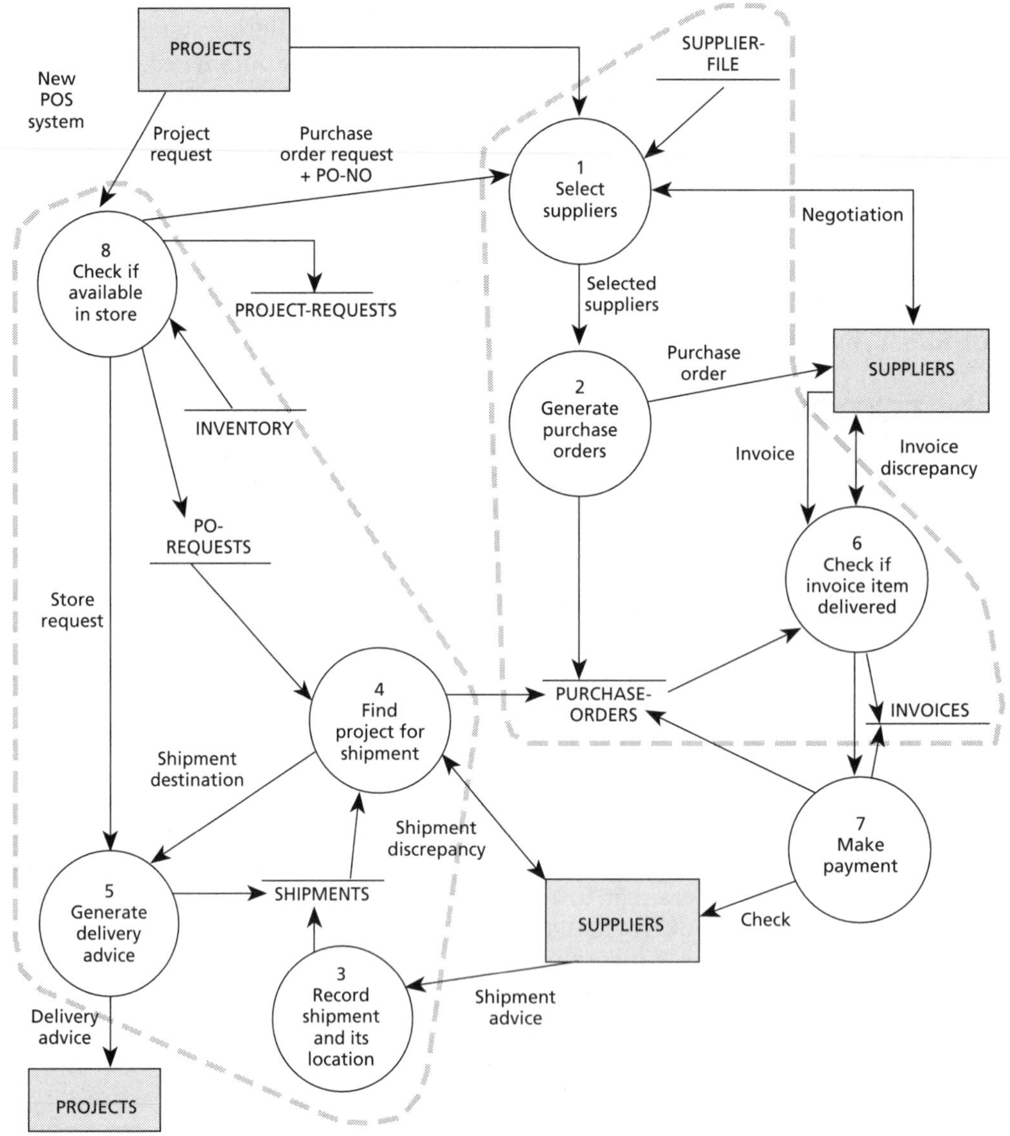

Figure 16.5 *Logical model*

into two parts. Processes 3.1, 3.2, and 3.3 form the part that takes the order and converts it to a warehouse requisition. They check whether the ordered parts are in the warehouse and create a warehouse requisition for them. Processes 3.4 and 3.5 form the part of the business process that delivers the parts requested in the warehouse requisition. They remove the requisitioned parts from the warehouse, package them, and check whether they meet the order before delivery to the customer.

After computer subsystems are chosen, detailed process specifications are developed for each DFD process in each subsystem. Then the DFD processes are converted to program modules, which are shown on a **structure chart**. The program modules are grouped into load modules during implementation.

Structure chart Program modules and their interconnection.

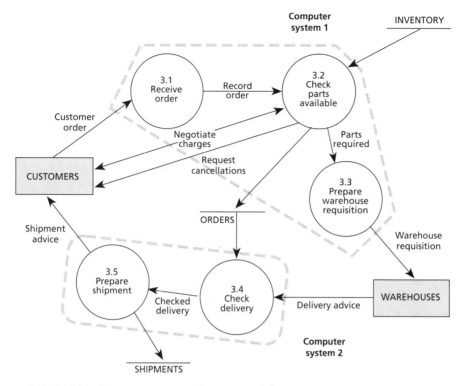

Figure 16.6 *Identifying computer subsystems with the automation boundary*

STRUCTURE CHARTS

Structure charts are one of the most commonly used methods to describe module architectures. In a structure chart, each program module is represented by a rectangular box. Modules at the top level of the structure chart call the modules at the lower levels. The connections between modules are represented by lines between the rectangular boxes. The connections describe data flows between the called and calling modules. Figure 16.7 illustrates a simple structure chart. This chart is made up of four modules. The top module is called COMPUTE-SALE-TOTAL. This module calls three lower-level program modules to accomplish its task. It calls

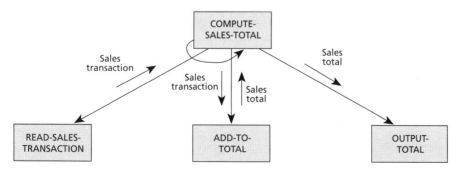

Figure 16.7 *A structure chart*

module READ-SALES-TRANSACTION to read individual sales transactions. It then calls module ADD-TO-TOTAL to sum the amount in each transaction. Finally, it calls module OUTPUT-TOTAL to output the sum.

STRUCTURE CHART CONVENTIONS

Structure charts use a number of conventions to describe system operations. The most important conventions specify the execution sequence and parameter passing between modules.

PARAMETER PASSING

The calling module passes a set of values to the called module and receives a set of values in return. These values are passed as parameter values. The parameters are shown in the structure chart next to the connection. Thus, in Figure 16.7, a value of 'sales transaction' is passed from module READ-SALES-TRANSACTION to module COMPUTE-SALES-TOTAL. Module COMPUTE-SALES-TOTAL then passes the value of 'sales transaction' to module ADD-TO-TOTAL and gets a value of 'sales total' in return. The value of 'sales total' is then passed from module COMPUTE-SALES-TOTAL to module OUTPUT-TOTAL.

EXECUTION SEQUENCE

By convention, modules are executed from left to right. Thus, in Figure 16.7, module READ-SALES-TRANSACTION is called before module ADD-TO-TOTAL. Module OUTPUT-TOTAL is the last module to be called.

Certain conventions are also used to represent decisions and repetition. Decisions occur whenever a calling module has to decide to call only one of a number of modules. Repetition, on the other hand, occurs when some modules are called repetitively by the calling module.

Repetition is modeled by a looping arrow. As an example, in Figure 16.7, module COMPUTE-SALES-TOTAL calls modules READ-SALES-TRANSACTION and ADD-TO-TOTAL any number of times.

Decisions are modeled by a diamond symbol. An example of decisions is given

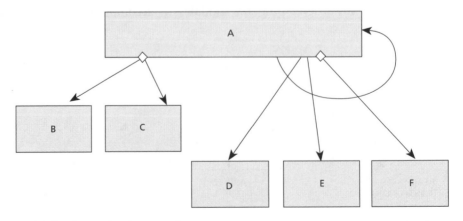

Figure 16.8 *Procedural annotations*

in Figure 16.8. Here, module A calls either module B or module C, then executes a loop that calls module D, E, and sometimes F.

STRUCTURE CHARTS AND STRUCTURED DESIGN

Structure charts are developed by a process called *structured design*. The objective of structured design is to produce structure charts with properties commensurate with good programming practice. A number of properties are used to judge how well structure charts satisfy these objectives. These properties are usually known as *module coupling* and *cohesion*. Module cohesion is also sometimes called *module strength*.

Coupling describes the nature, direction, and quantity of parameters passed between modules; cohesion describes how system functions are coded into modules. A goal of structured design is to minimize the complexity of coupling between modules. Structure charts with simple coupling are said to have low coupling. Another goal of structured design is to represent a well-defined system function by one module. When this happens, we have high strength. Structure charts with low coupling and high strength result in greater independence between modules and easier maintenance, because one module can be changed independently of other modules.

The proponents of structured design have devised measures of coupling and module strength. A number of terms are used to describe coupling and cohesion. These terms are ordered into a range starting with the least desirable to the most desirable.

MODULE COUPLING

Module coupling measures the quality of the connections between modules in the structure chart. The objective is to design structure charts that pass only data and not control information between program modules. However, a first-cut design may not meet this objective, and the structure chart may exhibit other kinds of coupling. There are a number of ways to describe coupling. The best coupling is data coupling, followed by control coupling, common-environment coupling, and content coupling.

CONTENT COUPLING

Two modules are content-coupled if one module makes a direct reference to the contents of another module. This kind of coupling allows the calling module to modify a program statement in the called module or refer to an internally defined data element of the called module. It also allows one module to branch into another module. Content coupling should be avoided at all costs, and structure charts should ensure that the only way to pass information between modules is by parameter values passed during subroutine calls.

COMMON-ENVIRONMENT COUPLING

Two modules are common-environment-coupled if they refer to the same data structure or data element in a common environment.

Examples of common-environment coupling include common areas in user programs or shared files. The effect of common-environment coupling is that modules that appear unrelated in a structure chart are coupled through their use of common data.

CONTROL COUPLING

Two modules are control-coupled if one module passes a control element to the other module. This control element affects the processing in the receiving module. Typical examples of control coupling are flags, function codes, and switches.

Control coupling violates the principle of information hiding. Passing a control element from a calling module to a called module implies that the calling module must know the method of operation of the called module. Any changes made to the called module can then require changes in the calling module.

DATA COUPLING

Two modules are data coupled if they are not content-coupled, common-environment-coupled, or control-coupled. Only data elements are passed as parameters between two data-coupled modules.

Data coupling is seen as the most desirable form of coupling. The calling module passes data values by parameters to the called module and expects some computations to be made on these values. The results of the computation are then returned as parameter values to the calling module. The calling module need not be aware of how the program does the computation.

Common examples of data coupling are calls to input or output modules. A calling module may require some input; it calls another module, INPUT, to provide this input. The calling module does not care how the called INPUT module obtains the input—it is concerned only with the data that it receives.

Calls to output modules are of a similar nature. The calling module passes the data that is to be output to module OUTPUT; it is not concerned with how the called module OUTPUT outputs the data.

MODULE STRENGTH

Module strength measures reasons why code appears in the same module. Many writers use the following six levels of module strength (listed from the least to the most desirable):

* coincidental
* logical
* temporal
* procedural
* communicational
* functional.

A brief outline of these six levels follows. More detailed descriptions can be found in Yourdon and Constantine's *Structural Design* (1979).

COINCIDENTAL STRENGTH

Coincidental strength exists if there is no meaningful relationship between the parts in a module. It often occurs when existing code is modularized. Modularization often proceeds by searching existing code for multiple occurrences of sequences of commands and replacing these sequences by modules. Often such modules are not related to well-defined system functions, but result from techniques that have been used to write the program.

LOGICAL STRENGTH

Logical strength occurs when all elements in a module perform similar tasks. For example, modules that include all editing, or modules that include all accesses to a file.

Considerable duplication can exist in the logical strength level. For example, similar edit checks may be made on more than one data item—for instance, more than one data item in an input transaction may be a date. Separate code would be written to check that each such date is a valid date. A better way is to construct a DATE-CHECK module and call this module whenever a date check is necessary.

TEMPORAL STRENGTH

Temporal strength is very similar to logical strength. All functions related to time are grouped into one module. Typical examples are INITIATION and TERMINATION modules.

Temporal strength is generally regarded as stronger than logical strength, but it still has some undesirable features as far as change is concerned. For example, adding a new file to a system will result in changes to both the INITIATION and TERMINATION modules, and to those modules directly concerned with operations on the new file.

PROCEDURAL STRENGTH

Procedural strength often results when a flowchart is divided into a number of sections, each represented by one module. This division may not be ideal, as the flowchart can represent one well-defined system function; division distributes this self-contained function among a number of modules.

COMMUNICATIONAL STRENGTH

Communicational strength occurs when processes that communicate with each other are included in the same module. Thus, all actions concerned with a file may be included in the same module. This module will read the file, process it, and write the output back to the file.

The most often quoted problem with communicational strength is the interdependence of processes in the module. For example, we may include code that uses the input from a file in the same module as the code that reads the file. However, the file 'read' and the use of the information from the file may occur in different time frames and may share common buffers. A change that allows concurrent 'reads' and 'writes' to the file may result in unexpected problems.

FUNCTIONAL STRENGTH

A module that has functional strength carries out one well-defined function. This module does not have those properties that characterize coincidental, logical, temporal, procedural, or communicational strength.

In addition, some writers define sequential strength and informational strength. Sequential strength occurs where outputs from elements in a module serve as inputs to other elements in a module. In terms of a flowchart or data flow graph, sequential strength combines a chain of successive transformations to the data. Information strength exists where modules perform multiple functions, with each function represented by a different entry point to the module. Each such entry point has the characteristics of a module of functional strength. Sequential and informational strengths fall between communicational and functional strengths in the scale of module strength.

MORPHOLOGY

Coupling and cohesion are criteria applied locally to a few modules. Apart from such local criteria, structure charts can also be evaluated by considering their total structure. Some criteria used in this evaluation are span of control, fan-in, scope of control, and scope of effect rules.

SPAN OF CONTROL

Span of control comprises the number of immediate subordinates of a module. In Figure 16.9, modules B and C are the immediate subordinates of module A. Hence, the span of control of module A is two. Ideally, the span of control should not exceed seven.

FAN-IN

The fan-in is the number of modules that call a particular module. In Figure 16.9, module A has a fan-in of one and module C has a fan-in of two. Ideally, structure charts should have a high fan-in. This means that self-contained functions that can be used at a number of places have been identified.

SCOPE OF CONTROL

The span of control comprises all the subordinates of a module. It includes the immediate subordinates of a module, their immediate subordinates, and so on. Thus, the scope of control of the MAIN module in Figure 16.9 is all the modules A, B, C, D, E, F, and G. The scope of control of module A is modules B and C.

SCOPE OF EFFECT

The scope of effect of a decision consists of all modules whose processing is conditional on the outcome of the decision. In Figure 16.9, for example, suppose that a decision (d1) is made in module B. If this decision affects processing in

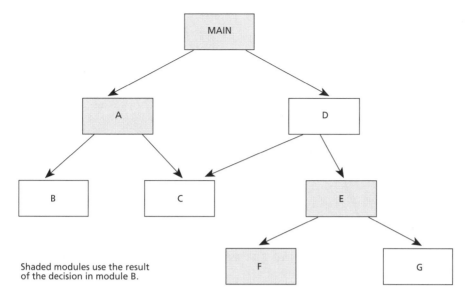

Shaded modules use the result
of the decision in module B.

Figure 16.9 *Scope of effect*

modules A, E, and F, then modules A, E, and F are said to be in the scope of effect
of module B.

Good design calls for the effects of a decision to be confined to as few modules
as possible. If this is done, then the tests based on one decision will not be
unnecessarily repeated, or results of a decision will not be passed through an
excessive number of modules. In Figure 16.9, for example, the result of decision
d1 (in module B) would need to be passed through modules A, MAIN, and D to
be useful in modules E and F. To eliminate such passing of control parameters, the
scope of effect of a decision should be within the scope of control of the module
where the decision is made. The decision need not then be returned to calling
modules. Indeed, it is desirable that the decision be made as close as possible to
the modules that use the decision. One way to achieve this is to have only immediate
subordinates within the scope of effect of a decision.

SOME COMMON STRUCTURES

Structure charts are often characterized by some constructs that tend to reappear in
many applications. Two such important constructs are transform-centered and
transaction-centered structures.

TRANSFORM-CENTERED STRUCTURES

Transform-centered structures receive an input that is transformed by a sequence
of operations, each operation being carried out by one module. Figure 16.10 is a
transform-centered structure where:

• the MAIN module calls the GET-X module to get input parameter X. Module
  GET-X calls the GET-Y module to read a value from the input device and pass it
  back to GET-X as a value of parameter Y. GET-X then calls the CHANGE-Y
  module to compute a value of X from the value of Y

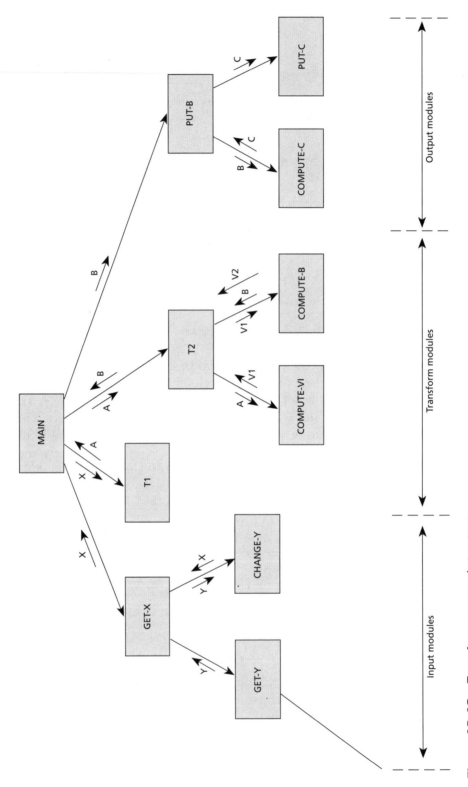

Figure 16.10 *Transform-centered structure*

- a module (T1) is then called to compute a value of A from the value of X
- another module (T2) is then called to compute a value of B from the value of A. To do this, module T2 calls its subordinate modules, COMPUTE-VI and COMPUTE-B
- the MAIN module then calls the PUT-B module to output the result of the computation. PUT-B first calls the COMPUTE-C module to compute a value of C and then calls PUT-C to output that particular value of C.

TRANSACTION-CENTERED STRUCTURES

A transaction-centered structure describes a system that processes a number of different types of transactions. It is illustrated in Figure 16.11. The MAIN module controls systems operations. Its function is to:

- call the INPUT module to read a transaction
- determine the kind of transaction and select one of a number of transaction modules to process that transaction
- output the results of the processing by calling the OUTPUT module.

Often transaction- and transform-centered structures can be found in the same structure chart. For example, the transaction-centered structure reads a transaction and then calls another module to process that transaction. This called module can be a MAIN module of a transform-centered structure.

CONVERSION FROM DATA FLOW SPECIFICATION TO STRUCTURE CHARTS

It now remains to develop techniques to convert a system specification given in terms of a data flow model to logical program structures expressed as a structure chart. Most work in this area has been associated with data flow diagrams and their conversion to structure charts. The design techniques proposed for this conversion are called *transform analysis* and *transaction analysis*.

The origin of these techniques can be traced back to the work of Constantine at IBM; this work has been substantially elaborated upon in later works of Yourdon and Constantine (1979).

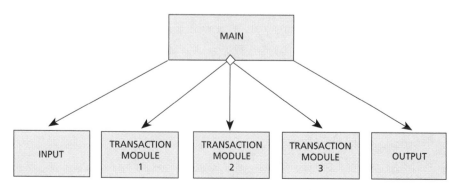

Figure 16.11 *Transaction-centered structure*

Both transform and transaction analysis start by generating an initial structure chart from a data flow diagram. The initial structure chart is then usually refined to yield the final structure chart.

Transform and transaction analysis convert each DFD process to one structure chart module and connect these modules in a way consistent with DFD data flows. The nature of these connections lies with the ideas of transaction- and transform-centered structures. We must identify flows in the DFD that correspond to transaction- and transform-centered structures and convert these flows to such structures.

TRANSFORM ANALYSIS

Transform analysis searches the DFD for a process that can be converted to a transform center in a structure chart. It looks for a central process together with well-defined input and output streams. One such process is Process F2 in Figure 16.12.

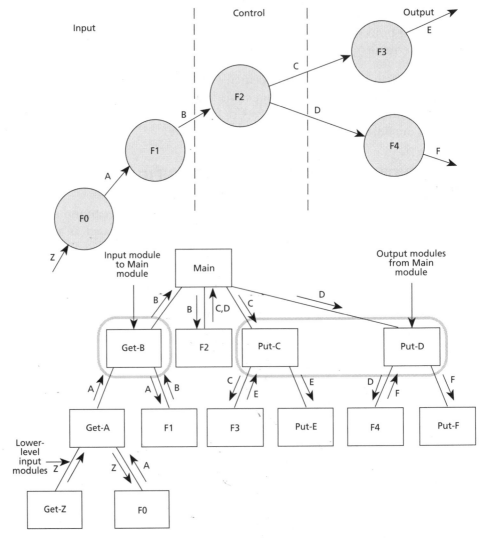

Figure 16.12 *Transform analysis*

The next step is to create a MAIN module and another module for the central process, then have the MAIN module call the central processing module (in this case F2).

Then we need to convert the processes that provide input to Process F2 to structure chart modules. These modules will provide the input to module MAIN, which in turn passes it to module F2. Thus, in Figure 16.12, input modules are created for DFD processes F0 and F1. Process F0 is converted to three modules. Module GET-A calls module GET-Z to read a value and pass it in parameter Z. It then calls module F0 to compute a value of A from the value of Z and passes the value of A to module GET-B. Module GET-B and module F1 are derived from process F1. Module GET-B first calls module GET-A to get a value of A and then calls module F1 to compute a value of B from the value of A. B is then passed to module MAIN.

Finally, the DFD processes that take the outputs from F2 are converted to output modules that obtain outputs from the MAIN module and deliver them to an output device. In Figure 16.12, DFD Process F3 is converted to modules PUT-C, F3, and PUT-E. Module PUT-C is called by module MAIN, and a value of C is passed to module PUT-C. Module PUT-C calls module F3 to compute a value of E from the value of C. It then calls module PUT-E to output the value of E. Similarly, Process F4 is converted to modules PUT-D, F4, and PUT-F.

TRANSACTION ANALYSIS

Transaction analysis revolves around identifying those parts of the DFD that can be converted to transaction-centered structure charts. In the DFD we look for an input stream that is split up into a number of input streams by a process. One such example is Process T1 in Figure 16.13. Process T1 receives input P and produces three output data flows labeled Q, R, and S. Each of these data flows is transformed and then recombined into a single data flow, W, by Process T5.

The data flow diagram in Figure 16.13A is converted to the structure chart shown in Figure 16.13B. Process T1 is converted to the MAIN transaction module in Figure 16.13B. Each of the three transaction processes T2, T3, and T4 are converted to structure chart modules T2, T3, and T4. These three modules are called by module MAIN.

The input process TO is converted to the input modules GET-P, GET-N, and TO. The MAIN module calls module GET-P to get the value of B. It then determines the kind of transaction in B. Depending on the transaction, it will activate one of the three modules T2, T3, or T4. The output from these transaction modules is sent back to module MAIN, which then initiates output by calling module PUT-W. Module PUT-W has been derived from Process T6. Process T6 has been converted to modules PUT-W, T6, and PUT-X. Module MAIN passes the value of W to module PUT-W. Module PUT-W calls module T6 to compute a value of X from the value of W and then calls Module PUT-X to output the value of X.

Again, we remind readers that both transform analysis and transaction analysis are the first step in a structure chart. Considerable refinement is often necessary to produce structure charts that have well-coupled and highly cohesive modules.

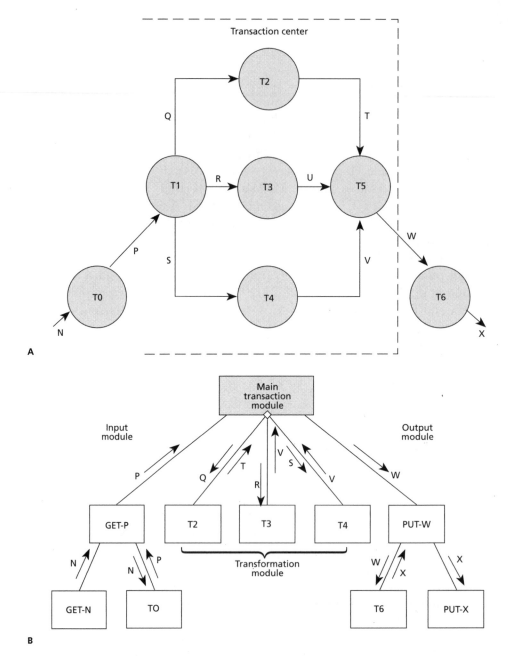

Figure 16.13 *Transaction analysis*

DESIGN USING OBJECT MODELING

System specifications in most object-oriented methods produce object classes together with either use cases or scripts. As is the case with structured systems analysis, this again corresponds to a usage-level model and a subject-level model. The system level specifications are usually converted to component diagrams. The components then become classes in the implementation if the implementation uses

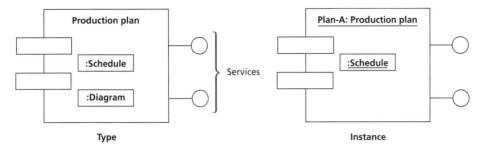

Figure 16.14 *Components*

an object-oriented system. If the implementation uses procedural programming, then the components can become modules in structure charts.

COMPONENT MODELS FOR OBJECT DESIGNS

Object-oriented methodologies that follow UML standards use component diagrams to represent the logical architectural structure. The notation used for components is shown in Figure 16.14. There is a distinction between component types and instances. A component type is shown as a rectangular box with two small rectangles protruding from its side. The name of the component is placed inside the box. Services provided by the component are shown as circles attached to the box. A component instance has the same shape but now has the name of the instance underlined in the box. Thus, Figure 16.14 shows a component type called the production plan, with two parts, Schedule and Diagram. It also shows that an instance identified as 'Plan A' has a schedule but not a diagram.

The idea of a component is that it defines a physical unit of implementation with well-defined interfaces and can be used as a replaceable part of a system. A component may include a number of parts. Here Schedule and Diagram are the parts that make up the component. Component diagrams can be adapted to both the type of application and the way that problem-solving steps are combined in rapid application development.

WEB DESIGN

Object modeling methods are often used with rapid application development. Rapid application development processes can combine some of the problem-solving steps shown in Figure 16.1. Thus, for example the architectural and physical steps may be combined when building a prototype. It may then be worthwhile to use the flexibility suggested by UML and extend the component diagram to include some implementation considerations.

TEXT CASE C:
Universal Securities—Component Model
Figure 16.15 illustrates the component diagram derived from the system specification shown in Figure 13.13. There are five database components:

- Suppliers, which contains the details about the suppliers.
- Supplier contracts, which include the parts that can be ordered under the contract.

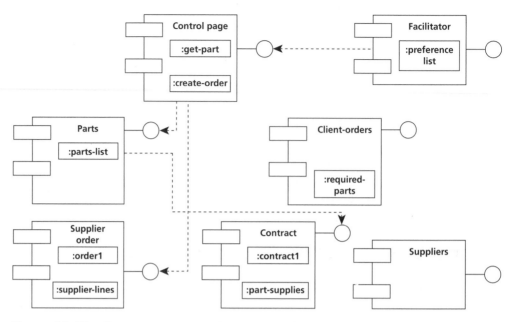

Figure 16.15 *Component diagram for Universal Securities*

* Client orders, including the parts required to fulfil the order.
* Supplier orders, including the parts ordered in the order.
* Parts, which contains a parts list.

In addition, there is a component called 'Control page', which contains the forms needed to access the server database through the Web. This component provides the services that allow the facilitator to find out where to get the needed parts and create an order for the parts. A facilitator component shows that the facilitator maintains a preference list used for guidance in selecting responses to quotes.

Contracts provide a service used by the parts component to find contracts that can be used to order the parts. Thus, the facilitator accesses the parts through the control parts component and then the contract component to find suppliers that can supply the needed part.

SUMMARY

This chapter describes architectural program design and how system specifications are converted to program modules. For structured systems analysis, conversion goes through a number of steps. The DFD is divided into subsets, and each subset eventually becomes a structure chart. The chapter describes how to go about developing structure charts and describes the difference between good and bad structure charts. Good structure charts exhibit good coupling and cohesion. The chapter concludes by suggesting that conversions from process specifications to well-structured code can be very straightforward.

The chapter also describes the conversion of system specifications expressed as object class diagrams into component diagrams. The components are later implemented as object classes.

EXERCISES

16.1   Consider the logical model in Figure 13.7 and see whether you can develop a different physical implementation from that shown in Figure 16.5. In this alternative physical implementation, inventory maintenance and supplier selection are to be done on the same machine. This machine will also process invoices. The idea is to integrate purchases for the store inventory with purchases for special project requests into one system. Would any changes to logical processing be required to implement such a system? Suggest any changes.

16.2   Look at the DFD and E–R model for Case 2 (travel arrangements). You should have developed this model as one of the problems in Chapters 8 and 9. Now it is time to propose some possible implementations for this system. Propose alternative automation boundaries for this system and comment on them. You may, for example, consider that some automation is impractical here. For example, keeping all hotel or airline schedules on computer file may simply call for too much data collection and therefore be impractical. Be careful to suggest only alternatives that can be practically implemented for automation.

16.3   What is the advantage of starting with a broad physical design in structured systems analysis? Discuss the role of the user interface in broad design. Does it illustrate what will be required of users in the new system?

16.4   What is the cohesion of modules shown in Figure 16.16? Figure 16.16 shows a brief process description of each module next to the module. Note that in Figure 16.16(a) it is necessary to sort by area before computing the delivery amount for each area. Furthermore, truck requirements can be estimated only after amounts are computed for each area. In Figure 16.16(b) the sort operations are independent of each other. In Figure 16.16(c) the same algorithm is used to achieve functional cohesion.

(a) Case A

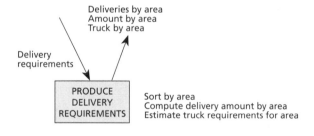

(b) Case B

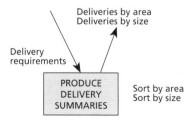

Figure 16.16   *Module cohesion*   (*continued . . .*)

(c) Case C

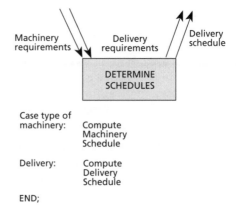

```
                    Case type of
                    machinery:      Compute
                                    Machinery
                                    Schedule

                    Delivery:       Compute
                                    Delivery
                                    Schedule

                    END;
```

Fig16.16   *(Continued)*

16.5   Identify any coupling problems in the structure chart shown in Figure 16.17. If you identify any problems, suggest a new structure chart where all coupling between modules is data coupling.

16.6   Use transform analysis to convert the data flow diagram in Figure 16.18 into a structure chart.

16.7   Use a combination of transform and transaction analysis to convert the data flow diagram in Figure 16.19 into a structure chart.

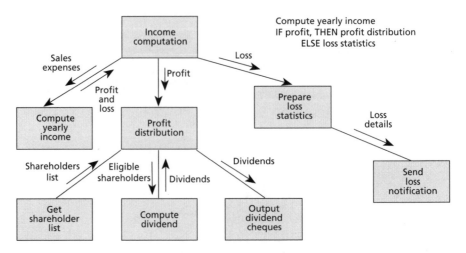

Figure 16.17   *Coupling*

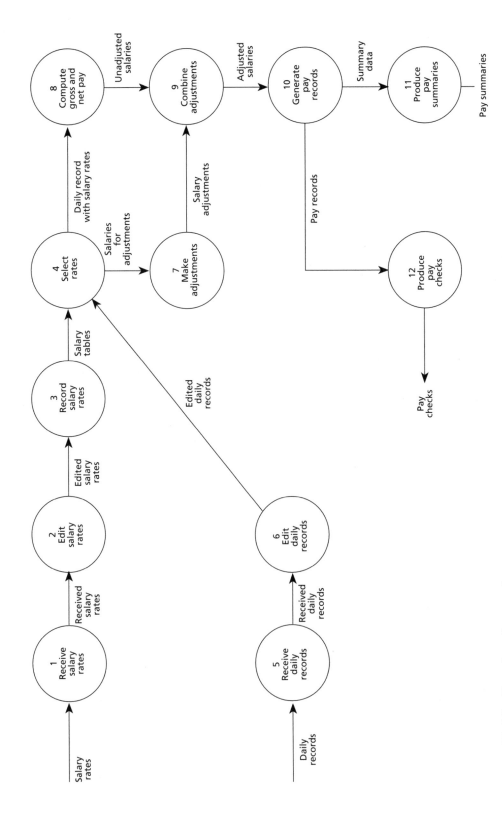

Figure 16.18 *Data flow diagram 1*

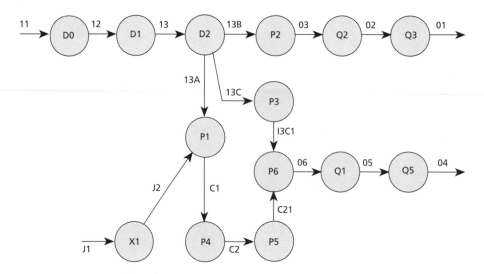

Figure 16.19 *Data flow diagram 2*

BIBLIOGRAPHY

Juliff, P. (1990), *Program Design* (2nd edn), Prentice-Hall, Sydney.

Meyer, B. (1988), *Object-Oriented Software Construction*, Prentice-Hall, New York, 1988.

Page-Jones, M. (1988), *The Practical Guide to Structured Systems Design* (2nd edn), Prentice Hall International, Englewood Cliffs, New Jersey.

Yourdon, E. and Constantine, L. L. (1979), *Structured Design*, Prentice Hall, Englewood Cliffs, New Jersey.

PHYSICAL DESIGN

CONTENTS

KEY LEARNING OBJECTIVES

How to convert relations to hierarchical and network structures

Data warehousing

Indices and file keys

What is done during file design?

Packaging programs

Deployment diagrams

INTRODUCTION

The previous chapters described architectural design and defined what must be done to implement the system. The next step is to describe how the architectural design will be implemented on physical systems (Figure 17.1). This often requires choosing among alternative ways to implement a particular architectural design. In many cases the selection is determined by the particular infrastructure chosen by an organization for its development. Thus, for example, an organization may have an organization-wide database management system or have a policy for implementing Web systems on a particular server. Others may use particular document management systems or workflow systems.

This chapter describes conversion to physical systems of both the programs and the database for some typical cases. It begins by showing how to convert database specifications to a number of the available implementation models. It then describes how program designs are organized into physical program structures. The next chapter will cover some development issues for specialized systems.

DATABASE IMPLEMENTATION MODELS

Databases are implemented using database management systems (DBMSs). DBMSs store information as sets of records with links between them. Each record type is made up of a number of fields. As an example, Figure 17.2 is a record type called PERSONS. Each record of this type has three fields, NAME, DATE-OF-BIRTH, and WHERE-BORN.

Records can be retrieved from DBMSs in a variety of ways. One way is to read records serially, starting with the first record and continuing until all the records have been read. Often, however, it is necessary to retrieve only one record from the database. For example, we may want to find Bill's record to see where Bill was born. To do this, we could start at the beginning and read records until we get to Bill's record. However, this can be time-consuming, because, in a large database,

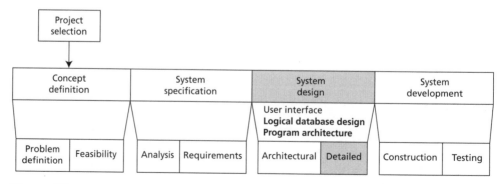

Figure 17.1 *Detailed physical design*

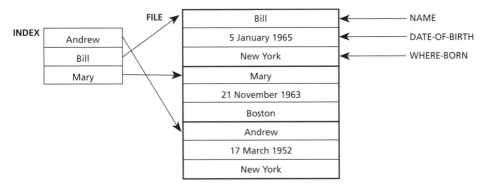

Figure 17.2 *File PERSONS*

many records may have to be read until the required record is found. What is needed is the ability to access one record directly rather than read the whole database until the required record is found.

Direct access is provided by indexes. An index uses one of the record fields as a key, and there is an entry in the index for each record in the database. This entry contains the value of the records key and a link from the index to the record. Thus, record type PERSONS in Figure 17.2 has an index whose key is NAME. It is made up of three entries with values 'Bill', 'Mary' and 'Andrew'. Each entry has a link to a record with that key value, and programs can access records through the index. A program specifies the value of the key, and file software will use the index to directly access the record with that key value.

Every DBMS supports a different kind of implementation model. A number of DBMSs support the relational model. These are known as relational DBMSs. There are also a number of DBMSs that support network structures. The network structure was once proposed as a standard, and a number of DBMS suppliers have followed this standard. DBMSs that support network structures are known as network DBMSs. Finally, a number of DBMSs support hierarchical structures. These are known as hierarchical DBMSs.

We will illustrate the structures supported by each of these three kinds of DBMSs. Figure 17.3 presents the DBMS record structures for the system, where:

- projects are assigned to departments
- projects are made of any number of jobs
- each project uses different part kinds.

DATABASE MANAGEMENT SYSTEMS

The relational model stores data as a set of tables or relations, as shown in Figure 17.3(a), where each record is a row in the table. In a relational DBMS, each such relation would be defined using the system's definition language. Commands provided by the DBMS would then be used to store and retrieve data. The next chapter will cover the relational model in greater detail.

Direct access
Retrieval of records based on keyword

(a) Relational model

DEPARTMENTS

DEPT-NO	MANAGER	DATE-ESTABLISHED

PROJECTS

PROJECT-NO	DEPT-NO	BUDGET

PARTS

PART-NO	COLOR	WEIGHT

PART-USE

PART-NO	PROJECT-NO	QTY-USED

JOBS

JOB-NO	PROJECT-NO	COST

(b) Network model

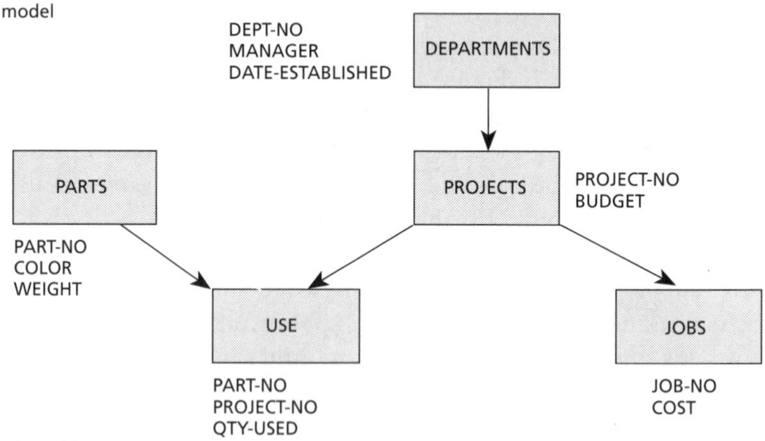

(c) Hierarchical model

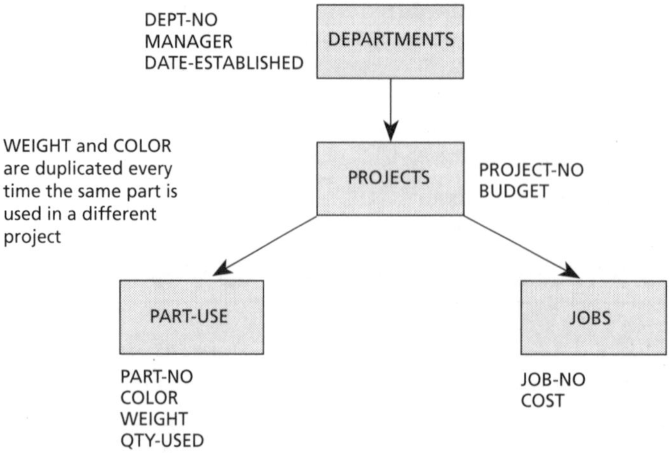

Figure 17.3 *Data models*

NETWORK DATABASE MANAGEMENT SYSTEMS

Network DBMSs store data as record types. Furthermore, parent–child relationships can be established between these record types. Such relationships are illustrated in Figure 17.3(b). Each DEPARTMENTS record will own any number of PROJECTS records. A DEPARTMENTS record will own PROJECTS records of projects that are assigned to that department. Each PROJECTS record owns the JOBS records of jobs that make up that project. Each PROJECTS record also owns those USE records that represent a project's usage of items. The parent–child links are called set types in network terminology.

In a network model, each record type can be a parent of any other record types. It can also have any number of parents. This is not, however, the case for hierarchical DBMSs.

HIERARCHICAL DATABASE MANAGEMENT SYSTEMS

The hierarchical data model differs from the network model because each record type can have only one parent. We can no longer have a record type such as USE, which had two parents in the network data model. The designer has to decide whether USE is to be modeled as a child of record type PARTS or a child of record type PROJECTS. Figure 17.3(c) represents the case where record type USE has become a child of record type PROJECTS. In that case, PARTS record fields are now combined with the USE record to make the PART-USE record, which is a child of the PROJECTS record. PART-USE records now store the details of each part used by each project. The disadvantage of this approach is that the COLOR and WEIGHT for a part are stored more than once. In fact, they are duplicated for each project that uses a part.

Alternative hierarchical representations of the same user data are possible. For example, we could make USE records a child of PARTS records and PROJECT records a child of PARTS records. In that case, PROJECT data would be duplicated for each part used by the project. Alternatively, we could store each USE record twice, once as a child of record PROJECTS and once as a child of record USE. Designers must choose between these alternatives.

CONVERSION TO DBMS STRUCTURE

During implementation, the logical record structure is converted to a **database definition** supported by a DBMS. This proceeds in two steps:

Database definition A definition of a database used as an input to a DBMS.

- Converting the logical record structure to a data model supported by the DBMS. The database is defied by a database definition, which is input directly into the DBMS and sets up the database record structures.
- Choosing the ways to access data in the database.

The conversion method depends on the type of data model supported by the DBMS.

CONVERSION TO RELATIONS

The simplest conversion is where the DBMS supports the relational model. Each logical record type becomes a relation in the relational DBMS. Thus, the logical record structure in Figure 15.3 would be converted to three relations, PROJECTS, PARTS, and USE, shown in Figure 17.4. The logical record fields would become the relation columns. Thus, relation PROJECTS would contain three columns, PROJECT-NO, START-DATE, and BUDGET. The fields of records in each relation are shown next to the relation in Figure 17.4. Thus, records in relation PARTS have four fields: PART-NO, WEIGHT, COLOR, and PRICE.

CONVERSION TO NETWORK DATABASE MANAGEMENT SYSTEMS

The logical record structure can be converted to a network model in a relatively simple way. Each logical record is converted to a record type, and each logical record link becomes a set type. An example of such a conversion is shown in Figure 17.5. The logical record structure to be converted is shown in Figure 17.5(a). This is converted to the network structure shown in Figure 17.5(b).

Each logical record becomes a record type in the network structure, and all the links in the logical record structure become links in the network representation.

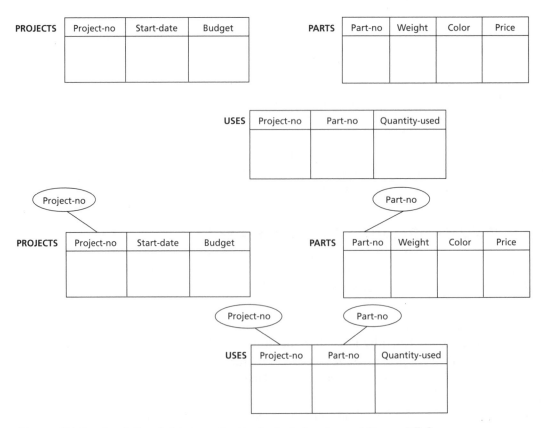

Figure 17.4 *A relational database for the logical structure of Figure 15.3*

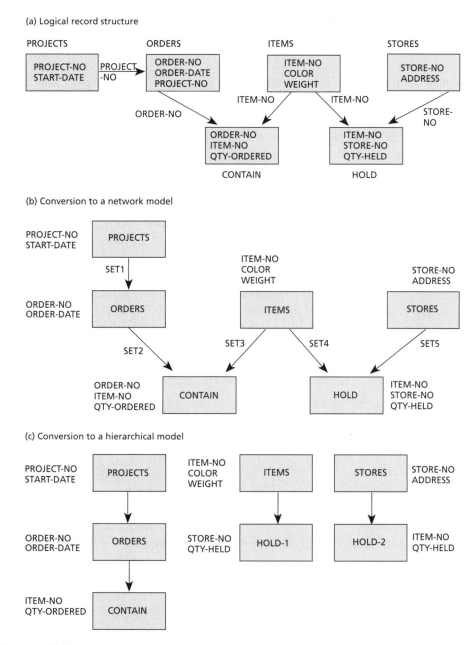

Figure 17.5 *Conversion to a data model*

CONVERSION TO HIERARCHICAL DATABASE MANAGEMENT SYSTEMS

Conversions to a hierarchical model impose an additional constraint. Each record type in a hierarchical model can have only one parent. This was not the case in network model design. For example, records CONTAIN and HOLD in Figure 17.5(b) have two parents in the logical record structure. After conversion, record types CONTAIN and HOLD in the network model also have two parents. In hierarchical systems, however, two parent records are not permitted for a record type. The designer must choose one parent out of a number of possible parents. Alternatively, a logical record structure (LRS) record may appear more than once in the hierarchical logical structure. It will appear once for each parent in the LRS.

Figure 17.5(c) illustrates a possible conversion to a hierarchical data model. Record type ORDERS is chosen as the parent of record CONTAIN. It is easier, in this design, to access the contents of each order rather than find the orders that contain a given item. The logical record HOLD, however, appears twice in the design, once as record HOLD-1 and once as record HOLD-2. It has ITEMS as a parent the first time and STORES as a parent the second time. The logical parent STORES is used to allow easy access to holdings in each store. The logical parent ITEMS is used to allow easy access to the location of items. However, some information, namely QTY-HELD, is now duplicated.

To eliminate such duplications, many hierarchical systems now offer logical links between physical databases. IBM's database management system IMS is a prime example here. These links allow designs that reduce data duplication by setting up links between separate physical databases.

PROVIDING WAYS TO ACCESS DATA

A number of additional things are done during physical design. Indexes to records are chosen, as are placements of logical records in physical files. Often we draw access diagrams to make these choices. Access diagrams tell us how the database is to be used to choose file keys during database design.

Information about quantitative data and access requirements is gathered during systems analysis. Quantitative data consists of:

* the size of data items
* the number of occurrences of record types.

The size of data items is usually given in a data dictionary. Record volumes can be added to the logical record structure by the method shown in Figure 17.6. The number of records is simply added into the logical record. Thus, Figure 17.6 shows that there are 100 PROJECTS records, 2000 PARTS records, and 5000 USES records.

SPECIFYING ACCESS REQUIREMENTS

Access requirements
Defining how databases are accessed.

Access requirements are initially picked up from user procedure specifications, which include statements about how users will access data. These statements become the database access requirements. During database design, access paths are plotted against logical record types for each access requirement. The access paths show how data is to be used and describe:

* the record types accessed by each access request
* the sequence in which the record types are accessed
* the access keys used to select record types
* the items retrieved from each record
* the number of records accessed.

There are many ways to draw access paths. One way is illustrated in Figure 17.6.

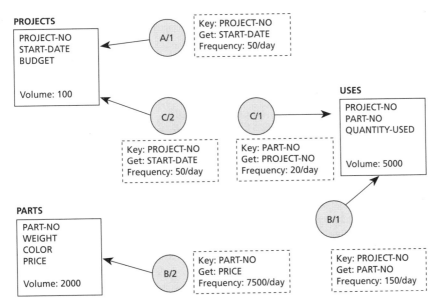

Figure 17.6 *Access paths to the logical structure level*

This figure shows three access requirements plotted against three logical record types. The three access requirements are:

A—find the START-DATE for a given project
B—find the PRICE of parts used on a project with a given PROJECT-NO
C—find the START-DATE of projects that use a given PART-NO.

There is one access path for each access requirement, and each access path can be made up of one or more access steps. Each access step has a label that is made up of the mnemonic that describes the requirement and a sequential number. The access step is labeled with an access key. The access keys are the names of fields whose values are known at the time of the access step. The access step also includes a description of the activity at the step.

Access requirement A is specified by one access step, which is labeled A/1 in Figure 17.6. This access step is used to access PROJECTS records. The access uses the value of PROJECT-NO as the access key and retrieves the value of START-DATE. It is made an average of 50 times a day and retrieves one record.

Access requirement B consists of two access steps, one to record type USES and the other to record type PARTS. Access step 1 (labeled B/1) finds the parts used by the project. It is made an average of 150 times a day, and each access retrieves an average of 50 records, because each project uses an average of 50 parts. The 50 is computed by dividing the number of USES records by the number of projects. Access step 2 (labeled B/2) finds the PRICE of these parts. Note that access step B/2 has a frequency of 7500 per day. The frequency is computed by multiplying the 150 times that access step B/1 is executed by the 50 PARTS records that are retrieved following each execution of access step B/1.

The final access requirement in Figure 17.6 is access requirement C. The first step (labeled C/l) finds the PROJECT-NOs of projects that use the given PART-NO. Step 1 of access step 3 is executed 20 times a day and retrieves, on average, 2.5 records (as each part is used by 2.5 different projects). Access step 2 (labeled C/2) finds the START-DATE of these projects.

The access requirements are now used to complete a database physical design by selecting appropriate physical structures. Usually, a number of alternative physical structures can be chosen to satisfy the access requirements. Part of the design is to select from these structures. An iterative process is used to make such choices, and performance estimates are made at each iteration to compare the alternatives.

SELECTING PHYSICAL STRUCTURES

Access paths are used to select access keys. As described earlier, each key in an access step becomes a file index. This rule holds for both files and structures supported by a DBMS. Apart from choosing keys, there are other choices to be made in physical design. Many of these depend on the type of DBMS and how to place physical records on DBMS files. For example, placing records that are often retrieved together close to each other can improve system performance.

IMPROVING ACCESS PERFORMANCE

The kinds of choices described above require some analysis using access paths. One way is to examine access steps and physically link records, which are accessed in successive steps. The number of access steps can also be reduced by combining fields, which are retrieved in successive access steps, into one record. For example, in Figure 17.6, the value of PRICE could be added to record USES. Access step C/2 now becomes unnecessary, and performance of access requirement C is improved. The performance improvement, however, is made at the price of losing normal form, because record type USES is no longer in normal form. Thus, redundancy is introduced in a controlled manner to improve performance. It is up to the designer to make the trade-off between improved performance and good data representation.

DESIGN TRADE-OFFS

Designers usually choose design options in an iterative manner, as shown in Figure 17.7. A design problem is identified and a solution is proposed. Usually the solution requires the designer to make some trade-off. The major trade-offs are:

- performance against storage use. The introduction of new physical links usually improves performance but requires additional storage to store the physical links
- performance against structure. The controlled introduction of redundancy improves performance but requires additional storage and introduces restructuring problems.

To make these trade-offs, designers first estimate any improvements of performance and the cost required to achieve these improvements. Usually a number of options are evaluated, and the trade-offs for each option are identified. Environmental criteria are then used to select the appropriate option. Typical criteria include:

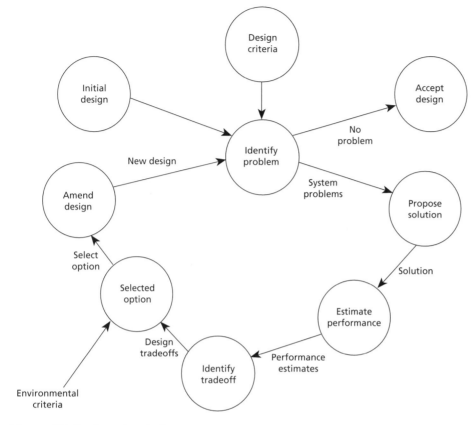

Figure 17.7 *Iterative design*

- the availability of storage
- the importance of improved performance for a given requirement
- the probability of restructuring a particular database part.

A particular option is then selected, given the criteria and the user environment. The user environment is usually important in such choices. In microprocessor systems, for example, where storage is at a premium, it is unlikely that designers would be prepared to sacrifice storage for marginal performance improvements. However, such a trade-off will be quite reasonable in large systems that are tightly integrated with user functions.

Performance estimates are made during the iterative process shown in Figure 17.7 to identify the trade-offs. The usual method for doing this is called **logical record analysis** (LRA). LRA uses access paths. Each access step is examined to determine the number of records accessed by that step.

Once this number of records is known, the number of disk transfers made can be estimated. Such an estimate of disk transfers considers the physical structure used to store the records. The numbers of disk transfers for different design options are then compared when making design trade-offs.

Logical record analysis A way of making performance estimates in design.

RECOVERY FROM ERROR

One important consideration in database design is to prevent faults or errors from destroying or corrupting the data in the database. Database errors can occur in many ways—computer system malfunctions, disk drive crashes, operating system faults, or power failures. Errors can also be caused by incorrect input data.

Recovery is usually provided by keeping back-up copies of the whole database or parts of the database. If the database is destroyed by a fault, it can be replaced by its back-up copy.

Back-up copies can be made in various ways. Early batch systems made a copy of the database at the end of each day. If an error occurred during the day, the corrupted database was replaced by the back-up copy, and all the daily transactions were run again.

On-line transaction systems use more sophisticated recovery methods. They maintain journal files. Every time a record is changed, the old record is copied to a journal file. The transaction that makes the change is also stored in the journal. If there is a system fault, the journal file is read backwards, and old record copies are restored back to the database. The transactions can then be rerun to update the file.

DATA WAREHOUSING

Designers of data warehouses must choose a data structure in the same way as described earlier in this chapter. However, because of the database size and importance, designers must also pay special attention to other factors. One of these is maintaining the quality of data. This usually means that special attention must be paid to processes that operate on the data warehouse. These include processes that update the database or transform its contents. It also requires special processes to check data consistency.

PROGRAM DESIGN

The goal of implementation in structured systems analysis is to convert the system model, specified as a structure chart, into a set of program modules. Figure 17.8 describes the steps in detail. A DFD specification has been developed, and each process has been described in structured English using the methods of Chapter 10. The description identifies all the conditions that can arise in the process and how they are treated. Thus, for example, Step 1 in Figure 17.8 defines that three things can happen if sufficient parts are not found in store to meet an order. The order can be cancelled, the order quantity can be reduced, or just that part can be deleted from the order. In addition, a structure chart has been developed showing the program modules that will implement the process. Two steps are now carried out in physical design:

• Packaging the structure chart modules

(a) Step 1. Produce detailed program specifications

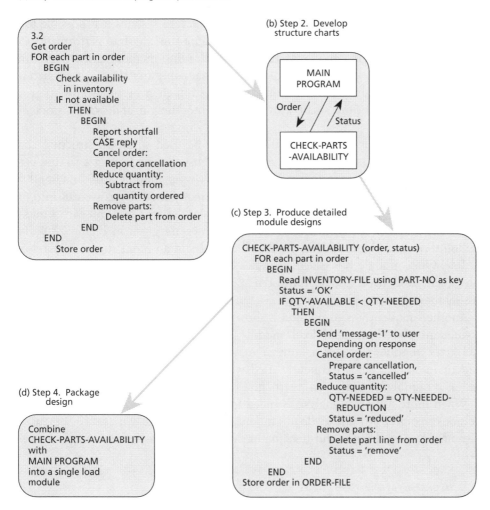

```
3.2
Get order
FOR each part in order
    BEGIN
        Check availability
        in inventory
        IF not available
        THEN
            BEGIN
                Report shortfall
                CASE reply
                Cancel order:
                    Report cancellation
                Reduce quantity:
                    Subtract from
                    quantity ordered
                Remove parts:
                    Delete part from order
            END
    END
        Store order
```

(b) Step 2. Develop
    structure charts

MAIN
PROGRAM

Order   Status

CHECK-PARTS
-AVAILABILITY

(c) Step 3. Produce detailed
    module designs

```
CHECK-PARTS-AVAILABILITY (order, status)
    FOR each part in order
        BEGIN
            Read INVENTORY-FILE using PART-NO as key
            Status = 'OK'
            IF QTY-AVAILABLE < QTY-NEEDED
            THEN
                BEGIN
                    Send 'message-1' to user
                    Depending on response
                    Cancel order:
                        Prepare cancellation,
                        Status = 'cancelled'
                    Reduce quantity:
                        QTY-NEEDED = QTY-NEEDED-
                            REDUCTION
                        Status = 'reduced'
                    Remove parts:
                        Delete part line from order
                        Status = 'remove'
                END
        END
Store order in ORDER-FILE
```

(d) Step 4. Package
    design

```
Combine
CHECK-PARTS-AVAILABILITY
with
MAIN PROGRAM
into a single load
module
```

Figure 17.8   *Developing program specifications*

- Completing the process by including things such as references to records, detailed error checks, I/O operations, and user dialogs.

PACKAGING INTO LOAD MODULES

Computer systems work by bringing a module into the memory when that module is called and then executing it. Some system time is used up every time a module is loaded into the memory.

Ideally, all the program modules should be brought into memory when the program starts. Thus, all modules would be in store at all times and would not need to be brought into memory when they are called. However, this ideal cannot be realized for large programs because of memory size limitations. The whole program will simply not fit into memory.

The next alternative is to make each structure chart module into a 'load' module.

This approach may have to be used if all the modules are relatively large and the amount of memory is small. However, in most practical systems it is possible to fit more than one module into the memory at the same time. System performance can be improved by bringing all closely related modules into memory at the same time. When these modules call each other, no additional computer time is used, because all the modules are already in memory. The question then becomes how to group modules so that the time spent moving 'load' modules in and out of storage is minimized.

There is no magic formula or rule that can be used to group structure chart modules into 'load' modules. Such grouping is usually made in a trial-and-error fashion and depends on factors such as memory size, module size, and module-calling frequency. Obviously, we want to put as many modules as we can into one 'load' module, while ensuring that such modules are related and call each other.

Some guidelines for constructing 'load' modules are given in Figure 17.9. The method shown in Figure 17.9(a) groups all modules in an input stream into a 'load' module. Once module B is called, modules C and D are also brought into memory. Subsequent calls to C and D do not require more 'load' modules to be brought into memory.

Another frequently used method is shown in Figure 17.9(b). All the lower-order modules are placed into one 'load' module. This method is useful only if that low-level 'load' module contains a number of shared modules and can be continually resident in storage. It is frequently used where common data access modules are used at a number of points on a structure chart.

When more elaborate analysis is called for, we need to look at iteration and the decisions required to make 'load' module choices. Take Figure 17.9(c) as an example. Here we have found that the most common decision outcome is to select

(a) Grouping by logical flow

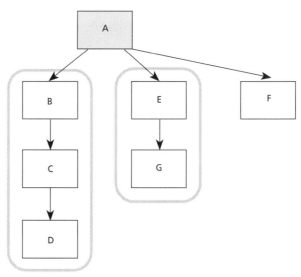

Figure 17.9   *Creating load modules*

(b) Grouping by level

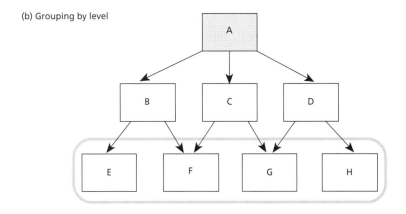

(c) Using decision and iteration structures

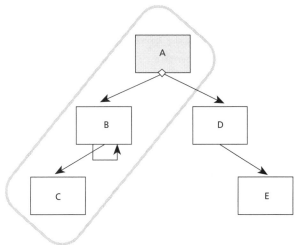

Figure 17.9 *(Continued)*

B rather than D. Furthermore, B calls C a number of times but calls E once only. It therefore makes sense to group A, B, and C into one 'load' module. If this is done, the outcome of the decision will, in most cases, not require a new 'load' module to be brought in or require continual loading of a module during each loop. Of course, E and D should also be included in the loaded module if there is sufficient space. However, in the example, we have assumed that such space is not available.

DEPLOYMENT DIAGRAMS

Object modeling methodologies based on the UML standard use deployment diagrams to define what is a close equivalent to load modules. They define the computing systems that will store the various objects. These diagrams show where different components are located. The notation used in deployment diagrams is shown in Figure 17.10. Here a node contains a component. In Figure 17.10, the node is called the central server and contains the production plans. The notation

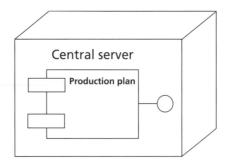

Figure 17.10 *A deployment diagram*

can also show the links between the nodes, thus identifying how information is moved between the different system nodes.

TEXT CASE C:
Universal Securities—Deployment Diagrams

Figure 17.11 shows the deployment diagram for Universal Securities. It shows how the components illustrated in Figure 16.15 are implemented on the system. Thus, the facilitator maintains the preference list on a private PC but interacts with both the regional server and central server. The regional server stores the regional customer orders, whereas parts, contracts, and supplier orders are stored centrally. The control page is also placed on the central server and is used to access the database components through the WWW. The facilitator uses the regional server to get requirements, and posts calls for quotes on that server. The facilitator then uses the responses to quotes to initiate supplier orders on the central server.

SUMMARY

This chapter describes physical design. It begins with the conversion of logical data models to a DBMS. Such conversion depends on the types of structures supported by the DBMS. DBMSs are often classified into three kinds: relational, network, and hierarchical. This chapter shows how logical record structures can be converted to logical relational, network, and hierarchical structures. It then describes the trade-offs made during physical design to get satisfactory database performance.

The chapter then describes the design of physical program modules.

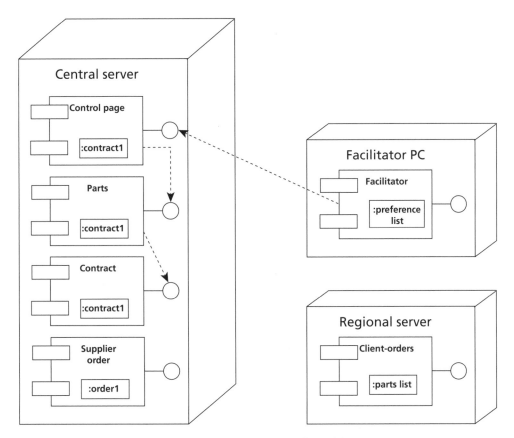

Figure 17.11   *Deployment diagram for Universal Securities*

EXERCISES

17.1   You are given the logical record structure shown in Figure 17.12. This LRS describes a delivery system. Each trip is made by one driver driving one vehicle. A number of deliveries can be made on each trip. Any number of part kinds can be delivered in each delivery. Draw access paths on the LRS for the following access requirements:

1.   List the NAMEs of drivers who made deliveries to a customer with a given CUSTOMER-NAME.

2.   List the NAMEs of drivers who used a vehicle with a given REGISTRATION-NO.

3.   List the REGISTRATION-NOs and MAKEs of vehicles used on a given TRIP-DATE. Assume that a trip never takes more than one day.

4.   List the QTY-DELIVERED of a given PART-NO on a given TRIP-DATE.

17.2   Convert the LRS in Figure 17.12 to a network and a hierarchical DBMS.

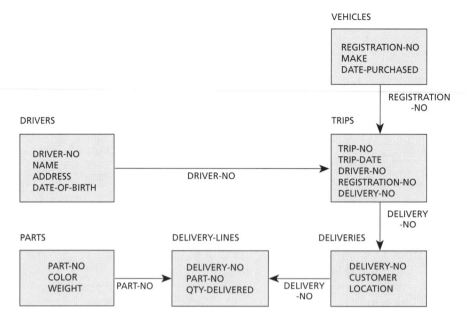

Figure 17.12 *A logical record structure*

BIBLIOGRAPHY

DeMarco, T. (1978), *Structured Analysis and System Specification*, Yourdon Press, New York.

Eaglestone, B. and Ridley, M. (1998), *Object Databases: An Introduction*, McGraw-Hill, New York.

Gane, C. and Sarson, T. (1979), *Structured Systems Analysis*, Prentice Hall, Sydney.

Inmon, W. (1996), *Building the Data Warehouse* (2nd edn), John Wiley and Sons, New York.

Jacobson, I., Christerson, P.J. and Overgaard, G. (1992), *Object-Oriented Software Engineering*, Addison-Wesley, Harlow, England.

Stonebraker, M. (1996), *Object Oriented DBMSs: The Next Great Wave*, Morgan Kanfman, San Diego.

DEVELOPMENT

CONTENTS

KEY LEARNING OBJECTIVES

Development platforms
Program development
Object development
Developing WWW sites
Workflow systems

CHAPTER 18

INTRODUCTION

The previous chapters have described how to develop system design models. First the architecture of the new system was defined. This was followed by defining how this architecture would be implemented by physical systems. The next step, as shown in Figure 18.1, is to actually develop computer systems that implement these models. Contemporary system development can be a very complex process. When you look back at Figure 1.3 in Chapter 1, you will see that there are many different kinds of systems. Each such system is often developed in different ways, and what is even more important is that all of these systems must often be integrated into organization-wide processes. Thus, developers not only develop particular systems but must also integrate them with other systems. No book can hope to cover all these processes. This chapter only introduces some ways to develop systems. You will need to consult specialist texts to find out the details.

Usually, developers are provided with a particular development platform and an environment for that platform. Typical development platforms can be:

- *a programming language and its environment, such as editors*
- *an object-oriented development and its environment*
- *enterprise resource programming (ERP) and a development environment such as SAP*
- *Web site development platforms*
- *workflow management systems.*

A development environment is also provided with the platform. Environments provide tools to help developers carry out their work and include programming languages, editors, and ways of managing configurations. Thus, there may be a development environment for languages used in the system or for a particular DBMS. Object-oriented environments also provide libraries of reusable object classes. The goal of using the environments is to reduce the development time. It helps developers to keep track of changes, in testing, and in the integration of modules.

The choice of a platform and development environment is usually a key infrastructure decision within an organization, as it affects the way systems are developed within the organization. It would not be a good idea to have each developer use a different environment. Such environments set a standard way of development for all developers and lead to better integration of software modules. All developers must thus be familiar

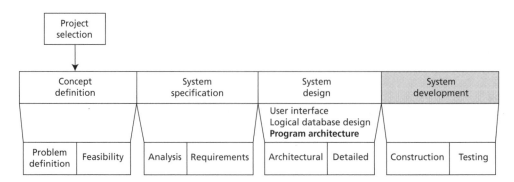

Figure 18.1   *Detailed physical design*

with both the development environment and the particular languages and tools within the environment. This chapter briefly outlines the different development methods. Development must include the development of data and programs. The kinds of methods used, however, depend on the technology available to the developers. The chapter begins by describing program development.

PROGRAM DEVELOPMENT

Irrespective of the development environment, it is necessary to write programs for the system to work. The traditional way to develop systems is to use a programming language to realize the program specification. Such programs are now developed using the programming languages provided by the development environment. The simplest view of writing a program is as a special exercise to implement a structure chart module or an object component using some programming language. Such programs must satisfy two requirements:

- They must correctly carry out the required system functions.
- They must be well written and easy to maintain.

Correctness requires correct interpretation of the design model. A good model with clear process definitions is an almost essential requirement here.

A good structure usually means that well-defined functions appear in the same section of program code, and changes to these functions affect only that section of code and not the entire program structure. Good structure makes it easier for people other than those who originally produced the code to later amend the program to accommodate the user requirements changes that always occur after a system is built.

It is now generally recognized that production of good code is realized by what is known as **structured programming**. Structured programming uses standard control structures to improve program clarity and maintenance. The control structures encourage top-down program development by orderly expansion of program blocks. Designers specify each top-level function by one program block, and the block is then expanded into more detailed components.

Structured programming A systematic way of writing programs.

To facilitate this top-down development, program blocks are made up of three main constructs, SEQUENCE, IF-THEN-ELSE, and REPEAT. Each block in the code should implement some well-defined function.

Good programming practice also calls for in-line comments to improve readability. Each block in the control structure should be defined, and its purpose should be described. Links to outside descriptions of code should also be included in the documentation.

It is relatively easy to use process specifications to create well-structured code. Process specifications use key words very similar to the constructs used in structured programming. Thus, conversion from process specification to structured code can be very straightforward. The key words of the process specification are replaced by the key words used in the programming language. The arithmetic or transformation statements used in the process specifications are replaced by the grammar used in the programming language. This improves the likelihood of a correct program.

DATABASE LANGUAGES

Most programs access data in databases. They do not need special programs to get data from a database but use languages such as SQL (Structured Query Language), which is available with most relational DBMSs. SQL simply states the conditions that retrieved rows must satisfy. As an example, consider the relations in Figure 17.3. To find out the projects in a department, all one needs to do is enter the statement:

SELECT PROJ-NO
FROM PROJECTS
WHERE DEPT-NO = 'Dep1';

A list of projects in 'Dep1' is then displayed on the screen. The SELECT-FROM-WHERE is the standard SQL clause used for data retrieval. It defines the data needed and conditions to be satisfied using the following syntax:

SELECT <data to be displayed>
FROM <relations that contain the data>
WHERE <conditions satisfied by the retrieved data>.

The user is free to select the data, relations and conditions at the time the query is input, and no special preprogramming is necessary. SQL supports a large variety of queries. Data displayed can include arithmetic statements, and conditions can include a large number of clauses separated by AND and OR and evaluated by using Boolean logic. One SQL statement can be used to retrieve data from more than one relation. For example, to find all the jobs in a department, we must first find all the department's projects from relation PROJECTS, and then find all the jobs for each project from relation JOBS. The jobs can be retrieved by the following SQL statement:

SELECT JOB-NO
FROM PROJECTS, JOBS
WHERE DEPT-NO = 'Dep1'
AND PROJECTS.PROJ-NO = JOBS.PROJ-NO;

The first part in the WHERE clause selects those rows in relation PROJECTS with the given value of DEPT-NO, 'Dep1'. The second part of the WHERE clause then matches the selected row in PROJECTS with rows in relation JOBS that have the same value of PROJ-NO. The value of JOB-NO in the matched row is then output. Functions can also appear in SQL statements. For example, the statement:

SELECT SUM(BUDGET)
FROM PROJECTS
WHERE DEPT-NO = 'Dep1';

will output the sum of the value of BUDGET for all rows with a given value of 'Dep1'. That is, the output is the total project budget of a department. SQL offers many additional facilities available, and its flexibility has now made it the standard language for relational systems.

REPORT GENERATORS

The idea behind report generators is somewhat similar to that for database languages. It is to get away from writing programs to generate reports and simply define what the report must contain and the report layout using a report definition language. Usually the contents of a report are defined using a statement from a query language. The report layout is defined using special commands, which may include specifications of report heading and footing, column names, any totaling or subtotaling, and the size of each column. Report generators provide the flexibility needed to change reporting requirements within a short time.

OBJECT DEVELOPMENT

The goal of object development is to provide environments that can directly implement object models. Once there is an object design model, the object classes in the model can be defined using the language within the environment. A number of languages and their environments have been developed for this purpose. Typical languages here include C++, Eiffel (Meyer, 2000), and recently Java. To describe these environments is outside the scope of this text.

Object development, however, has additional goals to those of writing programs. Object development methods were introduced after structured systems analysis was widely adopted. Their goal was to provide advantages over these earlier methods by improving productivity over that of existing systems. This improvement is achieved through the idea of **reuse**. The reuse goal is to build libraries of reusable object classes and to provide methods for integrating these object classes into systems. To achieve reusability, it is necessary to develop an object class library specifically for reuse. One of the earliest libraries developed concerns interface programming. As one example, object development environments based on the Java programming language, such as VisualCafe, provide a set of reusable objects, called Java Beans, especially for interface design.

Reuse Using an existing system or module for a new task.

Interface programming is a generic activity across many kinds of applications. Supporting more specific problem domains, such as banking or insurance, is more difficult. A good set of object classes requires careful analysis of the problem domain or the subject world. In problem domain analysis, it is necessary to identify the *meaningful objects* that are frequently used in a general sense in the problem domain. For financial systems they may be things such as investment, account, and transaction. The library would contain generalized **skeleton objects** that can be **customized** to particular applications or business processes. We can go even further and define *object patterns*. Object patterns are collections of objects and their connections that form skeletons for possible applications. One important aspect to library development is the idea of continually refining or specializing existing objects to meet new needs. The object paradigm supports this idea. The concept of inheritance leads to the possibility of continual refinement of objects. It allows new objects that meet new needs to be created from existing objects by changing or replacing some features of the existing objects.

Skeleton object An object that can be customized for more than one application.
Customization Adapting an object for use in an application.

Easy access to reusable objects is also important. It is usually found that if a programmer cannot find an object to reuse within about 20 minutes or so, they will simply give up the search and write their own. For this reason there has been considerable research on developing classification schemes that enable useful object classes to be quickly found. Once found, the object may be used in its original form or perhaps modified slightly to meet particular needs.

Figure 18.2 illustrates the kind of environment that encourages reuse. It includes a library of reusable modules. It also shows two development processes, one for application development and one for library development for a particular problem domain. The two processes are usually closely integrated. Part of the application development process is to examine the library to see whether it has modules that can be used in the development. The other is to identify possible new modules to be included in the library. Often the new modules are modifications to existing modules. In this case, we would gradually build up a library of modules that can be used again and again to build new applications.

WEB SYSTEM DEVELOPMENT

Web-based applications are now becoming popular because of the increasing trend toward electronic commerce. They can be standalone systems or be integrated with an organization's information systems and external clients through the Worldwide Web. Such Web sites will include interface components, discussed in Chapter 14. They also often require integration with existing legacy systems. Programming on the WWW is a relatively new area. What needs to be done depends on the kind of WWW site that is planned. There is a range of different kinds of sites, including:

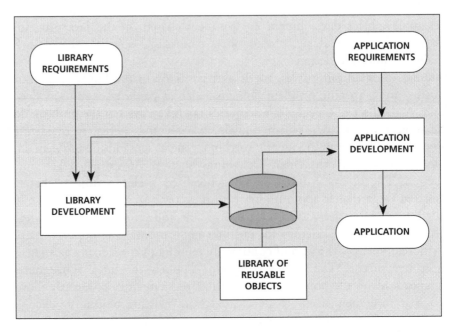

Figure 18.2 *System synthesis*

- sites that simply publish static information about an organization
- sites that support interaction with the organization to purchase products
- sites that support people working together.

Usually development evolves by starting with sites for publishing material about an organization and then continuing to include interactivity as experience grows with their use. Web site development also requires the choice of environment and language to develop any interactive features. Thus, careful planning is needed to start with an environment that supports such evolution.

Development must also fit in with the way information is transmitted across the Web. Figure 18.3 shows a server and a client. The simplest connection is to simply define Web pages that present information. Display programs simply present information in multimedia format to the user. The server contains all the multimedia pages written in HTML. The HTML pages are stored on a server and are accessed through a browser on the client workstation using what is known as the http protocol. The most popular browsers are Netscape Navigator and Microsoft Explorer. To access a page, it is necessary to know the server on which it is stored and the page address on that server. Thus, for example, the address to access my page is http://linus.socs.uts.edu.au/igorh.

Developers must first create the pages that present the interface and then add any interactive features. Developing the page itself is a relatively simple matter. It uses a language known as HTML, which can now be generated from a variety of editors, including the Netscape editor.

EXTENSION TO INTERACTIVE SITES

Special programs must be written to support interactive pages. Most languages used to develop these programs are object based. These programs reside on servers and are sometimes known as CGI (Common Gateway Interface) programs. These CGI

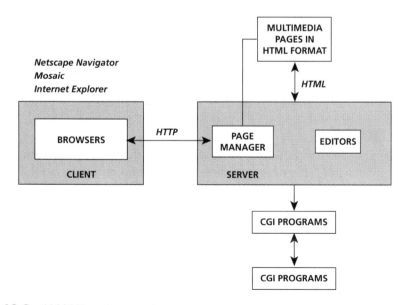

Figure 18.3 *WWW environment*

programs can be modules in a structure diagram. Early programs were written in a language known as Perl, but the trend now is toward using object-oriented languages such as Java.

The CGI program resides on the server and is activated from an HTML page through a special feature known as a FORM. The form provides input parameters, which are filled in by the users, captured, and used by the CGI program. The CGI program carries out any computation on the server and returns the results of this computation to the browser as an HTML page.

The way to develop interactive Web pages is continually changing. One goal of such improvement is to provide more options at the browser interface. HTML pages are static and do not provide features such as pop-up menus or multi-window displays. They also makes it difficult to provide video services, as these would have to be transmitted over telephone lines. The language Java provides an alternative to execution on the server. It allows part of the program to be executed on the client itself. Thus, the program, known as the applet, is automatically downloaded with the video and plays on the client computer. Alternatively, an applet can provide additional interface features such as multi-window displays or pop-up windows.

ACCESSING CORPORATE DATABASES

Web-based development is now becoming increasingly integrated with corporate systems. In these cases, the CGI programs must access the corporate database. Again, many vendors now provide tools for this purpose. For example, a program known as JDBC can accept commands from a Java program and convert them to database commands for the matching DBMS.

DOCUMENT MANAGEMENT AND WORKFLOWS

Another set of development tools and environments provides easy ways to manage and distribute documents or to define transaction-based workflows. Document management is now seen as the first step in knowledge development within an organization. The documents provide the basic inputs that may be used to develop new strategies, products, or services by an organization. Perhaps the most well known system here is Lotus Notes.

LOTUS NOTES

Creating applications with Lotus Notes centers on capturing information through document databases, distributing these documents through replication, and providing access to them through views. Accessed documents can be changed if needed. Document structures are defined as forms that are then used to create document instances. Buttons are included in the documents to initiate changes to the databases, or to change the states of documents. These changes will be replicated on other client sites, and the documents will be displayed in views of those clients if they satisfy the right conditions. Additional changes can then be made to the documents at those sites.

Databases can be defined by using an interactive forms definition facility to define document structures in terms of *forms*. The form defines a number of fields, some of which can be defined to be computed fields. Buttons can also be defined for a form. A program using Lotus script can be associated with each button. This program is activated when the button is selected and can be used to change documents of Lotus Notes databases. New documents can be created, or composed in terms of Lotus Notes terminology, using the defined form structure and stored in a selected database. A single form can be used to compose any number of documents. It is also possible to create what are known as responses to documents, and responses to these responses. Thus, a document can flow through a process with comments recorded as responses and attached to the document.

Document structures have many other features that are outside the scope of this book and are described in detail in the references. For example, documents can contain rich fields that can include information other than text—for example, graphics. Forms can also contain subforms. Lotus Notes also allows a response to be made to each document. This response is then stored and displayed with the document.

The complementary feature allows views that can be used to display the stored documents. Views specify conditions that must be satisfied by a document for display by a selected user. A different view can be provided to each user, depending on their role in the system, thus enabling workflows to be easily specified. A document can proceed through a number of states—say, 'new', 'checked' and 'mailed'—where a state is a document field. A different user operates on the document in each of the states. Thus, documents in state 'new' can appear in the view of a user who checks documents. When checking is complete, the state is changed to 'checked', in which case the document appears in the view of the user who mails it. It then enters the state 'mailed'.

KNOWLEDGE MANAGEMENT

Knowledge management goes beyond simple document repositories. It requires a more goal-oriented approach that brings experts together in a shared workspace and provides them with access to relevant information. Such workspaces are built on top of document repositories and allow experts to come up with new ideas, discuss them in the context of existing information, and develop new kinds of products or services.

WORKFLOWS—A WAY OF SUPPORTING PROCESSES

Workflow systems have been attracting considerable interest because they allow a user to easily define and construct many of the predefined process that are often found in networked business applications. They become the process tools that move documents created in the enterprise. Workflow systems support processes that follow a predefined set of steps. The idea of a workflow is illustrated in Figure 18.4. A requisition is made by a project engineer to get items from a supplier. The next step is for the manager to approve the requisition, followed by the order clerk making an order. The order is then sent to the supplier, and the items are received

Figure 18.4 *A workflow*

from the supplier. Each of these steps is carried out by a person assigned to the step. The process usually results in document flows, and consequently the process is often defined in terms of document flows between persons assigned to the process steps. Each person is required to take some action based on the document and pass it onto the next person.

Writing workflow problems using conventional programming languages has been quite a lengthy and costly process. Programs must be written to store the documents, pass the documents from person to person, and generate any output reports. Workflow systems provide a way to easily construct workflow processes using a high-level definition language. Here a process is defined as a process model with a high-level definition method. The users define the document structures and process flows, often visually. They generate the workflow system from this definition.

Workflows are defined either visually or by using a definition language. Many such languages use the idea of process state and define processes in terms of state transitions. Development based on a workflow system usually requires some modeling specific to the workflow and then directly converting the workflow model into the language specific to the workflow. Models are often used in re-engineering projects.

MODELING WORKFLOWS

Modeling usually starts with a graphical representation of the procedure, as in Figure 18.5. This diagrammatic representation shows:

- all the forms used in the procedure
- the flow control of the forms, shown by the circular structure
- the functions in the procedure
- the roles responsible for each function.

Workflows are often defined using a special language that defines:

- the system roles
- form structures by their fields
- conditions that indicate when a form is to be sent to a user
- the actions to be carried by a user.

The roles in Figure 18.5 are PM, the project manager, PO, the purchase officer,

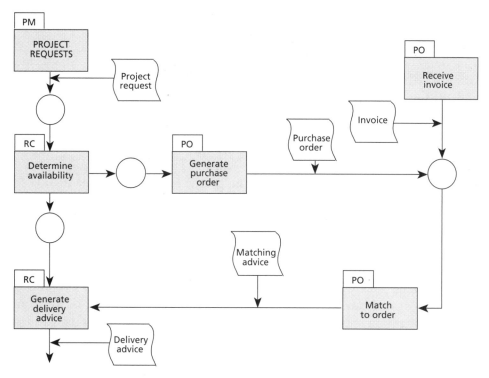

Figure 18.5 *Graphical procedure description*

and RC, the request coordinator. The exact syntax can change from system to system but includes definitions like:

ROLE Purchase-Officer: 'Jim' ... to define the role,
Purchase-Officer NEEDS Project-Request to 'Determine Availability'
    ... to specify action conditions for a role.

Logic can also be included in workflow specifications. For example,

Purchase-Officer NEEDS Purchase-Order and Invoice to Match Order
    ... this specifies that two inputs are needed before an action can commence.

Obviously such a tool would be very effective. We would do away with all the different kinds of modeling, the conversions, and other design activities. All we need do is specify the above structures and conditions and the system does the rest.

PROCESS IMPROVEMENT

Workflows are often used in re-engineering legacy transaction systems. Careful analysis and often simulation of workflows are used to test particular workflow designs. This is particularly the case with systems such as banking or insurance where many flows are large and predetermined, and savings of time in each flow multiply into large benefits to the organization. One example is a system known as ADONIS, developed at the University of Vienna and used widely in practice to re-engineer existing systems into workflows, including extensive system simulation. (For details, see http://www.boc.co.at/boc.)

WORKFLOW STANDARDS

Specifying workflows has further advantages in supporting alliances between businesses. This calls for standards to be used. The Workflow Management Coalition (http://www.aiim.org/wfmc/mainframe.htm) has been proposing standards for workflow definition and has developed a standard reference model, shown in Figure 18.6. The idea is that workflows defined using any process definition tool are translated to a standard enactment model.

The proposal thus identifies the major components of workflow management systems and defines interfaces between them. The interfaces are known as Application Program Interfaces (APIs). The goal is to identify standard APIs that can then be used to integrate components from different organizations and vendors into a workflow system. One of the important components is the process definition tool, which is used to define the workflow. Another is workflow enactment service, which executes the workflow as defined by the process definition tool. The idea is that a business can use any process definition tool that it finds appropriate to its way of working. This will generate a flow description in terms of API commands. Any enactment tool can then be used to execute the work process. It then becomes quite easy for two or more businesses to integrate their workflows through a shared enactment engine.

Another important aspect is to extend the workflow processes to the Internet so that the facilities can be used to support workflows across distance. It is again one of the objectives of the Workflow Management Coalition to develop such a standard.

There are further trends toward supporting workflows across the Internet, and a reference model based on objects is proposed by the Workflow Management Coalition.

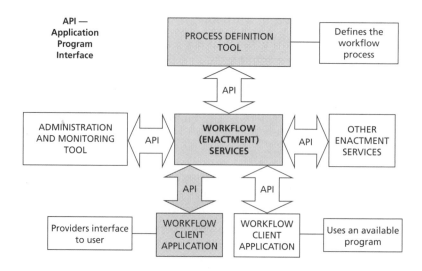

Figure 18.6 *Workflow reference model*

COMMERCIAL SYSTEMS

Workflow systems have been of interest to developers since the mid 1980s. Most of the early systems were built for specific uses or specific classes of use, but since 1990 there has been a prolific growth of systems and support tools for generating workflows.

One commercial product is ProcessWise, which was earlier marketed by ICL. ProcessWise is a system designed to define business processes that may cross organizational boundaries. It is based on the concepts of People, Roles, Systems, Services, Activities, Functions, Process, Task, Document, Object, and Product. It supports primarily asynchronous operation by participants at different locations. A graphical workbench is used to identify relationships between the people and their roles in activities, functions, and services. ProcessWise also includes facilities to monitor process performance.

STANDARD REFERENCE PROCESS ENVIRONMENTS

An increasingly used way to re-engineer processes is to use standard reference processes. These are based predominantly on the idea illustrated in Figure 18.2. The goal is to develop a library of objects that support generic processes and provide a workbench for developers to synthesize processes from these objects.

The kind of environment provided to developers is characterized by SAP R/3 and is described in detail by Bancroft, Seip, and Sprengel (1998). It is based on the three levels shown in Figure 18.7. Business unit objects can be customized to an organization's requirements and integrated into processes. The processes generate events, which are then used by a workflow to initiate events in other processes.

SAP provides a workbench to allow developers to construct applications from the business unit objects. It provides a variety of tools, including application debuggers, ways to access databases, code generators, report builders, and interface generators. It also includes CASE tools to model the various levels shown in Figure 18.7. It stores business process models defined using EPC, which was introduced in Chapter 13, and links them to the internal stored process and business

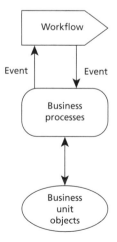

Figure 18.7 *An environment for reference processes*

unit objects. Developers must become familiar with the workbench and how to use it to construct business applications.

SUMMARY

This chapter gives a brief outline of the kinds of development platforms and environments that are used to build computer-based information systems. It describes how programs are written and then outlines the various ways of integrating programs into systems in object development environments and when building WWW sites. It also describes workflow management systems and their use in re-engineering and in environments that support standard reference processes.

DISCUSSION QUESTIONS

14.1   Which development platform would you suggest for a decision support system?

14.2   What do you understand by the term *development platform*?

14.3   What kind of software do you need to develop a WWW site?

14.4   What is the objective of workflow management systems?

BIBLIOGRAPHY

Arnold, K. and Gosling, J. (1998), *The Java Programming Language* (2nd edn), Addison-Wesley, Toronto.

Bancroft, N.H., Seip, H. and Sprengel, A. (1998), *Implementing SAP R/3*, Manning, Greenwich, London.

Brereton, P., Budgen, D., Bennett, K., Munro, M., Layzell, P., Macaulay, L., Griffiths, D. and Stannet, C. (December 1999), 'The future of software', *Communications of the ACM*, pp. 78–89.

Conradi, R. and Fuggetta, A. (July 1997), 'Assessing process-centered software engineering environments', *ACM Transactions Software Engineering Methodologies*, Vol. 6, No. 3, pp. 283–328.

Deitel, H.M. and Deitel, P.J. (1998), *Java: How to Program* (2nd edn), Prentice-Hall International, London.

Dwight, J. and Erwin, M. (1996), *Using CGI*, Que, Indianapolis.

Erikson, H.-E. (1998), *UML Toolkit*, John Wiley and Sons, New York.

Gunter, C., Mitchell, J. and Notkin, D. (December 1996), 'Strategic directions in software engineering and programming languages', *ACM Computing Surveys*, Vol. 28, No. 4, pp. 727–37.

Karagiannis, D., Junginger, S. and Strobl, R. (1996), 'Introduction to business process management concepts' in Sholz-Reiter, B. and Streickel, E. (eds), *Business Process Modeling*, Springer-Verlag, Berlin.

Larman, C. (1998), *Applying UML and Patterns: An Introduction to Object Oriented Analysis Design*, Prentice Hall International, London.

Larsen, G. (October 1999), 'Designing component-based frameworks using patterns in UML', *Communications of the ACM*, Vol. 42, No. 10, pp. 38–45.

Lie, H.W. and Saarela, J. (October 1999), 'Multipurpose Web publishing using HTML, XML, and CSS', *Communications of the ACM*, Vol. 42, No. 10, pp. 95–101.

Meyer, B. (2000), *Object-Oriented Software Construction*, (2nd edn), Prentice Hall, New York.

Prieto-Diaz, R. (May 1991), 'Implementation faceted classification for software reuse', *Communications of the ACM*, Vol. 34, No. 5, pp. 88–97.

Psankake, C.M. (October 1995), 'The promise and the cost of object technology: A five-year forecast', *Communications of the ACM*, pp. 32–49.

Schmidt, D., Fayad, M. and Johnson, R. (eds) (October 1996), 'Software patterns', *Communications of the ACM, Special Issue on Software Patterns*.

Schmidt, D.C. (October 1995), 'Using design patterns to develop reusable object-oriented communications software', *Communications of the ACM*, pp. 65–74.

Thomas, P. and Weedon, R. (1998), *Object Oriented Programming in Eiffel*, Addison-Wesley, Harlow, England.

Wall, L., Christiansen, T. and Schwartz, R.C. (1996), *Programming in Perl* (2nd edn), O'Reilly, Beijing.

Weske, M., Goesmann, T., Holten, R. and Striener, R. (1999), 'A reference model for workflow application development processes', Proceedings of the International Joint Conference on Work Activities Coordination and Collaboration, San Francisco, pp. 1–10.

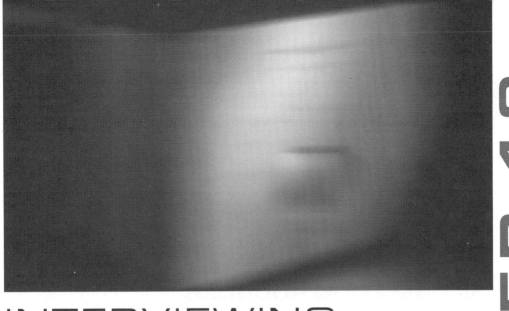

INTERVIEWING

CONTENTS

KEY LEARNING OBJECTIVES

The information sources in an organization
Developing an interview strategy
How to carry out an interview

CHAPTER 19

INTRODUCTION

Chapter 4 described a number of ways of determining user requirements. Whatever the method, however, some time must be spent in talking to or interviewing people in the organization. Such interviews may be needed only in the initial stages to identify the major issues, or they can proceed throughout the entire life of a project, perhaps mixed with other methods to supplement discussion.

Apart from the methods used in determining requirements, it is important to identify the information sources and to determine what, in the broad sense, must be analyzed. This chapter describes some alternative methods used to gather information. It starts by describing the sources of such information. It is often said that systems can be correct only if they are correctly specified at the beginning.

INFORMATION SOURCES

There are a variety of sources of information about a system. Each source usually yields a different kind of information and requires a different search method to get that information.

SYSTEM USERS

System users are the most important information source. From them it is possible to find out the existing system activities and to determine the user objectives and requirements. There are a number of ways of gathering information from users. One is through interviews. Another is to use questionnaires. A third is through observation of user activities and behavior. Interviews are one of the main methods used to gather information. (Interview techniques are described later in this chapter.) The interview approach may depend on the kind of user. For example, an overview is often sought with management, whereas detailed task information is sought from the users who carry out the tasks. Some users may cooperate more readily than others. Some users may be in favor of change and will provide considerable support, whereas others, who may have their positions to preserve, may not readily volunteer information. You might get someone who repeatedly exaggerates or who may even deliberately mislead you because they are not in favor of a study or any change. Later we describe some interviewing methods that can be used with different kinds of users.

FORMS AND DOCUMENTS

Forms and documents are useful sources of information about system data flows and transactions. There are many different kinds of such documents in any large system. They may include management information such as budgets or they may be detailed records of transactions within a business. It is important for the analyst to identify the complete list of such documents. Such a list is usually obtained from the system users. Analysts then go through the documents and analyze their contents. At this time it is usual to check for duplication of data and for naming consistency to ensure that the same data item does not appear under two names. Analysis should not begin by doing a detailed analysis of forms without first contacting users. You might just be unlucky and start with a form that is either out of date or no longer used. You should start by interviewing users and from them get the most recent and relevant forms.

COMPUTER PROGRAMS

Computer programs can be used to determine the details of data structures or processes. The search methods are often laborious and involve reading the program or its documentation, sometimes running the program with test data to see what it does, and examining the current user interface. Examination of existing programs is becoming an important part of analysis, because many systems now use computers, and an analyst must determine exactly what the current system does. The outcome of the design may be a proposal to redesign the existing computer system to meet new user requirements.

Of particular importance are the interactions with an existing computer system. The kind of information presented to users must also be examined during analysis. This information is often particularly useful, because it can identify specific shortfalls in the current system.

PROCEDURE MANUALS

Procedure manuals specify user activities in a business process. They can be used by analysts to determine detailed user activities, which is important in detailed system design. The search method is again a detailed examination of the contents of the manual. However, just as with forms and documents, it is necessary to make sure that the latest procedures are examined in a study. Many organizations are notorious for not updating their procedure manuals, so you should always check with a user any information obtained from a procedure manual.

REPORTS

This source indicates the kinds of outputs needed by users. It can be used as a basis for user interviews to determine any new output requirements that users may have.

It is very unlikely that one of these sources on its own will provide all the information that an analyst needs. If you start with an existing system, it is almost certain that you will have to obtain information about the organization from most, if not all, of the sources. The search procedure determines how to proceed through a search of the sources.

GATHERING INFORMATION THROUGH INTERVIEWING

Interviewing is the main approach used to analyze large structured systems. It is used by analysts to gradually build a subject world model and to understand any system problems. There are many important factors in successful interviewing. The first is to choose people to interview. The analyst must ensure that all key people within the study boundary are considered. The next important factor is finding the right way to conduct an individual interview. Good interpersonal relationships must

Interviewing
Gathering information by asking questions.

be considered, and the interviewer must establish some rapport with the interviewee to ensure the cooperation necessary to get all the relevant facts.

Information gathering through interviewing for a large and complex system can be an onerous task. Information must be gathered in an organized way to ensure that nothing is overlooked and that all system detail is eventually captured. All users must be consulted to ensure that every system problem and each user requirement is identified and that useful objectives are proposed. The search must avoid situations where different analysts seek the same information from the same user.

Before beginning a system study, an analyst or team of analysts must establish a **search strategy** for gathering the information needed to develop a model of a system. As illustrated in Figure 19.1, this search strategy is established by selecting the sources from which information is to be obtained and determining the methods to obtain the information from each source. These sources and methods are combined into a **search procedure**. The search procedure defines where to begin the search and how to continue. It also identifies the sequence in which sources will be searched and what information is to be gathered at each step.

INTERVIEW SEARCH PROCEDURES

Information gathering in large systems with many information sources must proceed in an organized way to ensure that all the relevant information needed to build the system is obtained. Analysts must determine what information needs to be gathered and the users who are to supply this information. They must then seek this information in an orderly way so that vital information is not neglected and people are not repeatedly bothered by being asked the same questions. To avoid such problems, it is necessary to develop a search procedure for gathering information. This procedure will define the steps to be followed in gathering information and

Search strategy
A selection of sources and methods to be used to gather information about a system.

Search procedure
The process followed to gather information about a system.

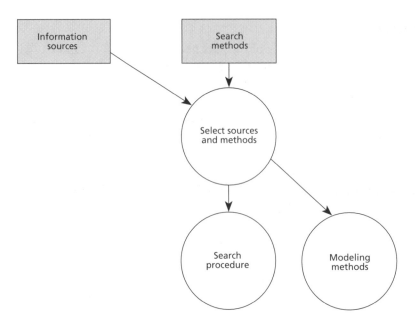

Figure 19.1   *Developing a search strategy*

the information to be obtained at each step. Such steps usually require the search to proceed in a top-down manner, objectives to be set for each step, and an appropriate search method to be chosen for each step.

The search procedure suggests what order is to be used to search information sources and which methods are to be used as the search proceeds. Thus, the search procedure becomes a plan stating which information is to be obtained from each source and what sequence is to be used to search the sources.

PROCEEDING IN A TOP-DOWN WAY

Most interview search procedures require all information sources to be examined to ensure that all information about the system is gathered. They suggest that this information be sought in a top-down way to gradually build up the system model. One should not proceed sequentially by first collecting all information about the system and then building the system model. Such an approach would be difficult to manage and may lead to errors. There would also be a large volume of seemingly unrelated information, which the analyst would have to sift through in order to try to reconcile inconsistencies and fill in missing pieces. In these circumstances, it is easy to overlook vital data, create incomplete models, or find that analysts are repeating interviews with disgruntled users.

A more orderly approach is therefore needed. The search procedures should specify the organizational level at which interviews should start, the personnel to be interviewed, and any other information sources to be used. They should also include an interview plan for users and allow time for examining detailed information. Such a plan must ensure that all users are considered in the study. Leaving out an important user may lead to later problems, as that user may feel that their work has not been considered and will then have less commitment to any new proposals or follow-up work. Many procedures follow a top-down approach, as shown in Figure 19.2. They begin with a set of initial interviews to find out what the system is about, usually by interviewing the higher levels of management, to get an overall picture.

Starting with management. The objective of interviewing management is to clearly identify the major components within the system and tasks within these components. These initial interviews identify functions inside or sometimes even outside the problem boundary. Often analysts will look at some of the more important reports or documents at this stage. They will then draw a top-level system model and verify this model during the following set of interviews.

There are also other reasons for starting interviews with management. First, management is aware of what goes on in the system and can give you leads as to how to go about getting information. They will suggest good sources of information and the people to interview to get this information, and they will introduce you to these people. Starting with management also ensures that management are on your side and will encourage cooperation from their section. Simply turning up and interviewing people in a section can have many undesirable effects. Management may feel that their activities are being investigaged without their knowledge, or that you are distracting their workers from important duties. You must therefore make

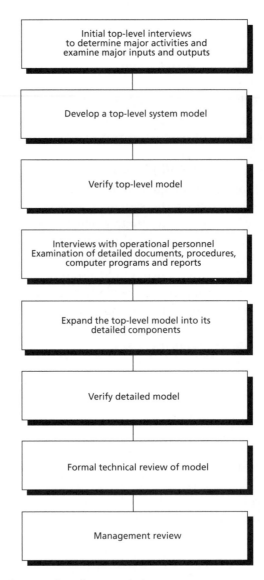

Figure 19.2 *Search procedure for an existing system*

management aware of your goals early in the study and of how your work will affect their section. It is important to establish confidence in management about your abilities. This can often be done by showing interest in their problems and describing how you can solve them in a systematic way. For example, a start may be to introduce yourself, and state the reason you are there, what you expect to get from the study, how long it will take, what steps you propose to follow, and what you will produce. You must ensure that management feels you are there to help them solve problems, not to find blame, and that you do not unduly disturb the operation of their section.

Considering detailed operations. Once the top-level model is validated, the next phase of the study may be to look at each of the major system components in turn.

The objective now will be to find the operation of the major business processes or functions of the system. These in turn will identify further detailed components and so on. As you go into detail, there may be less emphasis on interviewing and more time spent on looking at things such as forms, computer programs, or system reports. Analysts use this new, detailed information to expand the top-level model, which is then verified with the users. This should be done in an iterative manner.

The search for more detailed information usually begins by interviewing operational personnel. These interviews establish detailed sources of information, including computer programs, reports, and manuals. This procedure may be repeated a number of times as increasingly detailed information is sought. These subsequent interviews and searches become more important as the analyst learns more and more about the system. The analyst begins to identify problems in the system and, together with the users, to establish the objectives for the new system.

An analysis model is built during the interviews. This model is updated following each interview and checked in a follow-up interview. The analyst then searches for more detailed information and makes further updates to the model. Interviews at the detailed levels can thus be structured to fill in identified gaps or to obtain information that is needed to complete previously incompletely defined functions. These gaps will be identified as the analyst is developing the model. The iterations go on until the analyst is happy with the model. The iterations can be seen as part of the model validation process. Each iteration checks that model changes made in the previous iteration correctly describe the system. The model then proceeds through a more formal validation by being submitted for a number of formal reviews, starting with technical reviews to formally establish model correctness. Finally, it is submitted to a management review for agreement on system objectives and to obtain resources to develop the new system.

The kind of data sought at the lower levels depends on the system. If a computer system is to be improved, the analyst would look at existing computer systems and programs. If manual procedures are to be improved, the analyst would make a detailed examination of current procedures. In most cases, both the computer system and user procedures would be examined.

WHAT IF THERE IS NO EXISTING SYSTEM?

So far we have assumed that there is an existing system, parts of which are to be automated. But what if this is not the case, and there is no existing system? In that case we cannot examine what is currently happening or identify current user activities. In a sense, the search procedures are now simpler, because there are fewer sources of information. The procedures now emphasize user requirements and place less emphasis on the study of existing components. There are no reports or computer programs to go through and no manuals to examine. The whole procedure centers around interviews, but the thrust of these interviews is now different. The interviews do not need to search out how a system works but must determine users' expectations of the new system. They must then define the business processes and users' roles in these processes. Prototyping is often useful here. We can build up typical interfaces and outputs to get reactions from users on the kind of system behavior they would like.

Where totally new systems are proposed, analysts often look at sources outside the system for information. Often the new system is being suggested because someone has seen it somewhere else. Analysts can examine these external systems to see whether any of their features are applicable to the proposed new system.

THE INTERVIEW PLAN

The interview plan specifies:

- the users to be interviewed
- the sequence in which the users are interviewed
- the interview plan for each user.

The first step in developing an interview plan is to identify the users to be interviewed. Often an organization chart can be used to identify such users. This chart describes the organization's units, the positions in the units, and each position's occupant. The analyst uses the project's terms of reference to select the organizational units that fall within the boundary of the system study and are likely to be affected by any new system. Persons in these units then become candidates for interviewing. It is usually wise to begin interviewing at the top levels of the organizational areas to get support and cooperation from management before beginning to look into particular organizational activities or suggesting new solutions. Management may then often suggest other users that should be interviewed and are more likely to support any proposed changes.

There are also some common goals in each interview. Preparation for the interview is always essential. The analyst should have an idea of what information is needed from the interview and ask direct questions to get this information. The analyst should always endeavor to obtain leads to more information. If the current interviewee cannot answer, the analyst should ask for advice about where to go next.

The interview process thus follows a fairly structured path. You gain an appreciation of the overall system operation from management, then you go into detailed operations by interviewing system users at various levels of system operation.

You should not expect to obtain all the information required from one user in the course of one interview. There are usually two, three, or even more interviews with a given user. Usually the analyst begins with an initial interview to meet the users. This first interview is then followed by a number of fact-gathering interviews to gather all the major facts known to each user. Then there may be one or more follow-up interviews to verify these facts and any models developed by the analyst, and to gain additional information to complete the analyst's study.

Let us now consider the interview itself.

THE INTERVIEW

It is important to establish a good relationship with interviewees. This should start right at the beginning. Even some simple things are important here. For example, you should call and arrange an interview time and not simply barge in and expect the interviewee to drop everything to talk to you. You should also limit interviews to about 45 to 60 minutes and not expect people to give large amounts of their valuable time to you. If you are taking notes or recording the interview, ask for permission.

The interview should proceed in an organized and friendly manner. The interviewer should always be courteous and respect the user's needs and position. You should give the interviewee time to answer your questions, seek clarification if necessary, and occasionally summarize what you have learnt from them. You must give users confidence in your abilities and reassure them that you are there to help them solve some of their problems.

Be aware of some basic premises when conducting interviews. First, you must gain the confidence of your interviewees. To do this you must convince the interviewee of your own abilities and show that you are proceeding in an organized way and will not waste their time. You must also be sympathetic to their problems and not become aggressive and create the impression that you are there to apportion blame. It is not a good idea, for example, to begin by saying:

> 'I hear that there is something terribly wrong with the system and I am here to fix it.'

A better approach may be to begin by saying:

> 'You might have heard that a study of your system has been started to see how we can provide you with better support for your work. As a first step in this study, I am here to see exactly what you do and how we can better support you in your work.'

It is important not to force solutions upon users but rather to play the role of an advisor. Computer jargon should not be used to impress the user, though interviewers should explain the limitations of the computer in user terms and describe how it can help users in their work. You should not try to elicit the response you want by asking leading questions such as:

> 'You agree that the best solution is a computer solution?'

It is better to ask users for suggestions about what might be done to improve the system and then follow up these suggestions.

Interviewers should also take care to ensure that they obtain all the needed information from interviews. It helps to let the user know what information is required from the interview. The interviewer should then seek this information gradually and be precise and direct in their questioning. You should not get to a stage where you keep saying:

> 'I forget, did I ask you last time whether . . . ?'

It is better to obtain information gradually, and at a follow-up interview begin with:

> 'From last time, I understand that the following happens.'

You might then show the user a model you have developed and say:

'I have developed the following model from our last meeting. Can you verify whether this is correct?'

Then go through the model, explain it, and check it. The model may also be amended if it has any inaccuracies. The analyst is then ready to further elaborate on the model. The next step may be to say something like:

'Today I would like to follow up some of these points.'

The follow-up questions then find out more about the system. In this way, progress is apparent to both the analyst and the interviewee. The model and information about it grow at each interview. Analysts will be satisfied because their knowledge about the system is continually growing. The interviewees will be more cooperative because they can see that their time is being used productively.

THE INTERVIEW STRUCTURE

It is necessary to proceed in an orderly manner to get the best out of interviews. Figure 19.3 illustrates a procedure that has been found effective over time. It begins with some preliminaries. For example, analysts will have to introduce themselves at the first interview. Then some interview criteria may be identified. These include the length of the interview and what you expect to get from it, as well as obtaining

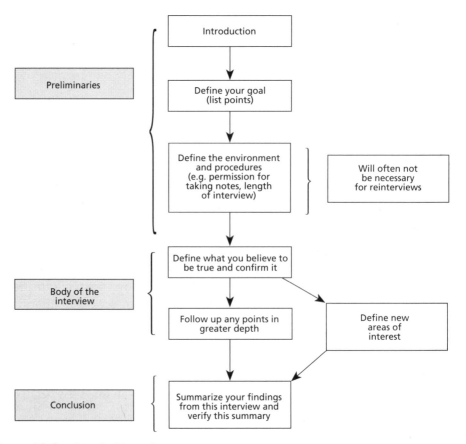

Figure 19.3 *A typical interview*

permission to take any notes or make recordings. It is a good idea to do this so that the interviewee will know what is expected from the interview and hence feel more able to contribute constructively.

Once the preliminaries are over, the main part of the interview can begin. It is usually a good idea to start by confirming any information you obtained from an earlier interview or any other investigation. At this stage you might show any models you have developed or confirm some of your beliefs about the system. This serves to put the interviewee in the picture and also helps you find any errors in your data. Once you reach agreement, follow up any relevant points in more detail.

The interview should be concluded by summarizing what you have found out in the interview and confirming it. Finally, it is usually a good idea to arrange a time for the next interview if necessary.

HOW TO GET INFORMATION FROM INTERVIEWS

One of the most important points about interviewing is which questions you need to ask. It is important to ask the right questions in the correct order to get the most out of interviews. Questions can be characterized by their subject content and type. Obviously, the subject content of a question will depend on the specific system study. However, the type of question can be generalized, and general guidelines can help you choose the most appropriate questioning method for the interview in question.

It is often convenient to make a distinction between three kinds of questions: *open questions*, *closed questions*, and *probes*. **Open questions** are general questions that establish a person's viewpoint on a particular subject. For example:

> 'What do you think of personal computers in your type of work?', or

> 'How relevant is the sales forecast to your activity?'

Open question
A question that requires the respondent to express a viewpoint.

Closed questions are more specific and usually require a specific answer. For example:

> 'Where do you send your summary of orders?', or

> 'How often do you need to get information about inventory levels from the warehouse?'

Closed question
A question that requires a direct answer.

A closed question often restricts an interviewee to some specific answer. This may be a number, an explanation of a report, or a reason for doing something. The answer to a closed question can then be followed with a probe to get more detail. **Probes** are questions that follow up an earlier answer. For example:

> 'Why do you send the summary of order to account?', or

> 'Why do you need this information so often?'

Probe A question that follows up an earlier answer.

Interviewers must often select a particular questioning method for an interview. A common method uses alternate sequences of closed and open questions. Some possible sequences are described below.

OPEN THEN CLOSED

It is often suggested that the first interview with a user should emphasize open questions. This enables analysts to identify the values a user places on the system and helps in gaining an impression of the user's likes and dislikes. This may then shape future interviews. Open questions also prompt the user to volunteer more detailed information, so that closed questions may be unnecessary. Variations on this sequence are possible. An analyst may spend all the interview time on open questions. Alternatively, for a small system it may be possible to get into closed questions very quickly.

OPEN, CLOSED, THEN OPEN AGAIN

It is possible to start with open questions, follow them with closed questions, and then finish up with some more open questions. This approach is used mostly in follow-up interviews. It is an easy way to gradually lead a user into detailed questioning. Some people when answering an open question may provide enough information to make future detailed questions unnecessary. It is also possible to start with detailed questions and finish with open questions. This can be useful when searching for user requirements. You might spend some time following up points about, say, placing orders, and then at the conclusion you might ask:

'What do you think would be a better way to place orders?'

Such a follow-up question will be very useful in getting an idea of what the user requires from the system. The user has expressed some ideas, then described some details and possibly problems with the system, and is probably now in the mood to state what is really needed.

CLOSED THEN OPEN

Interviews that start with closed questions are often used when interviewees, for one reason or another, are not ready to volunteer information. They may be shy and find it difficult to get into a discussion, or they may simply be uncooperative and not in favor of a study. Asking an open question makes it possible for the interviewee to avoid giving a useful answer. In this case it is usual to start with closed questions, follow up with some open questions, and then conclude with some closed questions. For example, one might start with a closed question such as:

'How do you choose a supplier?'

Following some further probes to find the detailed process, the interviewer may then ask:

'What other ways are there for choosing suppliers?'

There might be a discussion of the alternatives and you might settle on one. The interviewer might then ask:

'Suppose we had a computer system for choosing suppliers. Would you use their latest price for a given part, or would you consider the trend over the past few months?'

This might then be followed by a discussion of the requirements the user has for making a decision.

CHOOSING THE ALTERNATIVE

The type of approach you choose will often depend on the type of interviewee. It is often wise for a first interview to start with a series of open questions to develop rapport between the interviewer and a user. Detailed closed questions can follow later. With an uncooperative, reticent, or shy user who finds it difficult to offer opinions, it may be better to start an interview with closed questions. They are usually easier to answer and harder to avoid. Having answered the initial questions, these users may feel more at ease about giving their opinions.

FOLLOWING UP FOR DETAILED INFORMATION

Information can also be obtained from documents, forms, or other media. There are two aspects to examining detailed documents. One is simply to look at their content to determine what makes up the flows and stores in a system. The other is to analyze their content in detail to find out what in-depth relationships the system contains. Determining the contents of sources such as documents, reports, or computer programs tends to be relatively informal. Analysts must go through these sources, look at their content, and note it down on the system model. Often the search is guided by the modeling techniques provided by the design methodology. Thus, when we examine programs, reports, procedures, or documents, we compare them with the current system model. As we do this, we add any new information to the model and check for any inconsistencies. Alternatively, we could use the current model to determine whether there is any missing information and then go to the appropriate source to find it.

In-depth analysis may be more formal. For example, one might examine advertising budgets and compare them against sale variations to determine the effectiveness of advertising. Excessive inventory maintenance costs may also call for an analysis of inventory records and levels of holdings. Financial data are often used in such detailed analysis. For example, there may be a problem where the cost of providing items is too great. This will probably call for a detailed cost analysis to find out where the greatest cost overruns occur.

One aspect of in-depth analysis is that it is not possible to examine all relevant data. We cannot look at the cost of every item produced, or at all the items held in a warehouse. Analysts must choose samples of information for the analysis and use the correct sample selection technique to select the sample.

SUMMARY

This chapter concentrates mainly on one important method of gathering information, the interview. It describes a number of important points to remember when organizing a search for information about the system by interviewing. It considers how to plan a search strategy, and examines the differences between search strategies where there is an existing system and those where there is no existing system.

DISCUSSION QUESTIONS

19.1   What are the sources of information about a system?

19.2   What is meant by the term *search method*?

19.3   Why is interviewing one of the most frequently used search methods?

19.4   What are the important factors for devising a search strategy?

19.5   Would a search strategy for an existing system differ from a search strategy for a new system? How would it differ?

19.6   Why is an interview plan important?

19.7   Who are the important users that must be interviewed? How would you find them?

19.8   Describe how you would carry out an interview.

19.9   How would you go about establishing a good relationship with a user?

19.10   What is the difference between closed and open questions?

19.11   Describe some ways of questioning during an interview. Describe some typical users and the most appropriate questioning method for them.

19.12   What are some of the disadvantages of interviewing?

19.13   Describe some situations where division of labor occurs dynamically.

BIBLIOGRAPHY

Cash, C.J. and Stewart, W.B. Jr (1986), *Interviewing Principles and Practices* (4th edn), Brown Company Publishers, Dubuque, Iowa.

Dwyer, J. (1999), *The Business Communication Handbook* (5th edn), Prentice Hall, Sydney.

QUALITY ASSURANCE: REVIEWS, WALKTHROUGHS, AND INSPECTIONS

CONTENTS

KEY LEARNING OBJECTIVES

Quality assurance within the development process
The different kinds of quality assurance checks
When are inspections used?
What is a walkthrough?
How a walkthrough fits into the system development cycle
The structure of the walkthrough team
The process followed in a walkthrough

INTRODUCTION

Greater emphasis on quality in organizations requires quality assurance to be an integral part of information system development. It is important to keep in mind that quality assurance is not something that goes on separate from the development process. On the contrary, the development process must include checks throughout the process to ensure that the final product meets the original user requirements. Quality assurance thus becomes an important component of the development process. It is included in industry standards (IEEE 1993) on the development process. Chapter 5 described how a quality assurance process is integrated into the linear development cycle through validation and verification performed at crucial system development steps. The goal of the management process is to institute and monitor a quality assurance program within the development process. The QA program is an important stage in the process maturity levels described in Chapter 6, where maturity level 3 requires quality assurance activities to be in place. A quality assurance program includes:

- *validation of the system against requirements*
- *checking for errors in design documents and in the system itself*
- *checking for qualitative features such as portability and flexibility*
- *checking for usability.*

    Each of these objectives may be met by a different kind of QA activity. This chapter describes the reviews, walkthroughs, and inspections that often make up the QA program. The definitions and objectives of reviews, walkthroughs, and inspections are not yet standardized and may mean different things in different environments. One general view is that reviews are usually made to check whether project management goals have been achieved; walkthroughs are usually made to detect errors in the system; and inspections are made to evaluate its qualitative features. It is also usual to distinguish between reviews made about project resource use, and checks of whether a system model or technical proposal is correct. These are two different processes with different objectives, one to achieve correctness and the other to ensure effective use of resources.

    Most organizations now require that proper quality assurance be defined by a set of well-defined steps in the development process rather than being an ad hoc activity. Quality assurance thus requires special preparation. Because the goal of QA is to look at earlier work, designers are required to prepare for a review, inspection, or walkthrough, and be ready to carry out any checks or follow-up work needed. This chapter describes some of the methods used in quality assurance.

    There is also an important distinction between the quality control procedures needed in critical and non-critical systems. Critical systems are those where a fault or error can have dire consequences—for example, a nuclear plant failure or failure on an airplane. Special techniques are used to prove the correctness of such systems, including formal proofs of correctness using mathematical techniques.

    This is usually not the case with non-critical systems, where a less stringent approach is used, based usually on inspections of code, verification of models through discussion and evaluation, and testing of test cases derived from user specifications. In these systems, QA is usually carried out through reviews and inspections of products, in addition to the testing that takes place before product delivery.

IMPLEMENTING QUALITY ASSURANCE

So far, when we have discussed project tasks, it has been assumed that a person responsible for a task receives a request as an input and produces an output. Quality assurance mechanisms place additional requirements on a task—in particular, a review requirement followed by an approval before it can be used to initiate the next task. Reviews are carried out by team members not directly associated with a task to check that task outputs match requirements and satisfy best-practice criteria.

The approval is usually part of the management process and is included to ensure that all the necessary reviews have been completed. Reviews can take a number of forms—the most common are inspections and walkthroughs.

Two things are often considered important in achieving quality products. One is a precise set of user requirements, against which general quality is measured. The second is documentation. In a large project, documentation is the central source of all the information needed by the team members. If it is incorrect or out of date, then team members are working toward the wrong goals or using wrong inputs in their work. Management come in again here. Not only must they develop and monitor reviews of task outputs, but they must also ensure that any documents used in the tasks are the correct documents. This is often achieved by providing support based on a configuration management system and ensuring that all team members place their latest outputs into the configuration management system.

INSPECTIONS

Perhaps the simplest checks to describe are **inspections**, which are done in most reviews. It is usual to allocate roles to people involved in an inspection and outline a procedure for them to follow. Some common roles are the **producer**, whose product is under review, the **inspector,** who evaluates the product, and the **moderator**, who controls the review process. There is also a **reader**, who may guide inspectors through the product. It is important in such reviews that the people doing the inspection are not those that produced the product. Apart from their desire to ensure that their product passes the test, people who have worked on a product may not be aware of some of its shortfalls.

Some software engineers suggest that there be a phased program of inspections running parallel with system development, with an objective for each phase and an evaluation made against the objective for that phase. Inspections themselves would be formal, with reports made at defined life-cycle phases. Each of the roles in an inspection team would have well-defined responsibilities within the inspection team. Fagan (1986) suggests a procedure made up of five steps:

1. *Overview*, where the producers of the work explain their work to inspectors.
2. *Preparation*, where the inspectors prepare the work and the associated documentation for inspection.
3. *Inspection*, which is a meeting moderated by the moderator and guided by a reader, who goes through the work with the inspectors.
4. *Rework*, which is any work required by the producers to correct any deficiencies.

Inspection An examination of a product to assure quality.

Producer A person who produced a product.

Inspector A person who examines a product to assure quality.

Moderator A person who controls the progress of a review.

Reader A person who guides the way a system model is examined.

5. *Follow-up*, where a check is made to ensure that any deficiencies have been corrected.

The important thing here is that the inspections are formal and have a report that must be acted on. It is also important that any recommendations made during inspections be acted on and followed up to ensure that any deficiencies are corrected.

WALKTHROUGHS

Walkthrough A quality assurance activity to detect errors.

The **walkthrough** is a procedure that is commonly used to check the correctness of models produced by structured systems analysis, although its techniques are applicable to other design methodologies. Such checking has always been necessary in systems analysis and design. Walkthroughs differ from earlier methods in that they recommend a specific checking procedure and walkthrough team structure. Furthermore, they allocate specific tasks to various members of the walkthrough team and require documentation to be produced during and after the walkthrough.

The team must check that the model:

* meets system objectives
* is a correct representation of the system
* has no omissions or ambiguities
* will do the job it is supposed to do
* is easy to understand.

How these checks are made depends on the kind of model, but whatever the model, it is important for the checks to proceed in an orderly way. Walkthroughs are one important method used in QA. Another important feature of the walkthrough is that no actual design or system alteration takes place during the walkthrough; problems are only noted for further action. The responsibility of following up these problems is assigned to the walkthrough team members. The problems are documented in an action list, which also specifies which members of the team are to be responsible for following up these problems.

WHEN ARE WALKTHROUGHS CARRIED OUT?

Walkthroughs can take place throughout system development. In structured systems analysis, they begin when the physical and logical models of the existing system have been completed. The first walkthrough checks the existing system models to detect omissions and inaccuracies in them.

Walkthroughs should also be carried out on the new logical design to detect flaws, weaknesses, errors, and omissions in the proposed design. The walkthroughs should be made first on the new logical model and later on the new physical model.

There may be more than one walkthrough in each project phase, and there are no set times for doing them. It is time for a walkthrough when you reach a point where you have done all you can on a model and you need to be sure that this correctly represents the system. It is necessary to discuss what you have done with others to verify it.

HOW ARE WALKTHROUGHS CARRIED OUT?

The procedure followed in a walkthrough is shown in Figure 20.1. Before the walkthrough begins, the producers should ensure that obvious problems have been eliminated from the models. It is no good bringing together a walkthrough team to detect obvious and simple problems of which the model producers should already be aware. A walkthrough team is brought together to apply the benefit of its combined knowledge to the whole system and to detect the less obvious system problems. It is not there to find simple modeling faults. Some people suggest that you should use a checklist on a structured model before you submit your documents to a walkthrough. The checklist includes the common kinds of errors that often occur in modeling and serves as a guide for detecting such errors.

Once the producers are satisfied with the model, it is time for a walkthrough. Two outcomes are possible from the walkthrough. One is that no errors are found in the model and it is accepted. In that case, review documents are prepared for a subsequent project review. The other outcome is where errors are detected in the model. In that case, an action list is produced. The model is then amended and later submitted to another walkthrough.

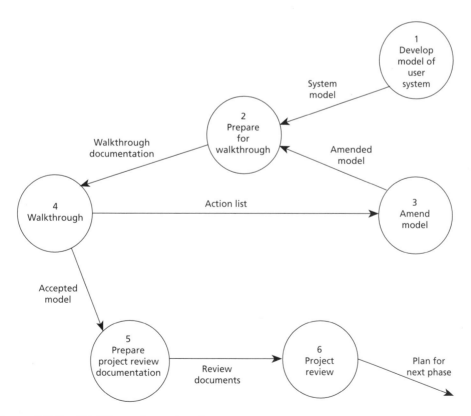

Figure 20.1 *Walkthroughs and reviews*

PREPARING FOR THE WALKTHROUGH

You must ensure that any model you have produced, and its associated documentation, does not have any simple problems before the walkthrough. The last thing you want to happen is to have the review find simple problems in your system. Hence, models developed during system analysis must be reviewed by the design team before they are submitted to a formal project review. As shown in Figure 20.2, the first preparation step is to assemble the walkthrough team and assign roles to each team member. The next step is to distribute relevant documentation to all team members. The distribution must be made early to give team members sufficient time to become familiar with the documentation. When this has been done, the team members can be called together for the walkthrough.

THE WALKTHROUGH

The procedure used during the walkthrough is shown in Figure 20.3. The person who developed the model actually tracks through the documentation. This may involve following the data flows in a DFD, describing data stores, or going through the logic of each process. Any omissions, ambiguities, or inaccuracies are noted in an action list during the walkthrough and followed up later.

WALKTHROUGH DOCUMENTATION

The outcome of the walkthrough is always documented. Usually only two documents are produced. One is a summary that describes the walkthrough outcome; the other is an action list. The action list shows all the issues raised during the walkthrough and, more importantly, who is to be responsible for resolving these issues. An example of an action list is shown in Figure 20.4.

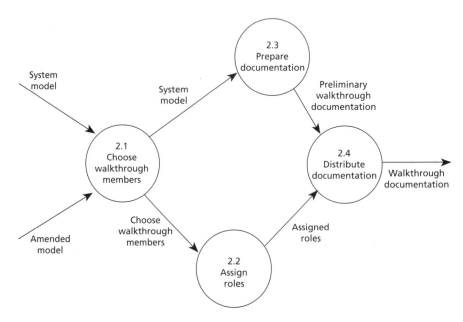

Figure 20.2 *Preparing for the walkthrough*

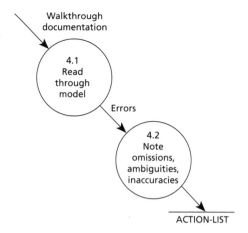

Figure 20.3 *The walkthrough*

FOLLOW-UP

The follow-up is based on the action list. The members who have been assigned responsibility for correcting errors amend the model. When this amendment is completed, the model is ready for the next walkthrough.

WALKTHROUGH TEAM COMPOSITION

There is no fixed optimum size for a walkthrough team. Team members should be selected in a way that will ensure that the material is adequately covered. The size of the team depends on the material to be covered and the skills and review experience of the potential participants. People with no knowledge of the system under review should not be in the walkthrough team. The number of participants will be somewhere between three and seven.

WALKTHROUGH ACTION LIST
  System Description:
  System Producer:
	Walkthrough Action List:
	Walkthrough Date:
	Walkthrough Leader:		

ISSUES RAISED	Diagram Reference	Assigned To	Amendment Completed
Walkthrough Secretary:			

Figure 20.4 *Action list*

Specific tasks are allocated to some of the walkthrough team members. Members are usually selected to take the roles of:

- the walkthrough leader
- the walkthrough secretary
- the walkthrough reader (or producer).

The remaining members of the team are usually called the participants. Specific roles of the team members are described below.

THE WALKTHROUGH LEADER

The job of the walkthrough leader is to ensure a good walkthrough, or to report why a good walkthrough was not achieved. A good walkthrough produces an accurate assessment of the product as it now stands. One reason for failing to achieve a good walkthrough may be that one or more members of the team were unprepared.

The leader, who should be technically competent to understand points raised in the walkthrough, has a number of functions to perform before, during, and after the walkthrough. When told by the producers that some work is ready for review, the walkthrough leader begins by collecting all the relevant materials, selects the people who are to attend the walkthrough, and distributes copies of the relevant materials to these people to ensure that they are well prepared. The leader must then set the meeting time and place and the length of the walkthrough. Finally, one of the team members must be appointed as the walkthrough secretary.

During the walkthrough, the leader must make sure that the meeting keeps to the relevant topics and that everybody contributes to the meeting. Finally, the leader must get agreement on the outcome of the walkthrough and make certain that the agreement is truly understood by all the participants.

After the walkthrough, the leader sees that accurate reports are produced promptly and checks that the producer has a reasonable basis for clearing up any issues requiring attention. All relevant people must receive the walkthrough report.

THE SECRETARY

Normally, the walkthrough leader chooses one of the participants to take on the role of secretary, whose function is to record the result of the walkthrough.

Before the walkthrough, the secretary should meet all the other walkthrough participants and be able to identify them by name. The secretary must also collect all the available materials necessary for keeping accurate records of the walkthrough.

During the walkthrough, the secretary must record all issues accurately and state each outcome explicitly, unambiguously, and neutrally.

After the walkthrough, the secretary prepares all reports promptly and gets all the participants to sign them. When this is done, the secretary distributes copies of the reports to all the relevant people.

THE READER (OR PRODUCER)

The reader's job is to describe the product under review. For structured systems analysis, this is usually a DFD together with any process description, data flow, and data models. The reader called the meeting, so it is the reader's job to get the most out of the people at the meeting.

It is the reader's responsibility to go through the documentation and bring out any points that caused difficulty or uncertainty during the development of the documentation. The general procedure is to follow data flows through the DFD. As each process is encountered, it may be elaborated (or leveled if necessary) to explain what it does. Any contentious issues about processes or flows should be raised in order to resolve them as soon as possible.

THE WALKTHROUGH PARTICIPANT

Each member of the walkthrough team is a reviewer of the product being walked through and has personal responsibility for the outcome. Experience has shown that each participant should follow a number of rules to help make the walkthrough a success. First, the participant must be well prepared. If you do not know what is going on, you will contribute little or, more seriously, waste the time of the meeting.

The participant should take a neutral and constructive stand on all issues raised in the walkthrough. Thus, discussions of style should be avoided, and participants should not become aggressive, or criticize or evaluate the producers.

Each participant should make at least one positive and one negative point. This guarantees that each participant will have an input. Participants should raise issues, rather than resolve them, and attempt to learn about unfamiliar parts of the system rather than unnecessarily criticize them.

SUMMARY

This chapter gives a brief introduction to the kinds of activities that make up quality assurance. It stresses the importance of processes that are objective in order to find modeling errors in an organized way. The chapter outlines one such process, walkthroughs.

DISCUSSION QUESTIONS

20.1   Why is it necessary for QA checks to be separated from system development?

20.2   Why is there a difference between checks carried out to detect errors and those to evaluate some of the more qualitative system features?

20.3   Why is it important for people to take formal roles during QA?

20.4   Define some common roles used in inspections.

20.5   What are the roles used in walkthroughs?

EXERCISES

20.1   Take a DFD that you have developed for one of the cases at the back of the text and walk through this diagram. The walkthrough should include a number of persons, each taking a walkthrough role. An action list should be created during the walkthrough.

20.2   Did you find that the walkthrough you just carried out was useful? Did any problems arise because you did not follow the walkthrough rules?

20.3   Why is a formal review of DFDs preferred to non-formal discussion?

20.4   Review once again the role of walkthrough team members in the light of your experience of a walkthrough.

20.5   Go through the walkthrough action list and, in retrospect, see whether you missed anything important in the walkthrough. If so, what do you think was the cause of the omission?

BIBLIOGRAPHY

Bach, T. (October 1995), 'The challenge of "good enough" software', *American Programmer*, Vol. 8, No. 10, pp. 2–11.

Card, D. and Glass, R. (1990), *Measuring Software Design Quality*, Prentice Hall, Englewood Cliffs, New Jersey.

Fagan, M.E. (July 1986), 'Advances in software inspections', *IEEE Transactions on Software Engineering*, Vol. 12, No. 7.

IEEE (1993), *Software Engineering, IEEE Standards Collection*, The Institute of Electrical Engineers, Inc., New York.

Kaplan, C. and Clark, R. (1995), *Secrets of Software Quality*, McGraw-Hill, New York.

Knight, J.C. and Myers, E.A. (November 1993), 'An improved inspection technique', *Communications of the ACM*, Vol. 36, No. 11, pp. 51–68.

Martin, J. and Tsai, W.T. (February 1990), 'N-fold inspection: a requirements analysis technique', *Communications of the ACM*, Vol. 33, No. 2, pp. 225–32.

Sanders, J. and Curran, E. (1994), *Software Quality: A Framework for Success in Software Development and Support*, Addison-Wesley, Reading, Massachusetts.

Seigel, S. (January 1992), 'Why we need checks and balances to assure quality', *IEEE Software*, pp. 102–3.

CASE 1—SALES/ORDER SYSTEM

GENERAL DESCRIPTION

A distribution organization sells its wares through personal contact. Its salespersons visit prospective customers with either sample wares or brochures and other descriptive material. Any customer orders are forwarded by the salespersons to either branch offices or the main office (depending on the salesperson's location). The orders may be forwarded by mail or by telephone. Each order contains the information shown in Table C1.1.

Table C1.1 *An order register*

Order no	Customer	Salesperson	Location	Status	Comment
075	K. Bloggs	Vicki	Store 7	Ordered	
092	B. Jog	Stan	In manufacture	Waiting	Expected August

Table C1.2 *Information in an order*

Sequence	Information
1	Customer name
2	Customer address
3	Date order taken
4	Salesperson
5	Any number of lines containing:
	• an item code (if the salesperson remembers it) and an item description
	• the quantity of items
	• the negotiated price.
6	Any special requirements

Branch offices and the main office have an order-processing clerk who receives orders from salespersons. The order-processing clerk keeps a register of all the orders.

A typical register is shown in Table C1.2. Part of the registration process is to give the order a unique number, identify any items not completely described, and allocate the appropriate item codes to items in the order. Any errors or inconsistencies detected by the order-processing clerk are often checked with the salesperson before final registration. The whole order is held over by the order-processing clerk until all inconsistencies have been resolved.

COMMISSIONS

The order-processing clerk is also responsible for computing salespersons' commissions. Salespersons send in their commission invoices at the end of every month, itemizing the commission for each individual order on the invoice. The order-processing clerk must check these invoices to determine their correctness and verify any discrepancies with the salesperson. The clerk then subtracts any commission for orders cancelled or lost through delivery delays. Usually a cancelled or lost order results in only 5% of the normal commission, with a limit of $5. The normal is 12.5% for sales generating up to $2500 sales revenue for the month and 15% for amounts that exceed $2500. When the commission amount is adjusted (if this is necessary), a commission check is ordered, and a payment advice is sent to the salesperson. Records of commissions paid to salespersons are kept by the order-processing clerk.

CANCELLATIONS

Occasionally customers cancel orders. These cancellations are telephoned through to the order-processing clerk by the salesperson. The clerk changes the order register to reflect the cancellation and also sends a cancellation advice to the expeditor.

ORDERS AND ORDER REGISTERS

Orders are normally kept by the order-processing clerk, whereas the order register is the responsibility of the expeditor. The order-processing clerk, however, has access to the register to make the initial order entry. The order is also sometimes removed from the order-processing area and sent to other areas if needed.

HOW ORDERS ARE FILLED

Orders can be filled in one of three ways:

- by obtaining the required items from a store (owned by the organization)
- by ordering the items from a manufacturer (or wholesaler)
- by manufacturing it.

It is the job of the order expeditor to choose one of these ways. The preferred way is to obtain the item directly from store. If the item is not available in store, the next preference is to order it from a wholesaler. The last resort is to manufacture it internally (if it is one of the items produced by the organization). If none of these methods is possible, the order must be rejected.

Each line of the order can be treated separately. Thus, items in one order line can be obtained from the store, and those from another line can be obtained by ordering from a manufacturer. In fact, it is also possible to split an order line so that part of it is obtained from one source and part from another.

OBTAINING AN ITEM FROM STORE

The usual approach here is to select one or more orders for processing and formulate an item request to store for all items in these orders. One copy of the item request is sent to the store and another is retained by the expeditor. Each item has a unique identifier.

The staff at the store will check whether the requested items are in store and uncommitted. If so, the store advises the expeditor accordingly by an availability note. A copy of the availability note is kept in the store. This commits the store to holding the item for a specified period. The availability note includes a commitment number and length of commitment. A subsequent order must quote the commitment number.

If the expeditor requires the items requested in the availability note, a store order is sent. The store checks each received order against its record of availability notes and, if the order quotes a current commitment number, that order is met. If there is no current commitment number for the order, then the store file is checked for item availability. If the item is available, it is issued; otherwise, the order is rejected.

PLACING WHOLESALER PURCHASE ORDERS

If the expeditor cannot get an item from store, they will attempt to buy it from a wholesaler. The expeditor will search a wholesaler file and then negotiate a price and delivery date with a selected wholesaler. Once the wholesaler is selected, a wholesaler order request form is prepared. A copy of this request form is stored by the expeditor in a purchase order request register.

Purchase order processing is done centrally. The wholesaler order request form (prepared by the expeditor) is sent to the main office for processing. At the main office, the purchase order request form is first keypunched and then read into the machine. The machine then generates the actual purchase orders, which are then sent to the wholesaler.

The wholesaler supplies the goods, together with a delivery advice. The delivery advice is checked against the purchase orders. If the delivery matches the purchase order, it is accepted; otherwise, a query is sent back to the wholesaler.

Advices about accepted orders are sent back to the expeditor from the main office, together with the delivered goods.

A computer system is currently available to keep track of orders sent to wholesalers. The system consists of a suite of programs made up of:

- INPUT—reads cards containing details of the wholesaler order as prepared by the expeditor (see Table C1.3). Edits the cards and creates an ORDER INPUT file

- PURCHASE-ORDER-GENERATE (POGEN)—generates purchase orders. Purchase orders are generated once a week to take advantage of grouping lines from different customer orders into larger purchase orders to obtain volume discounts. A file of generated purchase orders is also created.
- DELIVERY—receives delivery advices from suppliers and correlates them against generated purchase orders.
- READY—prepares advices about customer order lines that have been fulfilled by a delivery.

Table C1.3 *An expeditor order input to input 1*

Manufacturer/ wholesaler	Date	Item-code	Qty-needed	Customer-order-no	Order-line

INVOICES FROM WHOLESALERS

Wholesalers send invoices for provided goods. These invoices are received by the accounts department. They are checked against the purchase orders and, if correct, a check is issued. Any discrepancies between the invoice, purchase order, and delivery advice must be resolved before a check can be issued to the wholesaler.

DELIVERY TO THE CUSTOMER

The expeditor forwards the goods to the customer as soon as they are received from the store or wholesaler. The goods are sent, together with a goods provided advice. At the same time, the expeditor checks the status of the orders. If the receipt of some goods completes the order, then a completed order advice, together with this order, is sent to the invoice clerk.

OTHER ACTIVITIES

Customers may ring to inquire about the status of their orders. The enquiry is usually made to the salesperson, who then refers it to the order-processing clerk. The order-processing clerk attempts to answer the query by referring to the orders register and then locating the order together with any attachments.

Invoices are sent to a customer as soon as an order is filled. The invoice is prepared by the invoice clerk, who receives the complete order from the expeditor.

The store has a computer-based ordering system. This keeps track of standard lines and automatically generates orders for them (as soon as a reorder point is reached). Updates to the store database are made in batch. Cards are punched for any orders from the expeditor and are used to generate stores issues. An online terminal is used to look up the current stock quantities. The order-processing clerks are responsible for compiling monthly reports on sales volumes by item and area.

QUANTITY OF DATA

The quantity of data for this case is given in Table C1.4.

Table C1.4   *Quantitative data*

Data	Quantity
Number of orders	100/day
Average lines/order	3
Average value of an order	$250
Percentage of request met by the store	70%
Average time to get item from manufacturer	14 days
Average number of lost orders	15%
Average time from registration to store issues	2–5 days
Delivery time (expeditor to customer)	(sometimes longer)
Time to issue items from store to expeditor	2 days
Average duration for order registration and other processing	1–2 days
Number of order-processing points	15
Number of salespersons	200

CASE 2—TRAVEL ARRANGEMENTS

GENERAL DESCRIPTION

Customers come to a travel agency to arrange a variety of trips. A sales consultant deals with each customer for a particular trip. At the initial interview the sales consultant records all the customer's requirements.

The sales consultant will then advise the customer of possible trip alternatives and make any bookings on the customer's behalf. To do this, the sales consultant may refer to a variety of timetables, hotel locations, or tour brochures.

If the travel arrangements cannot be completed at the first visit, the sales consultant will follow up with further bookings on the customer's behalf and confirm these with either a phone call or during a subsequent visit.

Once an itinerary is completed, an itinerary schedule is prepared and sent to the customer.

As soon as the itinerary is agreed, the invoicing can begin. The method of invoicing is explained below.

INVOICING

A copy of the invoice sent to customers is illustrated in Table C2.1. It contains the REF-NO of each trip and details of any bookings made on the trip. The invoice is a multi-purpose one in which each invoice line is applicable to all kinds of bookings made by the agency. For this reason, the DESCRIPTION field is used to store some of the details associated with each booking. If only a deposit is required, then the amount of the deposit is included on the invoice.

Once a customer returns the invoice with the payment, the accounts clerk reconciles the payment with the invoice. First the TOTAL-PAYMENT-AMOUNT and DATE-RECEIVED of the payment are recorded and a receipt is prepared for the customer. Then each line of the invoice is checked against the payment. If some

Table C2.1 *Invoices*

CUSTOMER-ACCOUNT

NAME: JOE CAPONE				INVOICE-NO: 369	
ADDRESS:				DATE-SENT: 20 JULY 98	
REF-NO	HOTEL/AIRLINE/ TOUR AGENCY	HELD-UNTIL	DESCRIPTION	AMOUNT	DEPOSIT
3	QANTAS	3 Aug	Flight 203 to Melbourne	$230	$40
3	WINDSOR	6 Aug	4 night stay	$135	$40
5	VIVA	4 July	Hong Kong Tour B66	$950	$950
5	—	4 July	Extra China Trip	$143	$143
			Total owing	$1458	$1173

invoice lines are not paid for, the payment is called a part-payment; otherwise, the payment is in full. If the payment is in full, then all bookings are confirmed and are recorded as paid for. If a part-payment is made, then those bookings that are fully paid for, or for which a deposit has been paid, are confirmed, and the amount of payment (BOOKING-PAYMENT) is recorded against each booking.

Another action also takes place when a payment is returned. After the payment is received and recorded and a receipt is prepared for the customer, the accounts clerk prepares a payment advice (identified by PAYMENT-NO), together with a check (if necessary) to the hotel, airline, or tour agency. This payment advice includes the commission. Sometimes early payments are held over for a few days and combined with those received from other customers to be sent to the same provider. The payment advice is shown in Table C2.2.

Table C2.2 *Payments*

PAYMENT ADVICE

TO: WINDSOR			PAYMENT NO:	
ADDRESS:			DATE-SENT:	
OUR REF-NO	DESCRIPTION	AMOUNT	COMMISSION	ENCLOSED
3	4 night stay for JOE CAPONE from 6 Aug (deposit only, balance to be paid by customer)	40	13.50 '	26.50
11	3 night stay for MARY BLUM from 11 Sep	90	9.00	81.00
		130	22.50	107.50

Any confirmations received from providers following a payment are sent on to the customer. A record of the confirmation is also made at the travel agency, including any relevant provider reference (PROVIDER-REF).

Cancellations of bookings and packaged trips can be made. Unconfirmed and unpaid bookings are simply deleted. Where some payment has already been made to the provider, the refund is determined. If a refund applies, then the booking clerk advises the provider of the cancellation and forwards a cancellation advice. This, together with any travel agency commission, is returned to the customer. Each cancellation is for one booking only, or one packaged trip only.

DATA USED IN MAKING TRAVEL ARRANGEMENTS

The following facts are known about travel arrangements made by the travel agency:

1. The travel agency keeps a record of each customer. CUSTOMER-NAME is a unique identifier of the customer, and each customer has a CUSTOMER-ADDRESS and PHONE-CONTACT-NO.

2. Each trip is identified by a unique REF-NO. Each customer can make any number of trips, but each trip is for one customer only. The DATE-ARRANGED and COST are stored for each trip.

3. Trips may either be PACKAGED-TRIPS or SPECIALLY-ARRANGED-TRIPS.

4. A packaged trip is for one of the packaged tours that are regularly organized by tour agencies. The travel agency maintains the following information about each packaged tour: the ORGANIZATION-NAME of the tour agency, the TOUR-NAME, the START-DATE, and the BASIC-COST. The TOUR-NAME is the unique identifier of the packaged tour within an ORGANIZATION-NAME, and the same packaged tour may be available at more than one time. (Thus, there may be a Hong Kong tour organized by both Agency A and Agency B. The Hong Kong tour by Agency A may in fact take place twice, with one tour starting on 6 December 1998 and the other on 9 January 1999.)

5. Each packaged trip is for one packaged tour. A field called EXTRAS is provided for each packaged trip to record any special needs of the CUSTOMER. There may be any number of packaged trips taking the same packaged tour.

6. A specially arranged trip is one where the travel agency constructs a trip out of a number of bookings. The booking can be either a hotel booking or an airline booking. Each booking is given a unique LEG-NO within a trip REF-NO. A special description is stored for each specially arranged trip. Each booking has a DATE-MADE associated with it.

7. A hotel reservation is made with one hotel. The travel agency keeps a list of a number of hotels. Each hotel is identified by a unique HOTEL-NAME and has an ADDRESS and FAX-NO. Each hotel booking is made with one hotel. The data stored with a hotel booking includes FIRST-DAY, LAST-DAY, ROOM-TYPE, and DAILY-RATE.

8. An airline reservation is made with one airline. The travel agency keeps a list of airlines. Each airline is identified by a unique AIRLINE-NAME and has a PHONE-CONTACT-NO. Each airline reservation is made with one airline and is identified by a RESERVATION-NO. The reservation includes the STARTING-AIRPORT and DEPARTURE-TIME. The reservation also includes details of all the stopovers in the reservation. Each stopover includes the ARRIVAL-TIME, DEPARTURE-TIME, and STOPOVER-AIRPORT.

CASE 3—STANDING ORDERS SUPPORT

The following activities have been identified in an organization.

STANDING ORDER CREATION

An organization receives standing orders from customers (identified by CUSTOMER-ID). The standing order is made for items (identified by ITEM-NO) that are supplied to customers on a regular basis, in this case daily. It is ascertained by a detailed examination of this activity that:

- items are identified by ITEM-NO, and information about the weight of items is kept
- the address of customers is kept. Both the one MAIN-OFFICE-ADDRESS and any number of DELIVERY-ADDRESSes are stored for each customer. A manager is associated with each DELIVERY-ADDRESS
- standing orders are identified by an identifier, S-ID. There is a separate standard order for each ITEM-NO to be supplied to each customer and an agreed ITEM-PRICE for that standing order. The standing order contains the ITEM-QTY required for each DAY-OF-WEEK. The DATE-INITIATED and the TERMINATION-DATE are stored for each order.

ITEM DELIVERY

Deliveries are made on each day to satisfy standing orders for that day. Each uniquely identifiable delivery is to one DELIVERY-ADDRESS of one customer and can include items in more than one standing order for the one customer. Detailed analysis shows that:

- a delivery docket (identifiable by DELIVERY-NO) is made out at each delivery
- the delivery docket contains the DELIVERY-DATE, DAY-OF-WEEK, and QTY-DELIVERED of each ITEM-NO together with S-ID for that item.

INVOICING

Every week an invoice is prepared and sent to the organization's customers. The invoice consists of all the deliveries made up to Friday of the previous week to one customer.

The invoice is shown in Table C3.1.

Table C3.1 *Invoices*

	DELIVERY-NO	ITEM-NO	S-ID	TOTAL-QTY	AGREED-PRICE	COST
INVOICE-NO: 63	6	23	70	6	1.20	$7.20
CUSTOMER-ID: John	17	37	10	30	0.70	$21.00
WEEK-ENDING: 17 June 1996	33	87	16	10	2.30	$23.00
					TOTAL	$51.20

PAYMENT RECEIPT

Customers return their payments with the copies of one or more invoices. The payment is recorded against the invoice. Full or part-payments may be made for each invoice. The record consists of AMOUNT-PAID and DATE-RECEIVED. Each payment is allocated a unique number. This unique number is the same as the RECEIPT-NO of the receipt that is returned to the customer.

Note that one invoice may be paid in more than one payment.

CASE 4—INSURANCE CLAIMS SYSTEM

GENERAL DESCRIPTION

The system is to be developed to support business processes for providing insurance for income protection. Such policies enable clients to receive payments when they suffer an illness or disability that prevents them from continuing full-time employment. The system is required to maintain client policies, to manage claims made on those policies, and to manage any payments made on accepted claims to replace income.

The main processes supported are the establishment of policies, the collection of premiums on the policies, and the management of claims on those policies.

A number of use cases have been analyzed. The use case model is shown in Figure C4.1.

USE CASES

SET UP INSURANCE POLICY
- Client approaches salesperson with requirements.
- Salesperson looks up benefit tables and advises client.
- Client decides to set up policy, identifying the type.
- A policy account is set up for the policy.
- The client is notified of the account and payments schedule.

CLIENT PAYS PREMIUMS
- A premium is computed on acceptance of a policy and the client is notified of the amount.

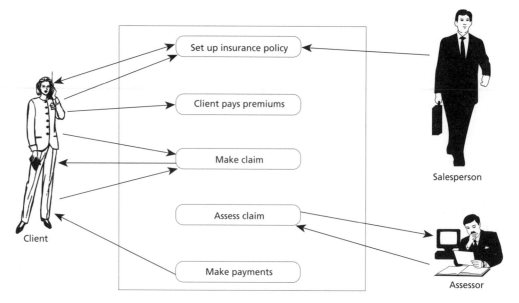

Figure C4.1 *Use case model*

- Client makes regular payments.
- The payment is recorded.
- A reminder notice is sent out when a premium is overdue.

MAKE AND ASSESS A CLAIM

- A client notifies the company of the intention to make a claim.
- A claim form is sent to the client and returned to the organization by the client once it is filled in.
- The claim is recorded and sent to the assessors.
- The assessment is recorded and the client is informed of the benefits.
- A payments record is set up to make the payments.

Draw a sequence diagram for each of these uses cases, then suggest an initial object class diagram for the system, showing attributes and methods.

There are two additional major activities, although use cases have not been written for them. These are making a claim assessment and making benefits payments that result from the claim.

Claims are then assessed, and a decision is made on whether a benefit is to be paid. In making the assessment, the claims assessor can refer to the client's policy and benefit tables and conditions. Once a claim is accepted, regular payments must be made to the claimant.

MAKING BENEFIT PAYMENTS

Payments can be manual or automatic. The check system issues the check and informs the policyholder of the actual amount paid.

A manual check is issued whenever fast payment is needed. In that case, a letter to the client is also produced, together with a check requisition form. Details from the requisition are used to issue the check.

SOME ADDITIONAL INFORMATION

Some additional information has also been found during the analysis. This especially concerns the following policy details.

New policies include details of the client, including address and date of birth. The policy also includes the name of the agent who sold the policy, the waiting period and expiry date, and a surrender value, if any. The start and expiry dates are included with each premium, together with the amount paid. New benefits can be added to the policy at any time, or refunds can be made on existing benefits, should the client so desire.

Policies can also include the following information:

- a waiting period
- an expiry date
- benefit type
- any rehabilitation expense
- minimum benefit for a specific condition
- benefit during the waiting period.

There are different conditions to be met for all of these benefits.

Policies also have no-claim bonuses that increase by a percentage over the first few years of a policy. Once a no-claim bonus period is completed, an escalating claim benefit is applied. Benefits increase with the consumer price index.

There is also some additional information about claims. A claim must include information about:

- the disability type and whether it was caused by sickness or accident
- income before disability
- any worker's compensation payments
- third-party insurance claims
- other legislative amounts.

You can use this additional information to construct a detailed object class diagram.

Access requirements. Defining how databases are accessed.

Accounts payable. The subsystem that keeps track of the moneys owed by the organization.

Accounts receivable. The subsystem that keeps track of moneys owed to the organization.

Analysis model. A description of the way in which a system works.

Appointments system. Keeping track of appointments.

Argumentation structure. A set of arguments used in making a decision.

Artifact. An object processed by a computer.

Association. A non-permanent relationship between two object classes.

Asynchronous cooperation. Cooperation where the participants refer to shared information at different times.

Attribute in a relational model. A column of the table or list of values.

Attribute in an E–R diagram. A property of a set in an E–R model.

Autonomous agent. A person or system that makes its decisions independently of other systems.

Batch system. A system that groups a number of transactions for later processing.

Binary relationship. A relationship that contains entities from at most two entity sets.

Block structured language A way of programming that clearly expressed process logic.

Brainstorming. Coming up with new ideas.

B2B. Carrying out business electronically between businesses.

B2C. Carrying out business electronically between consumers and businesses.

Bulletin board. A space that stores messages accessible to all members of a cooperating group.

Business alliance. One or more businesses working together for mutual benefit.

Business analyst. A person analyzing a business at the subject level.

Business process. A set of steps used to achieve a business goal.

Business process re-engineering. Changing an existing business process.

Business-to-Business networking. Conducting business across the Internet.

Business unit. A part of a business responsible for a well-defined business operation.

Cardinality. The number of relationships in which one entity can appear.

CASE tool horizontal integration. CASE tools integrated from same life-cycle stages.

CASE tool integration. The ability of one CASE tool to accept the input from another.

CASE tool vertical integration. CASE tools integrated at the adjacent life-cycle stage.

CASE tools. Computer Assisted Software Engineering tools used to keep track of system models.

Class. An object that describes a set of objects with the same features.

Class instance. An object that belongs to the class.

Client relations subsystem. That part of the system that interacts directly with the organization's clients.

Client relationship management. Finding ways to get, work with, and retain clients.

Client–server process. A process that describes how a server provides a service to a client.

Clients. People from outside an organization that deal with the organization.

Closed question. A question that requires a direct answer.

Collaboration. Two or more people deciding together on their future activities.

Collaboration diagram. A diagram showing exchange of messages between objects.

Communication. Interchange of information between people, machines, or both.

Computer network. A set of computers connected by communication lines.

Computer node. A computer in a computer network.

Computer operator. A person who operates computers.

Computer-based information system. An information system that uses computers.

Conceptual solution. A broad description of how a system will work.

Configuration. A set of documents used in a system development project.

Configuration management. Managing a configuration.

Containment. A permanent relationship between two object classes.

Context diagram. A diagram that shows the inputs and outputs of a system.

Controls. Checks to ensure that system inputs are correct and do not destroy system integrity.

Conversation. A sequence of speech acts.

Conversion technique. A method used to convert one system model to another system model that uses a different set of modeling constructs.

Cooperative design. Creating new artifacts by joint agreement between a number of participants.

Core business. The main business function of an organization.

Creativity. Coming up with new and innovative ideas.

Critical business process. A business process that is crucial to the survival of a business and supports the core business of the organization.

CSCW (computer-supported cooperative work). Systems that support groups of people working toward a common goal.

Customization. Adapting an object for use in an application.

Data dictionary. A document that contains the DFD and a description of all its components.

Data entry. Transaction input for later batch processing.

Data flow. Data flowing between processes, data stores, and external entities.

Data flow diagram (DFD). A method to illustrate how data flows in a system.

Data mining. Looking for patterns in databases.

Data store. A component of a DFD that describes the repositories of data in a system.

Data warehousing. The storage of large volumes of data for organizational use.

Database. An organized store of data.

Database definition. A definition of a database used as an input to a DBMS.

Decision support system. A system that supports decision making.

Dependent entity set. A set of entities whose existence depends on other entities.

Design. Creation of an artifact.

Design methodology. A collection of modeling methods and conversion techniques that start with a model of the user system and produce a computer system.

Design model. A description of the required system using system terms.

Design rationale. The reasoning behind making a decision.

Development process. A set of steps used to build a system.

Development world. The context in which technical development takes place.

Dialog. An interaction sequence between a user and a computer.

Direct access. Retrieval of records based on keywords.

Document configuration. The structure of documents within a project dictionary.

Domain of change. That part of a logical DFD of an existing system that will be changed in the new system.

Dropdown menu. A menu selection that provides more detailed selections.

E-mail. A way of using computers to exchange messages between people.

Economic feasibility. An evaluation to determine whether a system is economically acceptable.

Electronic commerce. Conducting business between organizations and consumers using the Internet.

Emic. An inside view of a system.

Empowerment. Giving people additional authority within an organization.

Encapsulation. Inclusion of many features in the one object.

Entity. A distinct object in a system.

Entity set. A component in an E–R diagram that represents a set of entities with the same properties.

Entity–relationship (E–R) model. A model that represents system data by entity and relationship sets.

Ethnography. Gathering information by observation.

Etic. An outside view of a system.

Event flow diagram. A diagram that shows the sequence of information flows between objects.

Evolutionary design. An experimental way of gradually building a system.

Expert system. A system for helping people make decisions.

Extends relationship. A use case that defines another use case in more detail.

External entity. An object outside the scope of the system.

Extranet. A network supporting exchanges between clients and an organization.

Feasibility analysis. An evaluation of whether it is worthwhile to proceed with a project.

Features. A characteristic of a class.

Federated database system. A set of databases managed independently but accessible in a unified way.

Feedback. Using variations from a system goal to change system behavior.

Fifth normal form. A set of relations that have no multivalued redundancy.

Financial services system. A system that keeps track of an organization's financial resources.

Formal interaction. A set of rules that define how people must interact.

Forward engineering. Re-engineering an existing system based on a model.

Function. A part that produces well-defined outputs from given inputs.

Functional dependency. Where one value of an attribute determines a single value of another attribute.

General accounts. A system that keeps track of funds within an organization.

Group support system. A system that supports decision making by groups.

Groupware. Software that assists workgroups.

Heterogeneous network. A computer network made up of different computers and software.

Homogeneous network. A computer network made up of the same computers and network.

Human resources subsystem. The part of a business that maintains personnel policy.

Identifier. A set of properties whose values identify a unique object in an object set.

Informal interaction. Working together without a set of prearranged rules.

Information system. A system that provides information to people in an organization.

Infrastructure. A basic structure for supporting a system.

Inheritance. Using the same features as another object.

Inspection. An examination of a product to assure quality.

Inspector. A person who examines a product to assure quality.

Instance. A unique occurrence of a type of object.

Integrated business process. A system that uses more than one business function.

Integrated databases. A set of databases managed by a single controlling system.

Interaction. The way people work together to achieve a goal.

Internet. A worldwide public network allowing global exchange of information.

Interviewing. Gathering information by asking questions.

Intranet. A network supporting information exchange within an organization.

Inventory. The business function that manages an organization's parts.

Knowledge-intensive organization. An organization that primarily produces knowledge.

Legacy system. An existing working computer system that is to be used in a new business process.

Leveling. Expanding a process into more detailed processes.

Library. A collection of objects that can be reused in many applications.

Linear cycle. A set of predefined steps for building a system.

Logical DFD. Describes the flow of logical data components between logical processes in a system.

Logical process. Describes any changes of values made by the processes on logical data.

Logical record analysis. A way of making performance estimates in design.

Logical record structure. A way of describing records at the system level.

Loyalty program. Providing rewards to clients for continuous support.

Management. People responsible for organizing and allocating resources.

Management process. The tasks required to manage a development process.

Management information system. A system for providing information to management.

Market research. Determining how to make products acceptable to customers.

Marketing subsystem. A system that determines what an organization is to produce and then publicizes its products.

Materials subsystem. The part of the business that keeps track of its material resources.

Mental model. The way a user sees a problem.

Menu. A set of alternative selections presented to a user in a window.

Merged subset. A collection of objects from more than one entity set.

Methods. A feature that describes programs within an object.

Middleware. Software that connects network services.

Mission. A reason for the existence of an organization.

Modeling construct. A representation that can be used to represent a system component in a system model.

Modeling method. A method used to construct a model of a system.

Modeling procedure. A set of steps provided by a modeling method to create a system model.

Moderator. A person who controls the progress of a review.

Monitoring a system. Checks made to see whether a system is meeting its goal.

Multimedia. Integrated storage of information in different media such as graphs, voice, video, and alphanumeric data.

Multiple relationship set. A relationship set where the same two entities can appear in more than one relationship.

Multi-user domain (MUD). An interface that supports more than one kind of user.

Multivalued dependency. Where one value of an attribute determines a set of values of another attribute.

N-ary relationship. A relationship that includes entities from more than two entity sets.

Network service. A technical system to support interaction between people.

Network manager. A person reponsible for a computer network.

Network programmer. A person developing programs to support networking.

Normal-form relations. A set of relations that describes the data in a system, but where each data component is a simple value.

Object class. See *class.*

Object class diagram. A diagram that shows object classes and relationships between them.

Object set. A generic term that includes entity sets, relationship sets, subsets, and dependent entity sets.

Occurrence diagram. A diagram that represents entities and relationships.

Office information system. A system that supports office operations.

On-line transaction. A transaction made through a terminal.

Open question. A question that requires the respondent to express a viewpoint.

Operational feasibility. An evaluation to determine whether a system is operationally acceptable.

Optimal normal form. Set of relations that have no single-value redundancy.

Organization chart. A chart that shows the business units of the organization.

Outsourcing. Arranging for computer processing to be done outside the organization.

Patterns. A collection of objects that can be adapted to an application.

Payroll subsystem. A business system for paying the organization's personnel.

Perlocutionary act. The effect of a speech act on a hearer.

Personnel development subsystem. A business system for maintaining people's skills.

Personnel subsystem. A business system for keeping information about people.

Phase. A step in the development process.

Physical DFD. Describes the flow of physical data components between physical operations in a system.

Physical process. Usually a physical device used to transform data—for example, a computer or person.

Planned work. Work where the sequence of tasks can be predefined.

Platform. A collection of computer services.

Polymorphism. Selecting the method appropriate for the class of object called.

Presentation. The layout of information on a computer screen.

Privacy. Ensuring that information remains accessible to one or a selected number of people.

Probe. A question that follows up an earlier answer.

Problem-solving cycle. A set of steps that start with a set of user requirements and produce a system that satisfies these requirements.

Process. A set of steps that define how things are done. In a DFD that describes how input data is converted to output data.

Producer. A person who produced a product.

Production subsystem. An organizational function that produces physical goods.

Productivity tools. Software systems that assist analysts and designers to build computer-based information systems.

Programmer. A person who writes computer programs.

Project dictionary. A record of all documents produced during system design.

Project management. An organized way of completing a project.

Project management tools. Tools to assist project management.

Properties. A feature that describes values stored within an object.

Prototyping. A method used to test or illustrate an idea and build a system in an explorative way.

Quality assurance. A process to ensure the development of quality products.

Quality service. A service that meets all client needs in a mutually satisfactory way.

Questionnaire. A form to gather information.

Rapid prototyping. Overlapping development phases to speed up system development.

Reader. A person who guides the way a system model is examined.

Re-engineering. Changing an existing system.

Relation. A table or list of values.

Relation key. A set of attributes whose values identify a unique row in a relation.

Relational model. A set of tables that describe the data in a system.

Relationship. One interaction between one or more entities.

Relationship between object classes. Indicates object classes that have meaningful interaction between their object instances.

Relationship set. A component in an E–R diagram that represents a set of relationships with the same properties.

Requirements model. A description of what users require the system to do.

Reuse. Using an existing system or module for a new task.

Reverse engineering. Developing a model for an existing computer-based information system.

Rich picture. A pictorial representation of a system.

Robustness. Ability to prevent interface errors from corrupting the system.

Role. The responsibility undertaken by a person.

Scenario. A description of a process in the usage world.

Screen. All the information presented to a user by a computer.

Script. A description of a process.

Seamless platform. A platform whose services are closely integrated.

Search procedure. The process followed to gather information about a system.

Search strategy. A selection of sources and methods to be used to gather information about a system.

Security. Ensuring that computer system faults do not destroy the information stored.

Sequence diagram. A diagram showing dynamic relationships between objects.

Situated work. Work where the next task is determined from the current situation.

Skeleton object. An object that can be customized for more than one application.

Skill profile. The skills possessed by a person or by an organization.

Soft-Systems Methodology. A development process centering on the user and subject worlds.

Software configuration. The documents used in a development process.

Software process. Another term for system life cycle but concentrating on software development.

Specialization. Taking on features in addition to those inherited from another object.

Speech act. An interaction between two people based on a single utterance.

Staged development. Building a system by parts.

State diagram. A diagram showing how objects evolve.

Statement of user requirements. A formal definition of what the new system must do.

Storyboard. A sequence of screens that illustrate how a system will work.

Storyboarding. Using a sequence of computer screens to describe how a system will be used.

Strategic planning. Determining the broad objectives for an organization.

Strategy. A broad objective for an organization.

Structure chart. Program modules and their interconnection.

Structured programming. A systematic way of writing programs.

Subject world. A system seen and described in well-defined business terms.

Subset. Some of the objects from one entity set.

Subsystem. A part of a system.

Supply chain. The movement of parts from the input through to the output of a business process.

Supporting process. A process to provide facilities needed by development teams.

Synchronous cooperation. Cooperation where the participants refer to shared information at the same time.

Synthesis. Building a system from existing modules.

System. A collection of components that work together to realize an objective.

System analyst. A person who analyzes the way the system works and its problems.

System boundary. The set of system components that can be changed during system design.

System component. An identifiable part of a system. Examples are computers, persons, documents, and data records.

System development cycle. Another term for the problem-solving cycle.

System development methodology. A predefined set of steps and a collection of tools used to design a system.

System directory. Another term for project dictionary.

System environment. Things outside the system study that can affect system behavior.

System life cycle. Another term for the problem-solving cycle.

System model. A model of a system created by using a modeling method.

System procedure. Defines actions to be taken to accomplish a system task.

System specification. A precise description of what the system must do.

System world. A system seen and defined in general system terms.
Systems analysis. Finding out what a system does and what its needs are.

Technical feasibility. An evaluation to determine whether a system can be technically built.
Template. A screen layout requesting a user to fill in values.
Terms. Words with specific meaning used to describe systems.
Testing. Checking to see whether a system does what it is supposed to do.
Transaction. A simple interaction with a computer database.
Transaction processing system. A computer system that manages transactions.
Transition diagram. Defines changes of state of an object.
Transparent access. Accessing data from a network independently of its location in the network.
Tuple. A row in a relation.

Usability. A term that defines how easy it is to use an interface.
Usability metrics. The things that can be measured to describe usability.
Usage world. A system seen and described in everyday terms.
Use case. A description of work in usage world terms.
Use case model. A collection of use cases that describe the whole system.
Uses relationship. A use case using another use case.
User-friendly. Describes a helpful interface.
User requirements. What users expect the system to do for them.

Validation. Checking whether a particular product satisfies user requirements.
Verification. Checks to ensure that a particular input has been converted correctly to an output.
Version. A variation of the same document.

Walkthrough. A quality assurance activity to detect errors.
Waterfall cycle. The same as a linear cycle.
Weak entity set. Another term for a dependent entity set.
Web designer. A person developing Web sites.
Window. An enclosed area on a screen.
WISIWIG. 'What I see is what I get' interface.
WISIWYS. 'What I see is what you see' interface.
Workflow. An instance of a workflow process.
Workflow process. A process made up of a predefined set of steps.
Workgroup. A group of people working to a common goal.
Workspace. The space and facilities provided for a user on a screen.
Workstation. A local computer.
Worldwide Web (WWW). A service supported on the Internet for the exchange of multimedia information.

INDEX